D1713208

# VHDL AND FPLDs
## IN DIGITAL SYSTEMS DESIGN, PROTOTYPING AND CUSTOMIZATION

# VHDL AND FPLDs
## IN DIGITAL SYSTEMS DESIGN, PROTOTYPING AND CUSTOMIZATION

by

**Zoran Salcic**
The University of Auckland

**KLUWER ACADEMIC PUBLISHERS**
**Boston / Dordrecht / London**

**Distributors for North, Central and South America:**
Kluwer Academic Publishers
101 Philip Drive
Assinippi Park
Norwell, Massachusetts 02061 USA

**Distributors for all other countries:**
Kluwer Academic Publishers Group
Distribution Centre
Post Office Box 322
3300 AH Dordrecht, THE NETHERLANDS

**Library of Congress Cataloging-in-Publication Data**

A C.I.P. Catalogue record for this book is available
from the Library of Congress.



*Printed on acid-free paper.*

Printed in the United States of America

# Table of Contents

# PREFACE

This book represents an attempt to treat three aspects of digital systems, design, prototyping and customization, in an integrated manner using two major technologies: VHSIC Hardware Description Language (VHDL) as a modeling and specification tool, and Field-Programmable Logic Devices (FPLDs) as an implementation technology. They together make a very powerful combination for complex digital systems rapid design and prototyping as the important steps towards manufacturing, or, in the case of feasible quantities, they also provide fast system manufacturing.

Combining these two technologies makes possible implementation of very complex digital systems at the desk. VHDL has become a standard tool to capture features of digital systems in a form of behavioral, dataflow or structural models providing a high degree of flexibility. When augmented by a good simulator, VHDL enables extensive verification of features of the system under design, reducing uncertainties at the latter phases of design process. As such, it becomes an unavoidable modeling tool to model digital systems at various levels of abstraction. When augmented by a good synthesizer, VHDL has been proven to be the most powerful standard tool for digital systems design. It enables synthesis of models for various target technologies, such as Application Specific Integrated Circuits (ASICSs) or FPLDs. As ASICSs are still the privilege of those having access to more complex tools and resources, the FPLDs have become an ideal target technology for both prototyping and volume production. Besides verification performed at the level of simulation, FPLDs provide more realistic in-circuit verification by downloading configuration bitstreams to target FPLDs and checking the system behaviour in a real environment. Moreover, the process of modeling and design is extended to the field of easy customization, which becomes a major feature of new generations of digital systems. Having VHDL models of standard or non-standard digital systems, and having flexible target technology, the designer can perform customization of an existing system and provide with ease application-specific solutions that are tailored for the needs of a specific application.

The book is divided into three parts, which cover to some extent main aspects of modeling, design and simulation with the primary purpose of synthesis and prototyping. Part 1 represents an introduction to VHDL with emphasis on basic features of the language and both basic modeling and design with VHDL. The aim was not to present all features of the language and its full syntax, but rather to introduce it through a large number of small examples. Most of them represent models or designs of real digital circuits which are found as building blocks in almost every more complex digital system. All types of VHDL models, including behavioral, structural and dataflow models, are present to emphasize that all of them are usually valid and lead to viable designs. The chapters of Part 1 cover elements of VHDL syntax, data types, and basic built-in objects used in modeling and design. Advanced data types give the designer facilities to extend basic types to ones that best suit a specific problem. Much attention is paid to concurrency features of VHDL, which differentiate it from programming languages. Although VHDL looks like a typical programming language, features that distinguish it from programming languages, such as concurrency and processes to describe parallel activities, signals and structural descriptions are emphasized. Relatively low attention is paid to aspects of VHDL that do not result in actual designs, and are primarily used for simulation purposes. Still, all of Chapter 8 is devoted to the simulation aspects and writing test benches as a unique tool to model both a system under test and test facilities in VHDL. Similarly, Chapter 9 is devoted to synthesis aspects and discussion of VHDL models that lead to viable combinational and sequential circuits including any type of finite state machines. The whole of Part 1 can be considered as a familiarization and preparation for more complex design tasks with the aim to seamlessly bring the reader to the point when he/she is able to make his/her own more complex models and designs. VHDL, as presented in this book, mostly follows IEEE 1987 standard, but also presents some of the features of the revised IEEE 1993 standard. The reasons not to stick just to the newer IEEE 1993 standard are that most of the current VHDL compilers still better compile the older standard, on one hand, and also that all key features of VHDL are present in IEEE 1987. Changes and improvements in the revised standard are very often in the domain of improved and “cleaner” syntax, and do not affect synthesis features of VHDL. The readers already familiar with VHDL may skip more or less of this Part, although they may find some of the chapters useful to read at least briefly.

Part 2 represents a bridge towards designing and prototyping using FPLDs as prototyping and implementation technology. It introduces FPLDs from the point of view of general features that include architecture of FPLD devices and their basic building blocks, mechanisms to implement both combinational and sequential logic functions, and types of programming technologies. Field Programmable Logic has been available for a number of years, but, by the recent increase of power and variety of devices, it is extending its role from that of simply being a convenient way of implementing the system "glue logic", to an increasing ability to implement very complex system functions, such as microprocessors or even complete microcomputers. The speed with which devices can be configured makes them ideal for prototyping and for education, and overall production costs make them competitive for small to medium volume productions. The reprogrammable devices are opening up sophisticated new applications and new hardware/software trade-offs and diminishing the traditional hardware-software demarcation line. Advanced design tools are being developed for automatic compilation of complex designs, and routes to custom circuits are now available. Aspects of reconfiguration of FPLDs, which include both custom-configuration and dynamic reconfiguration are pointed out as the keys to efficiently use VHDL as a modeling and design tool. In order to maintain links with the real world, illustrations of all key concepts are shown on examples of Altera's FPLDs, with emphasis on devices using SRAM cells as programming technology. Integrated design environments that represent the environment in which VHDL is usually used are introduced, and phases of modeling and design process explained and illustrated in an example of Altera's Max+Plus II design environment. Finally, specifics of VHDL and synthesis in a concrete design environment and for concrete target technology are tackled. Prototyping systems that represent extension of standard design environments are also introduced with examples of simple systems for embedded applications using standard microcontrollers and FPLDs, PC/FPLD environment and Altera's prototyping board known as UP1, which is especially suitable for educational purposes. The readers familiar with FPLDs may skip some of the chapters. Those who intend to perform design and prototyping using VHDL and Altera's FPLDs should read Chapter 13, although most of the examples presented in this chapter represent common digital circuits used as building blocks in more complex designs, with descriptions that depend little, or do not depend at all, on implementation technology.

Part 3 contains a number of examples and case studies that demonstrate the effectiveness of using VHDL and FPLDs in the design of real systems. Some of them are just the cores that can serve for further modifications and customization. Chapter 14 presents three examples. The first is a simple sequence recognizer and classifier that can be used as an initial point in many applications for recognition of sequences of bits or parallel presented data such as passwords, synchronization sequences etc. The second example is a traffic light controller that controls the traffic and pedestrian crossing on an intersection of a main and side road. Both these designs show how to approach the design of finite state machines of low or medium complexity. The third example demonstrates a different type of system which is found in digital modulation. This example shows the use of VHDL and FPLDs with embedded memory blocks in direct digital synthesis of a modulated signal, as well as in parts of modulator for the frequency shift keying (FSK) modulation. Chapter 15 presents an example of a high-speed switch for Asynchronous Transfer Mode (ATM) based systems. The ATM switch, called FlexSwitch, is a complex, high-performance switch for 8 input/output ports with an aggregate throughput of up to 6.4Gbits/s. It also demonstrates the feasibility of design of such a complex system with relatively small resources. This chapter is written in collaboration with my student Kuan Hua Tan, with whom I spent numerous fruitful discussions on the approach to the ATM switch in the last two years. Chapter 16 introduces a custom-configurable flexible instruction execution microcomputer, FLIX, which represents a core for application-specific microcomputers. The design enables change of the basic data path and control unit, and also customization by adding various types of "co-processors" to the core. The whole design fits into a single moderately complex FPLD and includes the whole microcomputer. This chapter may also be useful for those who teach and learn basic processor architecture, because the FLIX design is completely open giving freedom to be changed and modified with not only additional instructions but also addressing modes or new elements of its data path. It has been used as the core of a number of application specific processors currently being developed at the Auckland University including image enhancement, Kalman filter application specific processor and others. Finally, two application-specific co-processors for image enhancement are presented in Chapter 17. They represent hardware implementations of two algorithms, high-boost filter and global histogram equalization, and can be implemented as the co-processors to the FLIX core, or as co-processors in a standard computer environment such as PC/Windows platform to speed-up computationally intensive tasks. A part of this chapter represents results of research I have been involved in together with my colleague Jayanthi Sivaswamy. The main point of this Part is customization. It is the driving power for new applications of FPLDs, and also for new forms of computation that have been unknown until dynamically reconfigurable logic was made a reality. However, customization requires also new design tools, VHDL being one of the first suitable for that purpose. The reader will be able to feel the power of VHDL, but also its weaknesses, and the need for even more powerful tools which will enable further developments in the digital systems arena.

All design examples presented in this book are compiled with either Accolade Design VHDL Compiler (PeakVHDL) and verified using its simulator, or with Altera's VHDL compiler within Max+Plus II integrated design environment, verified using Altera's simulator and finally tested on one of the prototyping systems presented in Part 2. Gratitude goes to the Accolade Design Inc. for their support to writing parts of this book. Special gratitude goes to Altera Corporation for enabling me and my students to use their tools and prototyping systems, as well as supporting our research over a number of years, thus enabling many of our concepts to become reality. A special thanks goes to Altera Corporation for providing their Max+Plus II environment on CD ROM as part of my first book on design and prototyping of digital systems using FPLDs, and also for this one, making VHDL compiler and all other tools needed for the design and prototyping process accessible to a wide reader audience.

This book represents a continuation of the project I initially conceived by starting writing the book "Digital Systems Design and Prototyping using Field-Programmable Logic", which was finalized together with my old friend Asim Smailagic from Carnegie-Mellon University. It gave me the opportunity to open the field further and fill many gaps which remained after completing the first book. I hope that many of the questions, which naturally emerged when reading the first book, will be answered in this one. However, new questions are created and remain to be answered by the readers themselves.

The book contains a number of examples which have been refined throughout the last three years while I have been teaching at the University of Auckland's Department of Electrical and Electronic Engineering, and also while working on research projects with my colleagues and postgraduate students. None of the designs claims to be ultimate -further improvement and refinement is always possible. I will be happy for the interested reader to make changes and improvements in any respect, and I would be interested to hear of the outcome. That will be the real success of my aim. Special thanks go to all students who have been involved in my research projects. And to those numerous students who provided valuable feedback at various stages. As with any book, this one remains open for improvement, especially as the subject area is changing rapidly and is directing towards new exciting developments. My thanks go to the University of Auckland for providing me support which has enabled me to progress from virtually zero to the present level of attainment in only three years. My thanks also go to Carl Harris from Kluwer Academic Publishers for the right support at the right time, and to two friends from Altera Corporation whom I have never met, Stephen Smith and Joe Hanson, whose role as "invisible" supporters was invaluable.

I cannot finish without expressing my special thanks to my family for their support - my wife Dushka and sons Srdan and Goran for their help and patience during those days and nights of designing and modifying, writing and editing. And to those who provided "remote" support, my Mum, who will never see the fruits of her son's work she supported for many years, and my Dad.

Zoran Salcic
Auckland
New Zealand

# About the Accompanying CD-ROM

***VHDL and FPLDs in Digital Systems Design, Prototyping and Customization,*** First Edition includes a CD-ROM that contains Altera's MAX+PLUS II 7.21 Student Edition programmable logic development software. MAX+PLUS II is a fully integrated design environment that offers unmatched flexibility and performance. The intuitive graphical interface is complemented by complete and instantly accessible on-line documentation, which makes learning and using MAX+PLUS II quick and easy. MAX+PLUS II version 7.21 Student Edition offers the following features:

- ✓ Operates on PCS running Windows 3.1, Windows 95, and Windows NT 3.51 and 4.0.
- ✓ Graphical and text-based design entry, including Altera Hardware Description Language (AHDL) and VHDL.
- ✓ Design compilation for product-term (MAX 7000S) and look-up table (FLEX 10K) device architectures.
- ✓ Design verification with full timing simulation.

**Installing with Windows 3.1 and Windows NT 3.51**

Insert the MAX+PLUS II CD-ROM in your CD-ROM drive. In the Windows Program Manager, choose **Run** and type: *<CD-ROM drive>*: `\pc\maxplus2\install` in the *Command Line* box. You are guided through the installation procedure.

**Installing with Windows 95 and Windows NT 4.0**

Insert the MAX+PLUS II CD-ROM in your CD-ROM drive. In the Start menu, choose **Run** and type: *<CD-ROM drive>*: `\pc\maxplus2\install` in the *Open* box. You are guided through the installation procedure.

**Registration & Additional Information**

To register and obtain an authorization code to use the MAX+PLUS II software, to **http://www.altera.com/maxplus2-student**. For complete installation instructions, refer to the **read.me** file on the CD-ROM or to the ***MAX+PLUS II Getting Started Manual***, available on the Altera world-wide web site (**http://www.altera.com**).

This CD-ROM is distributed by Kluwer Academic Publishers with *ABSOLUTELY NO SUPPORT* and *NO WARRANTY* from Kluwer Academic Publishers.

Kluwer Academic Publishers shall not be liable for damages in connection with, or arising out of, the furnishing, performance or use of this CD-ROM.

# PART1

# Basic Modeling and Designing with VHDL

# 1 INTRODUCTION TO VHDL

VHDL (VHSIC Hardware Description Language) is a language used to express complex digital systems concepts for documentation, simulation, verification and synthesis. The wide variety of design tools makes translation of designs described in VHDL into actual working systems in various target hardware technologies very fast and more reliable than in the past when using other tools for specification and design of digital systems. VHDL was first standardized in 1987 in IEEE 1076-1987 standard, and an updated and enhanced version of the language was released in 1993, known as IEEE 1076-1993.

VHDL has had an enormous impact on digital systems design methodology promoting a hierarchical top-down design process similar to the design of programs using high-level programming languages such as Pascal or C++. It has helped to establish new design methodology, taking the designers away from low level details such as transistors and logic gates to a much higher abstract level of system description, just as high-level programming languages take them away from the details of CPU registers, individual bits and assembly level programming. Unlike programming languages, VHDL provides mechanisms to describe concurrent events being of crucial importance for the description of behavior of hardware. This feature of the language is familiar to designers of digital systems who were using proprietary hardware description languages, such as PALASM, ABEL, or AHDL, used primarily to design for various types of PLDs. Another important feature of VHDL is that it allows design entry at different levels of abstraction making it useful not only to model at the high, behavioral level, but also at the level of simple netlists when needed. It allows part of the design to be described at a very high abstraction level, and part at the level of familiar component level, making it perfectly suitable for simulation. Once the design concepts have been checked, the part described at the high level can be redesigned using features which lead to the synthesizable description. The designers can start using language at the very simple level, and introduce more advanced features of the language as they need them.

Having these features, the language provides all preconditions to change the design methodologies which result in such advantages as:

- shorter design time and reduced time to market
- reusability of already designed units

- fast exploration of design alternatives
- independence of the target implementation technology
- automated synthesis
- easy transportability to other similar design tools
- parallelization of the design process using a team work approach

By providing independence of the target implementation technology, VHDL enables the same design specification to be used regardless of the target technology, making it possible to implement the design in either ASIC or FPLD, for example. The power of VHDL goes even beyond this, enabling us to describe designs on such levels of abstraction as PCBs or MCMs which contain as their parts standard SSI, MSI, or LSI ICs, FPLDs and full-custom ICs. The designers are taken away from low-level details, and can spend more time on aspects of different architectures, design alternatives, and system and test issues. This becomes more important with the growth in complexity of FPLDs and ASICs exceeding the equivalent of 200,000 low-level gates. Top-down design methodology and hiding of the details at the higher levels of design hierarchy make readable and understandable designs possible.

## 1.1 What is VHDL for?

VHDL is a hardware description language which is now an industry standard language used to document electronic systems design from the abstract to the concrete level. As such it aims to model the intended operation of the hardware of a digital system. VHDL is also used as a standardized input and output from various CAE tools that include simulation tools, synthesis tools, and layout tools. VHDL is firstly used for design entry to capture the intended design. For this purpose we use a text editor. The VHDL source code can be the input to simulation in order to verify the functionality of the system, or it can be passed to synthesis tools which provide implementation of the design for a specific target technology. All examples of digital systems in this book are described in either IEEE 1076-1987 or 1076-1993 VHDL, and compiled and simulated with the Accolade PeakVHDL or Altera Max+Plus II compilers and simulation tools. Accolade's compiler and simulator are used for the conceptual stage of the design, and Altera's tools to provide a synthesis for FPLDs as a target technology. We have found that the best way of approaching VHDL, which is in its full extent a very difficult language, is to use a subset of the language initially, and, as more complex models and features are required, they can be subsequently learned and used.

VHDL consists of several parts organized as follows:

- The actual VHDL language as specified by IEEE standard
- Some additional data type declarations in the standard package called IEEE standard 1164
- A WORK library reserved for user's designs
- Vendor packages with vendor libraries
- User packages and libraries

A VHDL description lists a design's components and interconnections, and documents the system behavior. A VHDL description can be written at various levels of abstraction:

- Algorithmic or behavioral
- Register transfer
- Gate level functional with unit delay
- Gate level with detailed timing

Using top-down design methodology, a designer represents a system abstractly at first, and in more details later. Some design decisions can be left for the latter phases of the design process. VHDL provides ways of abstracting design, or "hiding" implementation details. A designer can design with top down successive refinements specifying more details of how the design is built.

A design description or model, written in VHDL, can be run through a VHDL simulator to demonstrate the behavior of the modeled system. Simulating a design model requires simulated stimulus, a way of observing the model during simulation, and capturing the results of simulation for later inspection. VHDL supports a variety of data types useful to the hardware modeler for both simulation and synthesis. These data types will be introduced throughout the book, starting with the simple ones and then presenting advanced types which make VHDL unique among hardware description languages.

Some parts of VHDL can be used with logic synthesis tools for producing physical design. Many VLSI gate-array or FPLD vendors can convert a VHDL design description into a gate level netlist from which a customized integrated circuit or FPLD implemented piece component can be built. Hence the application of VHDL is for:

- Documenting a design in a standard way. This guarantees support by newer generations of design tools, and easy transportability of the design to other simulation and synthesis environments.
- Simulating the behavior, which helps verification of the design often using a behavioral instead of a detailed component model. It has many features that enable description of the behavior of an electronic system from the level of a simple gate to the level of the complete microcontrollers or custom chips.The resulting simulation models can be used as building blocks for larger systems which use either VHDL or other design entry methods. Furthermore, VHDL enables specification of test benches which describe circuit stimuli and expected outputs that verify behavior of a system over time. They are an integral part of any VHDL project and are developed in parallel with the model of the system under design.
- Directly synthesizing logic. Many of the VHDL features, when used in system description, provide not only simulatable but also synthesizable models. After compilation process, the system model is transformed into netlists of low level components that are placed and routed to the chosen target implementation technology. In the case of designs and models presented in this book, the target technology are Altera's FPLDs, although the design can be easily targeted to FPLDs of the other vendors or ASICs.

Designing in VHDL is like programming in many ways. Compiling and running a VHDL design is similar to compiling and running other programming languages. As the result of compiling, an object module is produced and placed in a special VHDL library. A simulation run is done subsequently by selecting the object units from the library and loading them onto the simulator. The main difference is that VHDL design always runs in simulated time, and events occur in successive time steps.

There are several differences between VHDL and conventional programming languages. The major are the notions of delay and simulation environment, and also the concurrency and component netlisting, which are not found in programming languages. VHDL supports concurrency using the concept of concurrent statements running in simulated time. Simulated time is feature found only in simulation languages. Also, there are sequential statements in VHDL to describe algorithmic behavior.

Design hierarchy in VHDL is accomplished by separately compiling components that are instanced in a higher-level component. The linking process is done either by compiler or by simulator using the VHDL library mechanism.

Some software systems have version-control systems to generate different versions of loadable program. VHDL has a configuration capability for generating design variations. If not supported by specific simulators and synthesis tools, it is usually by default taken that the latest compiled design is one which is used in further designs.

## 1.2 VHDL Designs

Digital systems are modeled and designed in VHDL by the top-down approach of partitioning of the design into smaller abstract blocks known as components. Each component represents an instant of a design entity, which is usually modeled in a separate file. A total system is is then described as a design hierarchy of components making a single higher level component. This approach to the design will be emphasized and used throughout all examples presented in the book.

A VHDL design consists of several separate design units, each of which is compiled and saved in a library. The four source design units that can be compiled are:

1. **Entity**, that describes the design's interface signals and represents the most basic building block in a design. If the design is hierarchical, then the top-level description (entity) will have lower-level descriptions (entities) contained in it.
2. **Architecture**, that describes design's behavior. A single entity can have multiple architectures. Architectures might be of behavioral or structural type, for example.
3. **Configuration**, that selects a variation of design from a design library. It is used to bind a component instance to an entity-architecture pair. A configuration can be considered as a parts list for a design. It describes which behavior to use for each entity.
4. **Package**, that stores together, for convenience, certain frequently used specifications such as data types and subprograms used in a design. Package can be considered as a toolbox used to build designs. Items defined within

package can be made visible to any other design unit. They can also be compiled into libraries and used in other designs by a **use** statement.

Typically, a designer's architecture uses previously compiled components from an ASIC or FPLD vendor library. Once compiled, a design becomes a component in a library that may be used in other designs. Additional compiled vendors' packages are also stored in a library.

By separating the entity (I/O interface of a design) from its actual architecture implementation, a designer can change one part of a design without recompiling other parts. In this way a feature of reusability is implemented. For example, a CPU containing a precompiled ALU saves recompiling time. Configurations provide an extra degree of flexibility by saving variations of a design (for example, two versions of CPU, each with a different ALU). A configuration is a named and compiled unit stored in the library.

The designer defines the basic building blocks of VHDL in the following sections:

- Library
- Package
- Entity
- Architecture
- Configuration

In order to introduce intuitively the meanings of these sections, an example of a design unit contained in file my_design.vhd is given below:

```
package my_units is                                   --package--
          constant unit_delay:  time :=10 ns;
end my_units;

entity compare is                                     --entity--
          port  (a, b : in bit ;
                    c : out bit);
end compare;

library my_library;
use my_library.my_units.all;
```

```
architecture first of compare is                --architecture--

begin
        c <=not (a xor b) after unit_delay;
end first;
```

There are three design units in a design my_design.vhd. After compilation, there are four compiled units in library my_library:

- Package my_units - provides a shareable constant
- Entity compare - names the design and signal ports
- Architecture first of compare - provides details of the design
- A configuration of compare - designates first as the latest compiled architecture.

Each design unit can be in a separate file and could be compiled separately, but the order of compilations must be as it is shown in the example above. The package my_units can also be used in other designs. The design entity compare can now be accessed for simulation, or used as a component in another design. To use compare, two input values of type bit are required at pins a and b; 10 ns latter a '1' or '0' appears at output pin c.

Keywords of the language are given and will be shown in bold letters. For instance, in the preceding example, the keywords are **architecture**, **package**, **entity**, **begin**, **end**, **is**, etc. Names of user-created objects, such as *compare*, will be shown in lowercase letters. However, it should be pointed out, VHDL is not case sensitive, and this convention is used just for readability purpose. A typical relationship between design units in a VHDL description is illustrated in Figure 1.1.

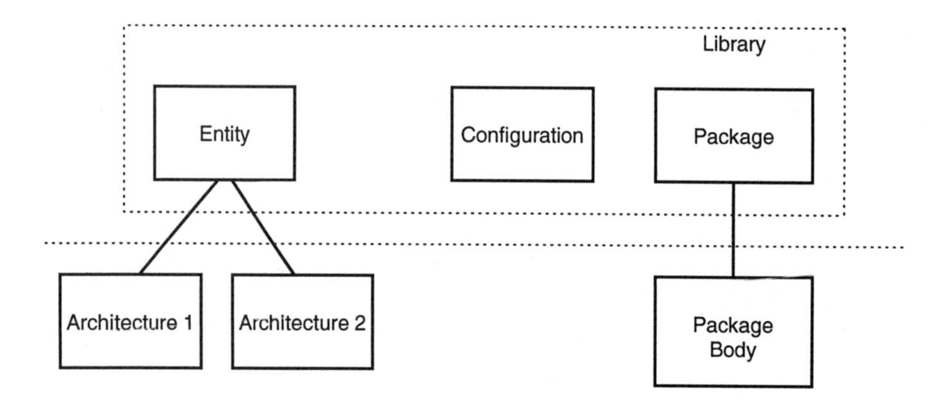


Figure 1.1 VHDL Design Units and Relationships

Basic VHDL design units are described in more details in the following sections.

## 1.3 Library

The results of a VHDL compilation are stored in a library for subsequent simulation, or for use in further or other designs. A library can contain:

- A package - shared declarations
- An entity - shared designs
- An architecture - shared design implementations
- A configuration - shared design versions

The two built-in libraries are WORK and STANDARD, but the user can create other libraries. VHDL source design units are compiled into WORK library unless a user directs it to another library.

To access an existing library unit in a library as a part of new VHDL design, the library name must be declared first. The syntax is:

**library** logical_name;

Now, component designs compiled into the specified library can be used. Packages in the library can be accessed via a subsequent **use** statement. If WORK library is used, it does not need to be declared.

Compiled units within a library can be accessed with up to three levels of names:

library_name.package_name.item_name

or

library_name.item_name

or

item_name if the WORK library is assumed

Units in a library must have unique names; all design entity names and package names are unique within a library. Architecture names need to be unique to a particular design entity.

In order to locate a VHDL library in a file system, it is sometimes necessary to issue the commands outside of the VHDL language. This is compiler and system dependent, and a user has to refer to appropriate vendor's literature.

## 1.4 Package

The next level of hierarchy within a library is a package. A package collects a group of related declarations together. Typically, a package is used for:

- Function and procedure declarations
- Type and subtype declarations
- Constant declarations
- File declarations
- Global signal declarations
- Alias declarations
- Attribute specifications
- Component declarations
- Use clauses

Package is created to store common subprograms, data types, constants and compiled design interfaces that will be used in more than one design. This strategy promotes the reusability.

A package consists of two separate design units: the package header, which identifies all of the names and items, and the optional package body, which gives more details of the named item.

All vendors provide a package named STANDARD in a predefined library named STD. This package defines useful data types, such as bit, boolean, and bit_vector. There is also a text I/O package called TEXTIO in STD.

A **use** clause allows access to a package in a library. No **use** clause is required for the package STANDARD. The default is:

```
library STD;
use STD.STANDARD.all;
```

Additionally, component or CAD tool vendors provide packages of utility routines and design pieces to assists design work. For example, VHDL descriptions of frequently used CMOS gate components are compiled into a separate library, and their declarations are kept in a package.

## 1.5 Entity

The design entity defines a new component name, its input/output connections, and related declarations. The entity represents the I/O interface or external specification to a component design. VHDL separates the interface to a design from the details of architectural implementation. The entity describes the type and direction of signal connections. On the other side, an architecture describes the behavior of a component. After an entity is compiled into a library, it can be simulated or used as a component in another design. An entity must have an unique name within a library. If a component has signal ports, they are declared in an entity declaration. The syntax used to declare an entity is:

```
entity entity_name is

        [generics]
        [ports]   [declarations {constants, types, signals}]
        [begin    statements]        --Typically not used
```

**end** [**entity**] entity_name;

An entity specifies the external connections of a component. In Figure 1.2 an AND gate (*andgate*) with two signal lines coming in, and one going out, is presented.

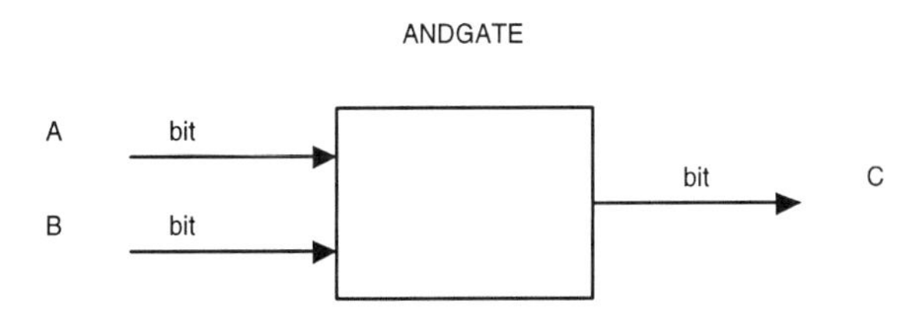


Figure 1.2 Example of AND gate

The diagram emphasizes the interface to the design. All signals are of the bit type, which mandates the usage; the andgate design only works on bit type data. VHDL declaration of this entity is:

```
entity andgate is
        port    (a, b: in bit;
                c: out bit);
end andgate;
```

In this example *andgate* is defined as a new component. The reserved word **is** is followed by the port declarations, with their names, directions (or mode in VHDL) and types.

Any declaration used in an entity port must be previously declared. When an entity is compiled into a library, it becomes a component design that can be used in another design. A component can be used without the knowledge of its internal design details.

All designs are created from entities. An entity in VHDL corresponds directly to a symbol in the traditional schematic entry methodology. The input ports in the preceding example directly correspond to the two input pins, and the output port corresponds to the output pin.

Optionally, the designer may also include a special type of parameter list, called a generic list, that allows additional information to pass into an entity. This information can be especially useful for simulation of the design model.

## 1.6 Architecture

An architecture design unit specifies the behavior, interconnections, and components of a previously compiled design entity. The architecture defines the function of the design entity. It specifies the relationships between the inputs and outputs that must be expressed in terms of behavior, dataflow, or structure. The entity design unit must be compiled before the compilation of its architecture. If an entity is recompiled, all its architectures must be recompiled, too.

VHDL allows the designer to model a design at several levels of abstraction or with various implementations. An entity may be implemented with more than one architecture. Figure 1.3 illustrates two different architectures of entity *alu*.

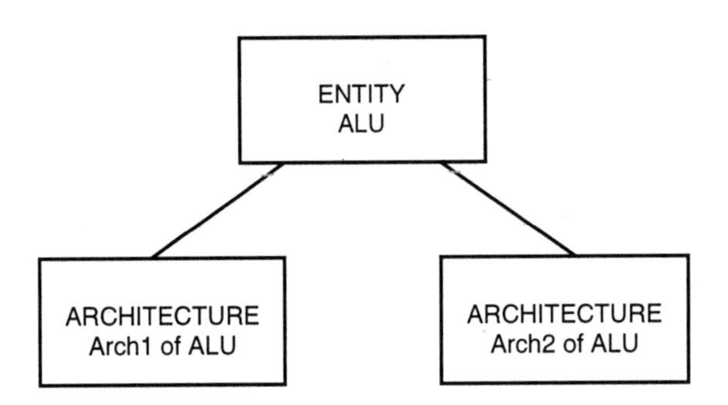


Figure 1.3 Entity with two different architectures

All architectures have identical interfaces, but each needs an unique architecture name. A designer selects a particular architecture of a design entity during configuration (for example *arch1*).

VHDL architectures are generally categorized in styles as:

- Behavioral - defines sequentially described process
- Dataflow - implies a structure and behavior
- Structural - defines interconnections of components

Different styles of VHDL designs actually represent different levels of abstraction of using the language. Generally speaking we can associate levels of abstraction with the architecture styles as in Table 1.1, although the boundaries between different styles are not strict, and often in the same model we can use a mix of these styles.

A design can use any or all of these design styles. Generally, designs are created hierarchically using previously compiled design entities. They can only be combined using structural style which looks like a list of components wired together (i.e., netlist).

Table 1.1 Architecture styles and levels of abstraction

| Architecture style | Level of abstraction |
|---|---|
| Structural | Physical information<br>Hierarchy of gates<br>Boolean Equations |
| Dataflow | Arithmetic operations<br>Concurrent signal assignment<br>Register transfers |
| Behavioral | State machines<br>Sequential descriptions<br>Test benches<br>High level specifications |

The architecture is defined in VHDL with the following syntax:

**architecture** architecture_name **of** entity_name **is**

[architecture_declarative_part]

**begin**

[architecture_statement_part]

**end** [**architecture**] [architecture_name];

The architecture_declarative_part declares items used only in this architecture such as types, subprograms, constants, local signals and components are declared.

the architecture_statement_part is the actual design description. all statements between the **begin** and **end** statement are called concurrent statements, because all of the statements execute concurrently.

The architecture can be considered as a counterpart to the schematic for the component in traditional designs.

### *1.6.1 Behavioral Style Architecture*

An example of an architecture called *arch1* of entity *andgate* is shown below.

```
architecture arch1 of andgate is

begin
        process (a, b);
        begin
                if a ='1' and b ='1' then
                c <='1' after 1 ns;
                else
                c <='0' after 1 ns;
                endif;
        end process;
end arch1;
```

It contains a process that uses signal assignment statements. If both input signals a and b have the value '1', c gets a '1'; otherwise c gets a '0'. This architecture describes a behavior in a “program-like” or algorithmic manner.

VHDL processes may run concurrently. The list of signals for which the process is waiting (sensitive to) is shown in parentheses after the word process. Processes wait for changes in an incoming signal. Process is activated whenever input signals change. The output delay of signal c depends upon the **after** clause in the assignment.

Parallel operations can be represented with multiple processes. An example of processes running in parallel is shown in Figure 1.4. The processes communicate with each other; they transfer data with signals. A process gets its data from outside from a signal. Inside, the process operates with variables. The variables are local storage and cannot be used to transfer information outside the process. Sequential statements, contained in process, execute in order of appearance as in conventional programming languages.

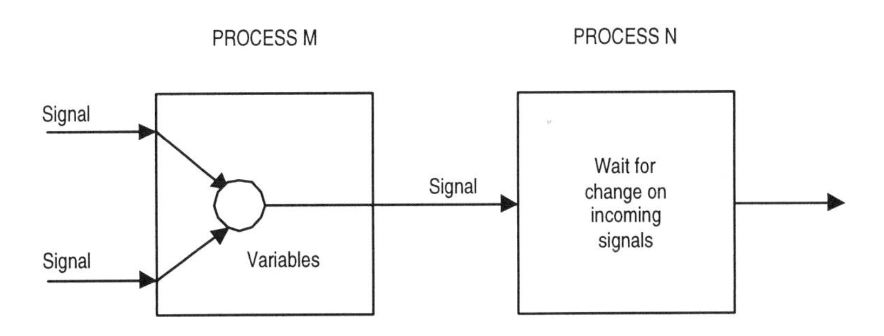


Figure 1.4 Process model

Process N in Figure 1.4 receives signals from process M. The running of one process can depend upon results of operation of another process.

In top-down design style, behavioral description is usually the first step; the designer focuses on the "abstract" behavior design. Later, the designer can choose the precise signal-bus and coding.

### *1.6.2 Dataflow Style Architecture*

A dataflow architecture models the information or dataflow behavior of combinational logic functions such as adders, comparators, multiplexers, decoders, and other primitive logic circuits. The following example defines the entity and architecture, in a dataflow style, of xor2, an exclusive-OR gate. xor2 has input ports a and b of type bit, and an output port c of type bit. There is also a delay parameter m, which defaults to 1.0 ns. The architecture dataflow gives output c exclusive-OR of a and b after m (1 ns).

```
entity xor2 is

        generic (m: time :=1.0 ns);
        port (a, b: in bit;
                c: out bit);
end xor2;

architecture dataflow of xor2 is
```

```
begin
        c <= a xor b after m;
end dataflow;
```

Once this simple gate is compiled into a library, it can be used as a component in another design by referring to the entity name xor2, and providing three port parameters and, optionally, a delay parameter.

### *1.6.3 Structural Style Architecture*

Top-level VHDL designs use structural style to instance and connect previously compiled designs. The following example uses two gates xor2 (exclusive-OR) and inv (inverter) to realize a simple comparator. The schematic in Figure 1.5 represents a comparator. Inputs in the circuit, labeled a and b, are inputs into first *xor2* gate. The signal wire I from *xor2* connects to the next component *inv*, which provides an output c.

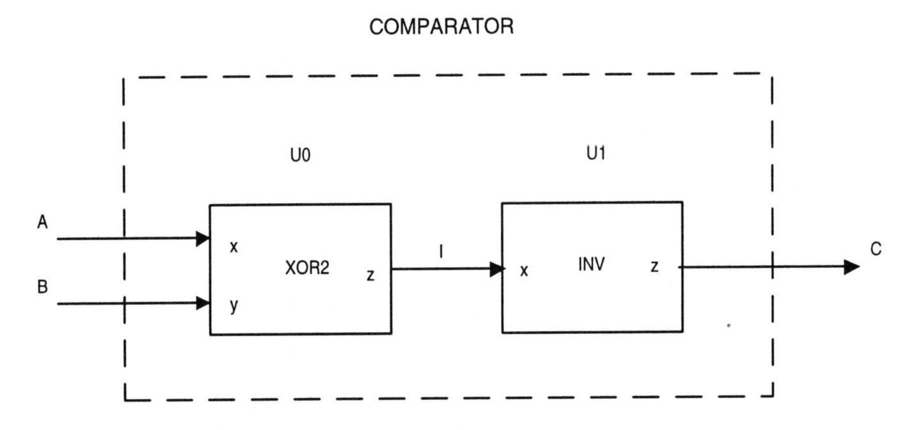


Figure 1.5 Schematic representation of comparator

```
entity comparator is
        port (a, b: in bit; c: out bit);
end comparator;

architecture structural of comparator is

signal i: bit;
component xor2 port (x, y: in bit; z: out bit);
```

```
end component;

component inv port (x: in bit; z: out bit);
end component;

begin
        u0:  xor2 port map (a, b, i);
        u1:  inv port map (i, c);
end structural;
```

The architecture has an arbitrary name structural. Local signal i is declared in the declaration part of architecture. The component declarations are required unless these declarations are placed in a package. Two components are given instance names u0 and u1. The port map indicates the signal connections to be used. The design entities xor2 and inv are found in library WORK, since no library is declared.

Which architecture will be used depends on the accuracy wanted, and whether structural information is required. If the model is going to be used for PCB layout purposes, then probably the structural architecture is most appropriate. For simulation purposes, however, behavioral models are probably more efficient in terms of memory space required and speed of execution.

## 1.7 Configuration

The configuration assists the designer in experimenting with different variations of a design by selecting particular architectures. Two different architectures of the entity andgate, called arch1 and arch2, have been illustrated in Figure 1.3. A configuration selects a particular architecture, for example arch1, from a library. The syntax is:

```
configuration identifier of entity_name is

        specification

end configuration identifier;
```

Different architectures may use different algorithms or levels of abstraction. If the design uses a particular architecture, a **configuration** statement is used. A configuration is a named and compiled unit stored in library. The VHDL source description of a configuration identifies by name other units from a library. For example:

```
configuration alu1_fast of alu is

for alu1;
for u0: comparator use entity work.comparator(dataflow);
```

In this example configuration alu1_fast is created for the entity alu and architecture alu1. The **use** clause identifies a library, entity, and architecture of a component (comparator). The final result is the configuration called alu1_fast. It is a variation of design alu. Configuration statements permit selection of a particular architecture. When no explicit configuration exists, the latest compiled architecture is used (it is called null configuration).

The power of the configuration is that recompilation of the whole design is not needed when using another architecture; instead, only recompilation of the new configuration is needed.

Configuration declarations are optional regardless of the complexity of a design. If configuration declaration is not used, the VHDL standard specifies a set of rules that provide the design with a default configuration. For example, if an entity has more than one architecture, the last architecture compiled will be bound to the entity.

## 1.8 Questions and Problems

1.1. Describe in your own words what are:

- Printed Circuit Board (PCB)
- Multi-Chip Module (MCM)
- Field-Programmable Logic Device (FPLD)
- Application-Specific Integrated Circuit (ASIC)

1.2. Explain the difference between schematic design entry and design entry using hardware description language. Can behavior be described using a schematic diagram?

1.3. How would you describe "reusability" of designs?

1.4. How would you define and describe modeling, design, prototyping, simulation, and synthesis?

1.5. What are the steps in the design and prototyping process?

1.6. What types of modeling are supported in VHDL? Use an example of a digital system for which you can describe all types of modeling.

1.7. What are the major similarities and differences between VHDL and high-level programming languages?

1.8. What are the VHDL design units? On an example of a digital system illustrate in parallel VHDL design units and equivalent descriptions using conventional tools for describing digital systems.

1.9. How are parallel operations described in VHDL?

1.10. What should you consider under "partial recompilation" in VHDL?

# 2 VHDL- LANGUAGE BASIC ELEMENTS

VHDL provides a variety of data types and operators in the package STANDARD that support the methodology of top-down design, using abstractions of hardware in early versions of design. Recent changes in language itself extended standards further, to help synthesis tool users and vendors by standard, portable data types and operations for numeric data, and by clarifying meaning for values in IEEE 1164 data types. In this Chapter we will concentrate on the basic language elements and features, and more advanced features will be introduced in subsequent chapters. Besides standard types and operations, it supports user defined data types that can be included in own user packages.

VHDL is a strongly typed language which assists designers to catch errors early in the development cycle. The compiler's analyzer is very exact and displays the errors for not using the correct data representation.

In many aspects VHDL is like a programming language. It draws its facilities from the familiar programming languages. In this Chapter we describe basic language elements for the reason of convenience.

## 2.1 Literals

A literal is an explicit data value which can be assigned to an object or used within expressions. Although literals represent specific values, they do not always have an explicit type. A scalar is a literal made up of characters or digits or it can be named. There are predefined scalar types, but the user can define other types. The predefined scalar types are:

- character
- bit
- real
- integer

- physical_unit

### *2.1.1 Character Literals*

A character literal defines 1-character ASCII value by using a single character enclosed in single quotes 'x'. Although VHDL is not case sensitive, it does consider case for character literals. Character literal can be any alphabetic letter a-z, digit 0-9, blank, or special character. Examples of character literals are

'X' 'x' '1' '$'

The data type of the object assigned these values dictates whether a given character literal is valid. The same value, for example ‘A’, is a valid literal when assigned to a **character** type object, but is not valid when assigned to a **bit** data type.

### *2.1.2 String Literals*

Literal character strings are collections of one or more ASCII characters and are enclosed in double quote characters. For example:

“the value must be in the range”

They can be assigned to arrays of single-character data types or objects of the built-in type **string**. They are useful when designing test benches around the synthesizable model.

### *2.1.3 Bit and Bit String Literals*

The value of the signal in a digital system is often represented by a bit. A bit literal represents a value by using the character literals '0', or '1'. Bit literals differ from integers or real numbers. Bit data is also distinct from Boolean data, although conversion functions may be implemented. Bit string literal is an array of bits enclosed in double quotes. They are used to represent binary, octal and hexadecimal numeric data values. When representing a binary number, a bit string literal must be preceded by the special character ‘B’, and may contain only the characters ‘0’ and ‘1’. When representing an octal number, the bit string literal must include only characters ‘0’ through ‘7’, and it must be preceded by the special character ‘O’.

When representing a hexadecimal value, the bit string literal may include only characters '0' through '9' and 'A' through 'F', and must be preceded by the special character 'X'. The underscore character '_' may be used within bit string literals to improve readability, but has no effect on the value of the bit string literal. Examples of bit string literals are:

B"011001"
B"1011_0011"
O"1736"
X"EB0F"

Bit literals are used to describe the value of a bus in a digital system. Most simulators include additional bit types representing unknowns, high impedance states, or other electrically related values. In VHDL standard 1076-1987, bit string literals are only valid for the built-in type **bit_vector**, but in 1076-1993 standard they can be applied to any string type including **std_logic_vector**.

### *2.1.4 Boolean Literals*

A Boolean literal represents a true or false value. It has no relationship to a bit. Relational operators like =, <=, >=, and /= produce a Boolean result. Boolean literals are

true TRUE True false FALSE False

A Boolean signal is often used to represent the state of an electronic signal or a condition on a bus.

### *2.1.5 Numeric Literals*

Two basic types of numeric literals are supported in VHDL, real literals and integer literals.

Real literals define a value with real numbers. They represent numbers from -1.0E+38 to +1.0E+38. A real number must always be written with a decimal point. Examples of real literals are:

+1.25

3.4
-2.5

Integer literals define values of integers in the range -2,147,483,647 to +2,147,483,647 (32 bits of precision, including the sign bit), but the instances can be constrained to any subrange of this one. It is not allowed to use decimal point in representing integers. Examples of integers are:

+5
-223
123

When the bit operations are needed, conversion functions from integers to bits must be applied. During design development, integers can be used as an abstraction of a signal bus or may represent an exact specification.

Numeric literals may include underscore character '_' to improve readability.

### *2.1.6 Based Literals*

Based literals are yet another form of integer or real values written in non-decimal form. The based literal is specified with a base specification (such as 2, 8 or 16) and inclose the non-decimal value with a pair or characters '#' as in following examples:

2#1110_1101#
8#3076#
16#AABC#
2#110.1#E12

### *2.1.7 Physical literals*

Physical literal represents a unit of measurement. VHDL allows the use of a number and unit of measure, such as voltage, capacitance, and time. Number must be separated from unit by at least one space. Examples of physical literals are:

1000 ps (picoseconds)
2 min (minutes)
12 v (volts)

Physical literals are most useful in modeling and representing physical conditions during design testing.

### *2.1.8 Range Constraint*

A range constraint declares the valid values for a particular type of signal or variable assignment. It must be compatible with the type it constrains, and be in compatible direction with the original declaration of the type. The syntax is

**range** low_val **to** high_val

In the example below a range constraint is used in port declaration:

**port** (b, a: **in** integer **range** 0 **to** 9 :=0)

### *2.1.9 Comments*

Comments in VHDL start with two adjacent hyphens ('- -') and extend to the end of the line. They have no part in the meaning of a VHDL description.

## 2.2 Objects in VHDL

VHDL includes a number of language elements, called objects, that can be used to represent and store data in the system being modeled. The three basic types of objects used in description of a design or a test bench are signals, variables and constants. Each object has its name and a specific data type, and a unique set of possible data values.

### *2.2.1 Names and Named Objects*

Symbolic names are used for objects in VHDL. A name (identifier) must begin with an alphabetic letter (a-z), followed by a letter, underscore or digit.

Other named objects are architecture names, process names, and entity names. VHDL has over 100 reserved words that may not be used as identifiers.

Examples of names in VHDL are:

my_unit
x_5 X23
my_unit.unit_delay

Examples of invalid names (reserved words) are:

**process**
**in**
**out**
**library**
**map**

Names are usually relative to a named entity and can be selected from a package or a library:

library_name.item_name
package_name.item_name

Named signal in VHDL represents a wire in a physical design. This signal is represented by a stored value during simulation. This allows us to observe changes in a signal value. Named objects are either constant (like fixed value of the signal) or varying in value.

Unlike programming languages, VHDL has two elements that can vary: the variable, which behaves just like a programming language variable, and the signal, which is assigned value at some specific simulated time. The type of variables and signals must be declared in VHDL. There are three object declarations in VHDL:

- constant_declaration
- signal_declaration
- variable_declaration

### *2.2.2 Indexed names*

Variables and signals can be scalars or arrays. Array references can be made to the entire array, to an element, or to a slice of an array. Examples are:

a Array

| | |
|---|---|
| a(5) | Element of array |
| a(1 **to** 5) | Slice of an array |

Arrays are especially useful in documenting a group of related signals such as a bus.

### *2.2.3 Constants*

A constant is name assigned to a fixed value when declared. Constants are useful for creating more readable designs, and make easier to change the design at a later time. If it is necessary to change the value of a constant, it is needed to change the constant declaration in one place. A constant consists of a name, a type, and a value. The syntax is:

**constant** identifier: type_indication [:=expression];

Examples of constant declarations are:

**constant** register: bit_vector (0 to 15):=X"ABCD"
**constant** v: real := 3.6;
**constant** t1: time := 10 ns;

Constants can be declared in a package, in a design entity, an architecture, or a subprogram. Frequently used or shared constants should be declared in a user-defined package. A constant specification can also be used to specify a permanent electrical signal in a digital circuit.

### *2.2.4 Variables*

A variable is a name assigned to a changing value within a process. It is used to store intermediate values between sequential VHDL statements. A variable assignment occurs immediately in simulation, as opposed to a signal that is scheduled in simulated time. A variable must be declared before its use. The syntax is:

**variable** identifier(s): type_indication[constraint][:=expression];

A variable can be given a range constraint or an initial value. The initial value, by default, is the lowest (leftmost) value of range for that type. Examples of variable declarations are:

```
variable alpha: integer range 1 to 90 :=2;
variable x, y: integer;
```

Variables are scalars or arrays that can only be declared in a process or a subprogram. They represent a local data storage during simulation of a process or subprogram. Variables cannot be used to communicate between processes. The important distinctions between variables and signals are covered in more detail in the later sections.

### 2.2.5 Signals

Signals connect concurrent design entities together and communicate changes in values within an electronic design. Signal assignments use simulated time to execute in VHDL. A signal must be declared before it is used. The syntax is:

```
signal identifier: type_indication [constraint] [:=expression]
```

Signals can be declared in an entity, an architecture, or in a package. If signal has to be initialized, it is indicated by literal in [:=expression]. The default initial value is the lowest value of that type. Examples of signal declaration are:

```
signal cnt: integer range 1 to 10;
signal gnd: bit :='0';
signal abus: std_logic_vector (7 downto 0):=(others=>'1");
```

The last signal declaration and initialization statement assigns all signals of the array abus initial value of '1'. Initialization values are commonly ignored by synthesis tools. However, they can be useful for simulation purposes.

Signal value changes are scheduled in simulated time. For example:

```
signal s: bit;
s <= '1' after 2 ns;
```

Signals cannot be declared in a process. If they are used within a process, unexpected results can be obtained because the value assignment is delayed until WAIT statement is executed. They provide global communication in an architecture or entity. Signals are usually used as abstractions of physical wires, busses, or to document wires in an actual circuit.

## 2.3 Expressions

An expression is a formula that uses operators and defines how to compute or qualify the value. The operators must perform a calculation compatible with its operands. Generally, operands must be of the same type. No automatic type conversion is done. In an expression, an operand can be a name, a numeric, or a character literal, but also a function call, qualified expression, type conversion, etc. The result of an expression has a type that depends upon the types of operands and operators.

A summary of VHDL operators is presented in Table 2.1. These operators create expressions that can calculate values. Logical operators, for example, work on predefined types, either bit or Boolean. They must not be mixed. The resulting expression has the same type as the type of operand. Relational operators compare two operands of the same type and produce a Boolean. The result of an expression formed with a relational operator is of type Boolean.

Table 2.1 Operators in VHDL

| Operations | Operators | Description | Operands |
|---|---|---|---|
| logical_operations | and | logical and | same type |
| | or | logical or | same type |
| | nand | complement of and | same type |
| | nor | complement of or | same type |
| | xor | logical exclusive or | same type |
| | not | complement | same type |
| | xnor | logical exclusive nor | |
| relational_operations | = | equal | same type |
| | /= | not equal | same type |
| | < | less than | same type |
| | <= | less than or equal | same type |
| | > | greater than | same type |
| | >= | greater than or equal | same type |

| | | | |
|---|---|---|---|
| arithmetic_operations | +<br>-<br>abs<br><br>+<br>-<br><br><br>*<br>/<br><br>mod<br>rem<br>** | unary plus<br>unary minus<br>absolute value<br><br>addition<br>subtraction<br><br><br>multiplication<br>division<br><br>modulus<br>remainder<br>exponentiation | any numeric<br>any numeric<br>any numeric<br><br>same type<br>same type<br><br><br>same type<br>same type<br><br>same type<br>same type |
| shift_operations (1076-1993 only) | sll<br>srl<br>sla<br>sra<br>rol<br>ror | logical shift left<br>logical shift right<br>arithmetic shift left<br>arithmetic shift right<br>logical rotate left<br>logical rotate right | one-dimensional integer array of bit or boolean |
| concatenation_operation | & | concatenation | same type |

Concatenation is defined for characters, strings, bits, and bit vectors and for all one-dimensional array operands. The concatenation operator builds arrays by combining the operands. For example:

"ABCDEF" & "abcdef" results in "ABCDEFabcdef"
"11111" & "00000" results in "1111100000"

in some cases operators are specifications for a hardware block to be built using logic synthesis tools. A plus (+) corresponds to an adder, and logical operators are models of gates. Table 2.2 lists precedence of operators. Each row represents operators with the same precedence. An operator's precedence determines whether it is applied before or after adjoining operators.

Table 2.2 Operators and their precedence

| Type | | | | | | | | Precedence |
|---|---|---|---|---|---|---|---|---|
| Logical | **and** | **or** | **nand** | **nor** | **xor** | | | Lowest |
| Relational | = | /= | | < | <= | > | >= | |
| Adding | + | - | & | | | | | |
| Unary | + | - | | | | | | |
| Multiplying | * | / | **mod** | **rem** | | | | |
| Misc. | ** | **abs** | **not** | | | | | Highest |

The default precedence level of the operators can be overridden by using the parentheses. More detailed insight to the use of VHDL operators will be covered in the later sections through a number of example designs.

## 2.4 Basic Data Types

VHDL allows the use of variety of data types, from scalar numeric types, to composite arrays and records, or file types. In the preceding chapters we have introduced the basic data types and objects supported by VHDL, particularly:

- signals, that represent interconnection wires that connect component instantiation ports together
- variables, that are used for local storage of temporary data visible only inside a process, and
- constants, that are used to name specific values

All these objects can be declared using a type specification to specify the characteristics of the object. VHDL contains a wide range of types that can be used to create objects. To define a new type, a type declaration must be used. A type declaration defines the name of the type and the range of the type. Type declarations are allowed in package declaration sections, entity declaration sections, architecture declaration sections, subprogram declaration sections, and process declaration sections.

The four broad categories of the types available in VHDL are

- Scalar types, that represent a single numeric value. The standard types belonging to this class are integer, real, physical, and enumerated types.
- Composite types, that represent a collection of values. There are two classes of composite types: arrays which contain elements of the same type, and records which contain elements of different types.
- Access types, that provide references to objects similar to the pointers used to reference data in programming languages.
- File types, that reference objects that contain a sequence of values (for example, disk files)

Each type in VHDL has a defined set of values. In most cases the designer is interested only in a subset of the possible values of specific type. VHDL provides a mechanism to specify a constraint in the declaration of an object. For example, declaration

```
signal data12: integer range 0 to 4095;
```

specifies that signal data12 can take values of unsigned positive integer values 0 through 4095..

Similarly, VHDL provides **subtype** mechanism for creation of an alternate data type that is a constrained version of an existing type. For example, the declaration

```
subtype data16 integer range 0 to 2**16-1;
```

creates a scalar type with a limited range. The subtype data16 carries with it all operations available for the integer base type.

Basic data types have been already introduced in an informal way. They belong to scalar types represented by a single value, and are ordered in some way that relational operators can be applied to them. Table 2.3 lists the built-in scalar types defined in VHDL Standard 1076.

Table 2.3 Built-in scalar types that allow application of relational operators

| Data Type | Example of Values | Comment |
|---|---|---|
| Bit | '0', '1' | Defined as an enumerated type. |
| Character | '0', '1', 'a', 'b', '#', etc. | ISO 8859-1 character set. |
| Boolean | False, True | Defined as an enumerated type. |
| Integer | 1, -56, 128 | Minimum range of -2147483647 to +2147483647. |
| Real | 1.28, -23.5E10 | Minimum range of -1.0E38 to +1.0E38 |
| Severity_error | NOTE, ERROR | Enumerated type used in severity clause report. |
| Time | 10 ms | A physical type. |

### *2.4.1 Bit Type*

The bit type is used in VHDL to represent the most fundamental objects in a digital system. It has only two possible values, '0' and '1', that are usually used to represent logical 0 and 1 values in a digital system. The following example uses bit data type to describe the operation of a 2-to-4 decoder:

```
entity decoder2to4 is

    port (a: in bit;
    b: in bit;
    d0: out bit;
    d1: out bit;
    d2: out bit;
    d3: out bit)
end decoder2to4;

architecture concurrent of decoder2to4 is
```

```
begin
        d0 <= 1 when b = '0' and a = '0' else '0';
        d1 <= 1 when b = '0' and a = '1' else '0';
        d2 <= 1 when b = '1' and a = '0' else '0';
        d3 <= 1 when b = '1' and a = '1' else '0';
end concurrent;
```

The bit data type supports logical and relational operations. The IEEE 1164 specification, which is now commonly used, describes an alternative to bit called std_ulogic. Std_ulogic is defined as an enumerated type that has nine possible values, allowing a more accurate description of values and states of signals in a digital system. A more detailed presentation of IEEE 1164 standard logic specification is given in the later sections of this Chapter.

### *2.4.2 Character Type*

This type is similar to the character types in programming languages. Characters can be used to represent actual data values in design descriptions, but more often they are used to represent strings and to display messages during simulation. However, characters in VHDL are defined as enumerated type and have no explicit value, and, therefore, they cannot be simply mapped onto numeric data types. In order to do that, type conversion functions are required. The character type defined in the 1076-1987 package is:

```
type character is (

NUL, SOH, STX, ETX, EOT, ENQ, ACK, BEL,
BS, HT, LF, VT, FF, CR, SO, SI,
DLE, DC1, DC2, DC3, DC4, NAK, SYN, ETB,
CAN, EM, SUB, ESC, FSP, GSP, RSP, USP,

' ', '!', '"', '#', '$', '%', '&', ''',
'(', ')', '*', '+', ',', '-', '.', '/',
'0', '1', '2', '3', '4', '5', '6', '7',
'8', '9', ':', ';', '<', '=', '>', '?',

'@', 'A', 'B', 'C', 'D', 'E', 'F', 'G',
'H', 'I', 'J', 'K', 'L', 'M', 'N', 'O',
'P', 'Q', 'R', 'S', 'T', 'U', 'V', 'W',
'X', 'Y', 'Z', '[', '\', ']', '^', '_',
```

```
'`', 'a', 'b', 'c', 'd', 'e', 'f', 'g',
'h', 'i', 'j', 'k', 'l', 'm', 'n', 'o',
'p', 'q', 'r', 's', 't', 'u', 'v', 'w',
'x', 'y', 'z', '{', '|', '}', '~', DEL);
```

The IEEE 1076-1993 specification extends the character set to the 256-character ISO 8859 standard.

### *2.4.3 Boolean Type*

The Boolean type is defined as an enumerated type with two possible values, True and False. It is a result of a logical test which is using relational operators or can be the result of an explicit assignment.

### *2.4.4 Integer Type*

The predefined integer type includes all integer values in range of -2147483647 to +2147483647, inclusive. New integer constrained subtypes can be declared using **subtype** declaration. The predefined subtype natural restricts integers to the range of 0 to the specified (or default) upper range limit, and predefined subtype positive restricts integers to the range of 1 to the specified upper limit:

```
type integer is range -2147483647 to 2147483647;
subtype natural is integer range 0 to 2147483647;
subtype positive is integer range 1 to 2147483647;
```

IEEE Standard 1076.3 defines an alternative to the integer type defining signed and unsigned data types, which are array types that have properties of both array and numeric data types. They allow to perform shifting and masking operations like on arrays, and arithmetic operations, like on integers. These types are presented in more details in subsequent sections.

In order to illustrate the use of integer data type consider example of 2-to-4 decoder as given below:

```
entity decoder2to4 is
port (x: in integer range 3 downto 0;
        d0: out bit;
        d1: out bit;
```

```
        d2: out bit;
        d3: out bit)
end entity decoder24;

architecture second of decoder2to4 is

begin
        d0 <= 1 when x=0;
        d1 <= 1 when x=1;
        d2 <= 1 when x=2;
        d3 <= 1 when x=3;
end architecture second;
```

In this example, the input port of the decoder is declared as constrained integer. The description of the decoder behaviour is simplified, and the checks of the input values are left to the VHDL compiler.

### *2.4.5 Real Types*

Real types have little use due to the fact that synthesizers do not support this type. They are primarily used for simulation purposes allowing to declare objects of this type and assign them real values in the specified range of -1.0E38 to +1.0E38. The real type supports arithmetic operations.

### *2.4.6 Severity_Level Type*

Severity_Level type is a data type used only in the report section of an **assert** statement. It is an enumerated type with four possible values that can be assigned to the objects of this type: note, warning, error and failure. These values may be used to control simulation process to indicate simulator to undertake an action if certain specific conditions appear.

### *2.4.7 Time Type*

Time data type is built-in VHDL data type used to measure time. Time has units of measure which are all expressed as multiples of a base unit, fentosecond (fs). The definition for type time might be as follows:

```
type time is range of  -2147483647 to +2147483647
        units
                fs;
                ps = 1000 fs;
                ns = 1000 ps;
                us = 1000 ns;
                ms = 1000 us;
                sec = 1000 ms;
                min = 60 sec;
                hr = 60 min;
        end units;
```

## 2.5 Questions and Problems

2.1. Explain what is a strongly typed language.

2.2. What are the literal types supported in VHDL? Explain the difference between the following literals:

1, '1', 1.0

2.3. What are the basic data types supported in VHDL?

2.4. Why is VHDL object-oriented language? What are the objects supported in VHDL?

2.5. What is the difference between variable and signal.

2.6. Two parts of a digital system communicate using TRANSFER signal that enables transfer of 20,000 different values. Show at least two data types that enable description of this signal.

2.7. What are the similarities and differences between bit and Boolean data type?

2.8. What is the physical data type useful for? Explain it on the example of time physical type.

2.9. Is the real type synthesizable? Explain it.

# 3 ADVANCED DATA TYPES

The advanced data types include enumerated types that allow for identifying specified values for a type and for subtypes, which are variations of existing types. There are composite types that include arrays and records. And there are predefined data types, text and lines, that facilitate text input and output operations. These types and functions are provided in package TEXTIO, which is contained in library STD. The advanced data types are:

1. Extended scalar types
   - enumerated types
   - subtypes
   - physical types
2. Composite types
   - arrays
   - records
3. Access types equivalent to pointers in typical programming languages
4. Other predefined types
   - files
   - lines

The extensible type facility of VHDL is helpful in modeling behaviour at a more abstract level and can also be used to specify very detailed behaviour.

## 3.1 Extended Types

As we have already seen, the VHDL language does not include many built-in types for signals and variables, but allows users to add new data types. The Package STANDARD, included in every implementation, extends the language to allow data types useful for description of hardware. These types include boolean, bit, bit_vector, character, string, and text. For example, declaration of bit type is:

**type** bit **is** ('0', '1');

It enumerates two possible values of type bit. However, in most environments, few more logical strengths, such as unknown, high impedance, weak 1, weak 0, are needed. Some vendors have up to 47 signal strengths.

To extend the available data types, VHDL provides a type-declaration capability and a package facility to save and deliver these new data types. VHDL also provides overloaded operators so that the use of new data types is natural and easy.

### *3.1.1 Enumerated Types*

As shown in preceding sections, enumerated types are used to describe many of the standard VHDL data types. They can be used to describe unique data types and make easier the description of a design. The enumerated type declaration lists a set of names or values defining a new type:

**type** identifier **is** (enumeration_literal {,enumeration_literal});

where enumeration_literal can be identifier or character_literal. This allows us to declare a new type using character literals or identifiers. For example, using identifiers we can declare:

**type** colors **is** (black, white, red};

The example identifies three different values in a particular order that define type colors. In subsequent declarations of a variable or signal designated type colors, assigned values could only be black, white, and red.

In some applications it is convenient to represent codes symbolically by defining own data type. In the following example data type fourval is defined using literals:

```
type fourval is ('0', '1', 'z', 'x');
```

If we declare type fourval in design, then we can declare ports, signals, and variables of this type.

Below is an example of a simplified CPU which is using enumerated type instruction_code to be the operation codes lda, ldb, sta, stb, aba, and sba. The CPU has two working registers a and b which are used to store operands and results of operations:

```
architecture behavior of simple_cpu is
type instruction_code is (aba, sba, sta, stb, lda, ldb);
begin process

variable a, b, data: integer;
variable instruction: instruction_code;
begin
          .............
          case instruction is
                    when lda => a:= data;
                    when ldb => b:= data;
                    when sta => data:=a;
                    when stb => data:=b;
                    when aba => a:= a + b;
                    when sba => a:= a - b;
          end case;
wait on data;
end process;
end behavior;
```

The only values that variable instruction can take on are enumerated values of the type instruction_code. Some extensions to VHDL allow to assign the numeric encodings, for example in the later stages of a top-down design.

Enumerated types provide through abstraction and information hiding a more abstract design style often referred to as object oriented. For example, they allow to observe symbolic type names during simulation process, or defer the actual encoding of the symbolic values until the time when the design is implemented in hardware.

### *3.1.2 Qualified Expressions*

If there is ambiguity in using the specific values in terms of its type, it is necessary to do typecasting to be explicit on the type of the value. The type is cast in VHDL by using a qualified expression. For example:

```
type name is (alpha, betta);
```

When a type has a shared value with the other types, the type can be clarified by using qualified expression with the following syntax:

```
type' (literal or expression)
```

for example

```
name'(alpha)
```

It is sometimes necessary to map one data type to another. A variable or signal can be converted by using conversion function. Assume that we have two types: incoming value of type threeval, and we want to convert it to outgoing value named value3:

```
type threeval is ('l', 'h', 'z');
type value3 is ('0', '1', 'z')
function convert (a: threeval) return value3 is
begin
        case a is
                when 'l' => return '0';
                when 'h' => return '1';
                when 'z' => return 'z';
        end case
end convert;
```

An example of the call to a conversion function is given below:

```
process (inp);
        variable inp: threeval;
        variable outp: value3;
begin
        outp := convert(inp);
end process;
```

### *3.1.3 Physical Types*

Physical types are used to represent physical quantities such as time, distance, current, etc. A physical type provides for a base unit, and successive units are then derived from the base unit. The smallest unit representable is one base unit; the largest is determined by the range specified in the physical type declaration.

An example of user-defined physical type follows:

```
type voltage is range 0 to 20000000
units
        uv;                 -- micro volts
        mv = 1000 uv;       -- milivolts
        v = 1000 mv;        -- volts
end units;
```

The type definition begins with a statement that declares the name of the type (voltage) and the range of the type in base units (0 to 20,000,000). The first unit declared is the base unit. After the base unit is defined, the other units can be defined in terms of the base unit or the other units already defined. The unit identifiers all must be unique within a single type.

## 3.2 Composite Types - Arrays

VHDL provides array as a composite type, containing many elements of the same type. These elements can be scalar or composite. They can be accessed by using an index. The only predefined array types in the Package STANDARD are bit_vector and string. New types have to be declared for real and integer arrays.

Access depends upon declaration. For example:

```
variable c: bit_vector (0 to 3);
variable d: bit_vector (3 downto 0);
```

In this example the indices for variable c are 0 for leftmost bit c(0) and 3 for the rightmost bit c(3); for variable d, 3 is the index for leftmost bit d(3), and 0 is the index for rightmost bit d(0). VHDL has no particular standard for the ordering of bits or the numbering scheme. One can number from 1 to 4, or 4 to 7, etc.

Examples of valid bit_vector assignments are given below:

```
c := "1100";
d := ('1', '0', '1', '0');
d := a & b & f & g;
```

In the last case a, b, f, and g must be 4 1-bit single variables concatenated by ampersand (&).

VHDL allows an access to the slice of an array that defines its subset. For example:

```
variable c: bit_vector (3 downto 0);
variable d: bit_vector (7 downto 0);

d(7 downto 4) := c;
```

Four bits of c are assigned to upper four bits of d. Any subrange or slice must declare subscripts in the same direction as initially declared.

### *3.2.1 Agregates*

An array reference can contain a list of elements with both positional and named notation, forming a typed aggregate. The syntax is:

```
type_name' ([choice=>] expression {, [others =>] expression})
```

where type_name can be any constrained array type. The optional choice specifies an element index, a sequence of indeces, or [**others**=>]. Each expression provides values for the chosen elements and must evaluate to a value of the element's type. An element's index can be specified by using positional or named notation. Using positional notation, each element is given the value of its expression:

```
variable x: bit_vector (1 to 4);
variable a, b: bit;

x := bit_vector' ('1', a and b, '1', '0');
x := (1 => '1', 3 => '1', 4 => '0', 2 => a and b);
```

An aggregate can use both positional and named notation, but positional expressions must come before any named [**choice** =>] expressions. If some values are not specified they are given a value by including [**others** =>] expression as the last element of the list. An example is given below:

```
variable b: bit;
variable c: bit_vector (8 downto 1)

c := bit_vector' ('1', '0', b, others => '0');
```

Eight bits on the right side come from various sources. The symbol => is read as "gets".

### *3.2.2 Array Type Declaration*

The syntax used to declare a new type that is an array is:

```
type array_name is array [index_constraint] of element_type
```

where index_constraint is:

```
[range_spec]
index_type range [range_spec]
index_type range <>
```

Examples of array type declarations are:

```
type byte is array (0 to 7) of bit;
type ram is array (0 to 7, 0 to 255) of bit;
```

After a new type is declared, it can be used for signal or variable declaration:

```
variable word: byte;
```

An enumerated type or subtype also can be used to designate the range of subscript values:

```
type instruction is (aba, sba, lda, ldb, sta, stb);
subtype arithmetic is instruction range aba to sba;
subtype digit is integer range 1 to 9;

type ten_bit is array (digit) of bit;
type inst_flag is array (instruction) of digit;
```

Hardware systems frequently contain arrays of registers or memories. Two-dimensional arrays can be useful for simulating RAMs and ROMs. VHDL allows multiple-dimensional arrays. A new array type must be declared before we declare own variable or signal array. For example:

```
type memory is array (0 to 7, 0 to 3) of bit;
constant rom: memory :=       ( ('0', '0', '1', '0'),
                              ('1', '1', '0','1'),
                              ('0', '0', '1','0'),
                              ('1', '1', '1','1'),
                              ('0', '0', '1', '1'),
                              ('0', '1', '1', '0'),
                              ('1', '0', '1', '0'),
                              ('1', '0', '1', '1'));
cont := rom(2, 1);
```

Multiple-dimensional arrays are not generally supported in synthesis tools, but can be useful for simulation purposes for describing test stimuli, memory elements, or other data that require tabular form. VHDL also allows declaration of array of arrays. Always array must be declared before a variable or signal of that type are declared.

Sometimes it is convenient to declare a new type (subtype) of an existing array type. For example:

```
subtype byte is bit_vector (7 downto 0);
```

Variable and signal declarations can now use the subtype:

```
variable alpha: byte;
```

### 3.3 Composite Types - Records

Records group objects of different types into a single object. These elements can be of scalar or composite types and are accessed by name. They are referred to as fields. Each field of record can be referenced by name. The pariod "." is used to separate record names and record element names when referencing record elements. An example of record declaration and use of record and its elements is shown below:

```
type two_digit is
        record sign: bit;
               msd:     integer range 0 to 9;
```

```
            lsd:     integer range 0 to 9;
        end record;

    process;
    variable cntr1, cntr2: two_digit;
    begin
        cntr1.sign := '1';
        cntr1.msd := 1;
        cntr1.lsd := cntr1.msd;
        cntr2 := two_digit' ('0', 3, 6);
    end process;
```

Records are not generally synthesisable, but they can be useful when describing test stimuli for simulation purposes.

### *3.3.1 Alias Declaration*

An alias is an alternate name assigned to part of an object, which allows simple access. For example, a 9-bit bus count has three elements: a sign, the msd, and lsd. Each named element can be operated using an alias declaration:

```
    signal count: bit_vector (1 to 9);
    alias sign: bit is count (1);
    alias msd: bit_vector (1 to 4) is count (2 to 5);
    alias lsd: bit_vector (1 to 4) is count (6 to 9);
```

The examples of accesses are:

```
    sign := '0';
    msd := "1011"
    count := "0_1110_0011"
```

## 3.4 Other Advanced Types

The other advanced types are briefly introduced in the following subsections.

### *3.4.1 Access Types*

Access types are rarely used by hardware designers. They are very similar to a pointer in a language like Pascal or C. It is an address, or a handle, to a specific object. Access types allow the designer to model objects of a dynamic nature, for example dynamic queues, stacks, fifos, etc.

Only variables can be declared as access types. By their nature, access types can be used only in sequential processing. They are currently not synthesizable. When an object is declared to be of an access type, two predefined functions, named **new** and **deallocate**, are automatically available to manipulate the object. Function **new** will allocate the memory of the size of the object in bytes and return the access value. Function **deallocate** takes in the access value and returns the memory to the system.

### *3.4.2 Text and Lines*

Two data types that are predefined in VHDL are type text and type line. Text and lines are used for input and output operations during simulation. These types are used in a file declaration. They are used with read, write, and end-of-file functions. Text files and lines are used in processes and subprograms. They allow reading and writing of ASCII text files. Both provide formatted input and output. Files of type text are treated as groups of lines.

In the library STD, a number of functions and types are predefined in Package TEXTIO. These functions allow to bring in and send out data between files and variables. The input procedures in the Package TEXTIO are readline, which reads a line of text out of a file, and read, which reads an item off particular line. The output procedures in the TEXTIO package are writeline and write. Examples of read operations are:

readline (f: **in** text; l: **out** line);

which reads a line from the file f into variable l, and

read (l: **inout** line; item: digit);

which reads an item out of line l into variable item.

Reading data from a file is a two-stage operation. For example:

```
a: process ...
file testvecotors: text is in "test.vec";
variable l: line;
variable av, bv : bit_vector (3 downto 0);
begin
        readline (testvectors, l);
        read (l, av);
        bv := av;
end process;
```

The logical file testvectors of type text that is called test.vec in file system is used within a process a. There is variable l of type line and four-bit variables av and bv. The operation after the begin reads a line from the testvectors file into the variable l. The subsequent read inputs the first four-bit value of line l and puts it into variable av.

Before using TEXTIO, it is needed to create a reference to this library package:

```
use std.textio.all
```

In addition, there are two functions which indicate end-of-file and end-of-line conditions by returning Boolean values:

```
endfile (filename) and
endline (filename).
```

Similarly, there is a write function that writes data contained in the object to the specified file:

```
write (filename, object)
```

In VHDL 1076-1993, file types and associated functions and procedures were modified to allow files to be opened and closed as needed.

## 3.5 Symbolic Attributes

VHDL has symbolic attributes that allow a designer to write more generalized code. Some of these attributes are predefined in VHDL, others are provided by CAD vendors. The designer can also define his own attributes. Attributes are related to arrays, types, ranges, position, and signal characteristics.

The following attributes work with arrays and types:

- aname'left returns left bound of index range
- aname'right returns right bound of index range
- aname'high returns upper bound of index range
- aname'low returns lower bound of index range
- aname'length returns the length (number of elements) of an array
- aname'ascending (VHDL '93) returns a Boolean true value of the type or subtype if it is declared with an ascending range

where character " ' " designates a separator between the name and the attribute. If the numbers that designate the lower and upper bounds of an array or type change, no change in the code that uses attributes. Only declaration portion should be changed. In the multirange arrays attributes are specified with the number of the range in the parentheses. For example, for array:

```
variable: memory  (0 to 10) of bit;

memory'right
```

will give value 10 because the second range has the index equal to 10.

Similarly array length attribute returns the length of an array:

```
a := memory'length ;
```

and a has a value of 11. The length of an array can be specified symbolically rather than with a numeric value.

An example of the use of function array attributes implementing an integer-based RAM device with 1024 integer locations and two control lines follows:

```
package package_ram is
        type t_ram_data is array (0 to 1023) of integer;
        constant x_val: integer := -1;
        constant z_val: integer := -2;
end package_ram;
use work.package_ram.all;
```

```
use work.std_logic_1164.all;

entity ram_1024 is
        port (data_in, addr: in integer;
         data_out: out integer;
         cs, r_w: in std_logic);
end ram_1024;

architecture ram of ram_1024 is

begin
process (cs, addr, r_w)
        variable ram_data: t_ram_data;
        variable ram_init: boolean := false;
        begin
                if not(ram_init) then
                 for i in ram_data'low to ram_data'high loop
                         ram_data(i) := 0;
                 end loop;
                        ram_init := true;
                end if;
                if (cs = 'x') or (r_w = 'x') then
                        data_out <= x_val;
                elsif (cs = '0') then
                        data_out <= z_val;
                elsif (r_w = '1') then
                        if (addr=x_val) or (addr=z_val) then
                                data_out <= x_val;
                        else
                                data_out <= ram_data(addr);
                        end if;

                else
                        if (addr=x_val) or (addr=z_val) then
                         assert false
                        report "writing to unknown address"
                                severity error;
                                data_out <= x_val;
                        else
                                ram_data(addr) := data_in;
                                data_out <= ram_data(addr);
                        end if;
```

```
                end if;
            end process;
        end ram;
```

This model contains an **if** statement that initializes the contents of the RAM to a known value. A Boolean variable ram_init keeps track of whether the RAM has been initialized or not. The first time the process is executed, variable ram_init will be false, and if statement will be executed, and the locations of the RAM initialized to the value 0. Setting the variable ram_init to true will prevent the initialization loop from executing again. The rest of the model implements the read and write functions based on the values of addr, data_in, r_w, and cs.

The range attribute returns the range of an object. The name'range and name'reverse_range are used to return the range of particular type in normal or reverse order. The best use of these attributes is when we actually do not know the length of an array, and varying sizes are provided.

Another use of symbolic attributes is for enumerated types. Enumerated type has the notion of successor and predecessor, left and right of the position number of the value:

- typename'succ (v) returns next value in type after v
- typename'pred (v) returns previous value in type before v
- typename'leftof (v) returns value immediatelly to left of v
- typename'rightof (v) returns value immediatelly to right of v
- typename'pos (v) returns type position number of v
- typename'val (p) returns type value from position value p
- typename'base returns base type of type or subtype.

Example below explains the usage of symbolic attributes in enumerated types:

```
type color is (red, black, blue, green, yellow);
subtype color_ours is color range black to green;
variable a: color;

a:= color'low
a:= color'succ (red);
a:= color_ours'base'right;
```

```
a:= color_ours'base'succ (blue);
```

Assignment statements assign to variable a following values:

- red
- black
- yellow
- green

respectively.

Signal attributes work on signals. they provide information about simulation time events:

- signalname'event returns true if an event occured this time step
- signalname'active returns true if a transaction occured this time step
- signalname'last_event returns the elapsed time since the previous event transaction
- signalname'last_value returns previous value of signal before the last event transition
- signalname'last_active returns time elapsed since the previous transaction occurred

Signal attributes allow designer to do some complicated tests. For example:

```
entity dff is
        port (d, clk: in std_logic; q: out std_logic);
end dff;

architecture dff_1 of dff is

begin
        process (clk)
        begin
                if (clk = '1') and (clk'event)
                and (clk'last_value = '0') then q <= d;
                end if;
```

```
        end process;
    end dff_1;
```

the process tests if clk is '1' and clk'event, which means the clock is changed to '1'. If the last previous value of clock is zero, then we have a true rising edge.

Attribute 'last_event is very useful for implementing timing checks, such as setup checks, hold checks, and pulse-width checks. consider example of a setup time and hold time in Figure 3.1. the rising edge of signal clock is the reference edge to which

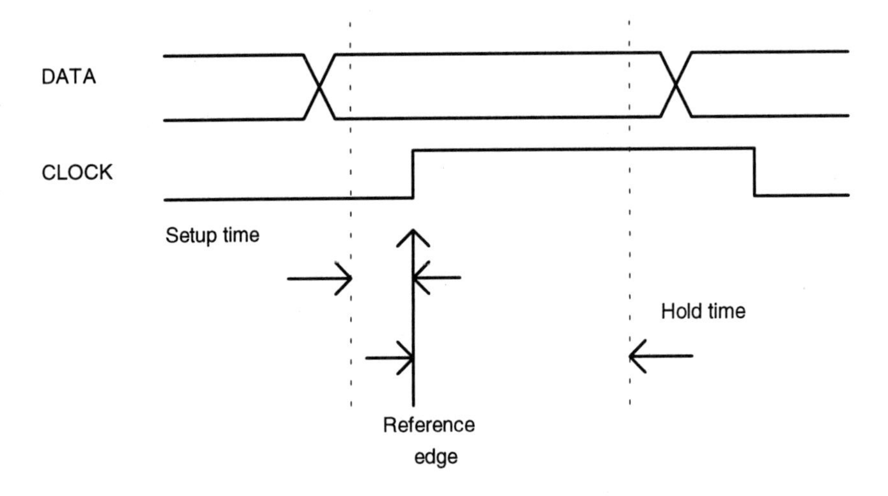


Figure 3. 1 Setup and hold time illustration

all checks are performed. a setup-time check will guarantee that the data input does not change during the setup time, and the hold-time check will guarantee that the data input does not change during the time equal to the hold time after the reference edge. This will ensure correct operation of the device. An example of the setup-check using 'last_event follows:

```
library ieee;
use work.std_logic_1164.all;

entity dff is
        generic (setup_time, hold_time: time);
        port (d, clock: in std_logic;
                q: out std_logic);
```

```
begin
setup_check: process (clock);
begin
if (clock = '1') and (clock'event) then
        assert (d'last_event >= setup_time)
        report "setup violation"
        severity error;
end if;
end process setup_check;
end dff;

architecture dff_1 of dff is

begin
        dff_process: process (clock)
        begin
                if (clock = '1') and (clock'event) then
                        q <= d;
                end if;
        end process dff_process;
end dff_1;
```

The setup_check procedure is contained in a passive process in the entity for the dff model. In this way it can be shared among any architecture of the entity. The passive process is executed for each event on signal clock. When the clock input has a rising edge, the **assert** statement will be executed and perform the check for a setup violation. Attribute d'last_event will return the time since the most recent event on signal d. If the time returned is less than the setup time (generic one), the assertion will fail and report a violation.

It should be noted that all event oriented attributes, except 'event, are not generally supported in synthesis tools.

Another group of signal attributes create special signals that have values and types based on other signals. These special signals can be used anywhere in the design description where a normally declared signal could be used. Special kind signal attributes are:

- aname'delayed(time), that creates a delayed signal that is identical in the waveform to the signal the attribute is applied to.

- aname'stable(time), that creates a signal of type Boolean that becomes true when the signal is stable (has no events) for a given period of time
- aname'quiet(time), that creates a signal of type Boolean that becomes true when the signal has no transactions (scheduled events) for a given period of time.
- aname'transaction, that creates a signal of type bit that toggles its value whenever a transaction or actual event occurs on the signal the attribute is applied to.

There are two additional attributes that return value and can be used to determine information about blocks or attributes in a design. The 'structure attribute returns true if there are references to the lower-level components, and false if there are no references to lower-level components. The 'behavior attribute returns false if there are references to lower-level components; otherwise it returns true. The prefix to both these attributes must be an architecture name.

VHDL 1076-1993 adds three new attributes that can be used to determine the configuration of entities in a design description. For more information about these attributes, refer to the IEEE VHDL Language Reference Manual.

## 3.6 Standard Logic

After 1076 standard, two other IEEE standards, 1164 and 1076.3, were introduced adding important capabilities for both simulation and synthesis.

### *3.6.1 IEEE Standard 1164*

One of the serious limitations of the first release of VHDL was the lack to provide multiple values (for example high-impedance, unknown, etc.) to be represented for a wire. These metalogic values are important for accurate simulation. To solve this problem, simulation vendors invented their own proprietary data types using enumerated types. Those proprietary data types were having four, seven or even thirteen unique values. IEEE 1164 is a standard logic data type with nine values as shown in Table 3.1.

Table 3.1 IEEE 1164 standard logic data type

| Value | Description |
|---|---|
| 'U' | Uninitialized |
| 'X' | Unknown |
| '0' | Logic 0 (driven) |
| '1' | Logic 1 (driven) |
| 'Z' | High impedance |
| 'W' | Weak 1 |
| 'L' | Logic 0 (read) |
| 'H' | Logic 1 (read) |
| '-' | Don't-care |

Having these nine values, it becomes possible to accurately model the behavior of a digital system during simulation. However, the standard is also valuable for synthesis purposes because it enables modeling of circuits that involve output enables, as well as to specify don't care logic that is used to optimize the combinational logic.

There are many situations in which it becomes useful to use IEEE 1164 standard logic. For example, if we want to observe during simulation behavior of the system when we apply to the inputs other values than '0' and '1', or if we want to check what happens when the input with an unknown or don't care value is applied. The resolved standard logic data types can be used to model the behavior of multiple drivers in a circuit. More about resolved types and resolution functions is given in Chapter 7.

However, the most important reason to use standard logic data types is to provide portability between models written by different designers, or when moving models and designs between different simulation and synthesis environments.

Two statements are added to the beginning of source VHDL files to describe that standard logic types will be used. Those two statements are found in the most of our previous examples:

```
library ieee;
use ieee.std_logic_1164.all;
```

If the source file contains several design units, the use clause has to be placed prior to each design unit. The exception is architecture declaration. If the corresponding entity declaration includes a use statement, then the use statement need not to be used before architecture declaration. These two statements are used to load the IEEE 1164 standard library and its contents (the std_logic_1164 package).

## Standard Logic Data Types

The std_logic_1164 package provides two fundamental data types, std_logic and std_ulogic. These two data types are enumerated types defined with nine symbolic values. The std_ulogic type is defined in the IEEE 1164 standard as:

```
type std_ulogic is ('U', -- Uninitialized
                    'X', -- Forcing Unknown
                    '0', -- Forcing 0
                    '1', -- Forcing 1
                    'Z', -- High Impedance
                    'W', -- Weak Unknown
                    'L', -- Wcak 0
                    'H', -- Weak 1
                    '-', -- Don't care
                     );
```

The std_ulogic data type is an unresolved type. It does not allow for two values to be simultaneously driven onto a signal of type std_ulogic. If two or more values can be driven onto a wire, another type, called std_logic, has to be used. The std_logic data type is a resolved type based on std_ulogic and has the following definition:

```
subtype std_logic is resolved std_ulogic:
```

Resolved types are declared with resolution functions, which define behavior when an object is driven with multiple values simultaneously. In the case of multiple drivers, the nine values of std_logic are resolved to values as indicated in Table 3.2.

Table 3.2 Resolution of std_logic data type

| | 'U' | 'X' | '0' | '1' | 'Z' | 'W' | 'L' | 'H' | '-' |
|---|---|---|---|---|---|---|---|---|---|
| 'U' | 'U' | 'U' | 'U' | 'U' | 'U' | 'U' | 'U' | 'U' | 'U' |
| 'X' | 'U' | 'X' | 'X' | 'X' | 'X' | 'X' | 'X' | 'X' | 'X' |
| '0' | 'U' | 'X' | '0' | 'X' | '0' | '0' | '0' | '0' | 'X' |
| '1' | 'U' | 'X' | 'X' | '1' | '1' | '1' | '1' | '1' | 'X' |
| 'Z' | 'U' | 'X' | '0' | '1' | 'Z' | 'W' | 'L' | 'H' | 'X' |
| 'W' | 'U' | 'X' | '0' | '1' | 'W' | 'W' | 'W' | 'W' | 'X' |
| 'L' | 'U' | 'X' | '0' | '1' | 'L' | 'W' | 'L' | 'W' | 'X' |
| 'H' | 'U' | 'X' | '0' | '1' | 'H' | 'W' | 'W' | 'H' | 'X' |
| '-' | 'U' | 'X' | 'X' | 'X' | 'X' | 'X' | 'X' | 'X' | 'X' |

Both these standard logic types may be used as one-to-one replacement for the built-in type bit. The following example shows how std_logic type may be used to describe a simple 2-to-4 decoder coupled to an output enable:

```
library ieee;
use ieee.std_logic_1164.all;

entity decoder is
        port (a, b, oe: in std_logic;
                y0, y1, y2, y3: out std_logic);
end entity decoder;

architecture arch1 of decoder is
        signal s0, s1, s2, s3: std_logic;
begin
        s0 <= not(a) and not(b);
        s1 <= a and not(b);
        s2 <= not(a) and b;
        s3 <= a and b;
        y0 <= s0 when oe='0' else 'Z';
        y1 <= s1 when oe='0' else 'Z';
        y2 <= s2 when oe='0' else 'Z';
        y3 <= s3 when oe='0' else 'Z';
end architecture arch1;
```

In addition to the single-bit data types std_logic and std_ulogic, IEEE standard 1164 includes array types corresponding to each of these types. The std_logic_vector and std_ulogic_vector are defined in the std_logic_1164 package as unbounded arrays similar to the built-in type bit_vector with the following definitions:

```
type std_ulogic_vector is array (natural range <>) of std_ulogic;
type std_logic_vector is array (natural range <>) of std_logic;
```

In actual models or designs, the user will use an explicit width or will use a subtype to create a new data type on std_logic_vector or std_ulogic_vector with the required width. The following example shows the use of a new subtype (defined in an external package) to create a 16-bit array based on std_logic_vector:

```
library ieee;
use ieee.std_logic_1164.all;

package new_type is
        subtype word is std_logic_vector(15 downto 0);
end package new_type;

use ieee.std_logic_1164.all;
entity word_xor is
        port(a_in, b_in: in word; oe: in std_logic;
                c_out: out word)
end entity word_xor;

architecture arch1 of word_xor is
        signal int: word;
begin
        int <= a_in xor b_in;
        c_out <= int when oe='0' else 'ZZZZ_ZZZZ_ZZZZ_ZZZZ';
end architecture arch1;
```

In this example a new subtype word is defined as 16-element array of std_logic_vector. The width of the word_xor circuit is defined in the package new_type, and easily can be modified. There is no need to modify the rest of description of the circuit.

If the designer needs to simplify operations on standard logic data, for example to use 3-, 4-, or 5-valued logic, the std_logic_1164 package contains the following subtypes:

```
subtype X01 is resolved std_ulogic range 'X' to '1'; --('X','0','1')
subtype X01Z is resolved std_ulogic range 'X' to 'Z'; --('X','0','1','Z')
```

```
subtype UX01 is resolved std_ulogic range 'U' to '1'; --('U','X','0','1')
subtype UX01Z is resolved std_ulogic range 'U' to 'Z'; --
                                           -- ('U','X','0','1','Z')
```

## Standard Logic Operators and Functions

Standard logic data types are supported by a number of operators defined as:

```
function "and" (l: std_ulogic; r: std_ulogic) return UX01;
function "nand" (l: std_ulogic; r: std_ulogic) return UX01;
function "or" (l: std_ulogic; r: std_ulogic) return UX01;
function "nor" (l: std_ulogic; r: std_ulogic) return UX01;
function "xor" (l: std_ulogic; r: std_ulogic) return UX01;
function "xnor" (l: std_ulogic; r: std_ulogic) return UX01; -- only
                                           -- standard 1076-1993
function "not" (l: std_ulogic) return UX01;
function "and" (l. r: std_logic_vector) return std_logic_vector;
function "and" (l. r: std_ulogic_vector) return std_ulogic_vector;
function "nand" (l. r: std_logic_vector) return std_logic_vector;
function "nand" (l. r: std_ulogic_vector) return std_ulogic_vector;
function "or" (l. r: std_logic_vector) return std_logic_vector;
function "or" (l. r: std_ulogic_vector) return std_ulogic_vector;
function "nor" (l. r: std_logic_vector) return std_logic_vector;
function "nor" (l. r: std_ulogic_vector) return std_ulogic_vector;
function "xor" (l. r: std_logic_vector) return std_logic_vector;
function "xor" (l. r: std_ulogic_vector) return std_ulogic_vector;
function "xnor" (l. r: std_logic_vector) return std_logic_vector; -- only
                                                   --1076-1993
function "xnor" (l. r: std_ulogic_vector) return std_ulogic_vector; -- only
                                                   --1076-1993
function "not" (l. r: std_logic_vector) return std_logic_vector;
function "not" (l. r: std_ulogic_vector) return std_ulogic_vector;
```

The strength stripping functions convert the 9-valued types std_ulogic and std_logic to the 3-, 4-, and 5-valued types, converting strength values 'H', 'L', and 'W' to their '0' and '1' equivalents:

```
function To_X01 (s: std_logic_vector) return std_logic_vector;
function To_X01 (s: std_ulogic_vector) return std_ulogic_vector;
function To_X01 (s: std_ulogic) return X01;
```

```
function To_X01 (b: bit_vector) return std_logic_vector;
function To_X01 (b: bit_vector) return std_ulogic_vector;
function To_X01 (b: bit) return X01;
function To_X01Z (s: std_logic_vector) return std_logic_vector;
function To_X01Z (s: std_ulogic_vector) return std_ulogic_vector;
function To_X01Z (s: std_ulogic) return X01Z;
function To_X01Z (b: bit_vector) return std_logic_vector;
function To_X01Z (b: bit_vector) return std_ulogic_vector;
function To_X01Z (b: bit) return X01Z;
function To_UX01 (s: std_logic_vector) return std_logic_vector;
function To_UX01 (s: std_ulogic_vector) return std_ulogic_vector;
function To_UX01 (s: std_ulogic) return UX01;
function To_UX01 (b: bit_vector) return std_logic_vector;
function To_UX01 (b: bit_vector) return std_ulogic_vector;
function To_UX01 (b: bit) return UX01;
```

The edge detection functions rising_edge() and falling_edge() provide a concise way to describc thc bchavior of an edge-triggered device such as a flip-flop:

```
function rising_edge(signal s: std_ulogic) return boolean;
function falling_edge(signal s: std_ulogic) return boolean;
```

The following functions can be used to determine if an object or literal is don't-care, which is in this case defined as any of the five values 'U', 'X', 'Z', 'W' or '-':

```
function is_X (s: std_ulogic_vector) return boolean;
function is_X (s: std_logic_vector) return boolean;
function is_X (s: std_ulogic) return boolean;
```

### *3.6.2 IEEE Standard 1076.3 (The Numeric Standard)*

IEEE Standard 1076.3 provides numeric data types and operations to help synthesis and modeling. It defines the numeric_std package that allows the use of arithmetic operations on standard logic data types. The numeric_std package defines the numeric types signed and unsigned and corresponding arithmetic operations and functions based on std_logic data type. Two numeric types declared in numeric_std package, unsigned and signed, are defined as follows:

```
type unsigned is array (natural range <>) of std_logic;
type signed is array (natural range <>) of std_logic;
```

Unsigned represents unsigned integer data in the form of an array of std_logic elements. Signed represents signed integer data in two's complement form. The leftmost bit is treated in both these types as the most significant bit.

The following example illustrates how the type unsigned may be used to simplify the description of a 16-bit up-down counter:

```
library ieee;
use ieee.std_logic_1164.all;
use numeric_std.all;

entity counter is
        port (clk, load, clr, up, down: in std_logic;
                data: in std_logic_vector(3 downto 0);
                count: out std_logic_vector(3 downto 0)
        );
end entity counter;

architecture count4 of counter is
        signal cnt: unsigned(3 downto 0);

begin
        process(clr, clk)
        begin
          if clr='1' then                    -- asynchronous clear
                cnt <= '0000';
          elsif clk'event and clk='1'then
                if load='1' then
                  cnt <= unsigned(data); -- type conversion
                elsif up='1' then
                  if cnt='1111' then
                        cnt <='0000';
                  else
                        cnt <= cnt+1;
                  end if;
                elsif down='1' then
                  if cnt='0000' then
                        cnt <='1111';
                  else
                        cnt <= cnt-1;
                  end if;
                else
```

```
                cnt <= cnt;
            end if;
        end if;

        count<= std_logic_vector(cnt); -- type conversion

    end process;

end architecture count4;
```

The type unsigned is used in this example within the architecture to represent the current state of the counter. IEEE 1076.3 standard describes the add operation ('+') and subtract operation ('-') for type unsigned, so the counter can be easily described. Conversion between unsigned and std_logic_vector is straightforward because these two types are based on the same element type std_logic.

## Numeric Standard Operators and Functions

### *Arithmetic Operators*

```
function "abs" (ARG: signed) return signed;
function "-" (ARG: signed) return signed;
function "+" (L, R: unsigned) return unsigned;
function "+" (L, R: signed) return signed;
function "+" (L: unsigned; R: natural) return unsigned;
function "+" (L: natural; R: unsigned) return unsigned;
function "+" (L: integer; R: signed) return signed;
function "+" (L: signed; R: integer) return signed;
function "-" (L, R: unsigned) return unsigned;
function "-" (L, R: signed) return signed;
function "-" (L: unsigned; R: natural) return unsigned;
function "-" (L: natural; R: unsigned) return unsigned;
function "-" (L: integer; R: signed) return signed;
function "-" (L: signed; R: integer) return signed;
function "*" (L, R: unsigned) return unsigned;
function "*" (L, R: signed) return signed;
function "*" (L: unsigned; R: natural) return unsigned;
function "*" (L: natural; R: unsigned) return unsigned;
function "*" (L: integer; R: signed) return signed;
function "*" (L: signed; R: integer) return signed;
```

**function** "/" (L, R: unsigned) **return** unsigned;
**function** "/" (L, R: signed) **return** signed;
**function** "/" (L: unsigned; R: natural) **return** unsigned;
**function** "/" (L: natural; R: unsigned) **return** unsigned;
**function** "/" (L: integer; R: signed) **return** signed;
**function** "/" (L: signed; R: integer) **return** signed;
**function** "rem" (L, R: unsigned) **return** unsigned;
**function** "rem" (L, R: signed) **return** signed;
**function** "rem" (L: unsigned; R: natural) **return** unsigned;
**function** "rem" (L: natural; R: unsigned) **return** unsigned;
**function** "rem" (L: integer; R: signed) **return** signed;
**function** "rem" (L: signed; R: integer) **return** signed;
**function** "mod" (L, R: unsigned) **return** unsigned;
**function** "mod" (L, R: signed) **return** signed;
**function** "mod" (L: unsigned; R: natural) **return** unsigned;
**function** "mod" (L: natural; R: unsigned) **return** unsigned;
**function** "mod" (L: integer; R: signed) **return** signed;
**function** "mod" (L: signed; R: integer) **return** signed;

*Numeric Logical Operators*

**function** "not" (L:unsigned) **return** unsigned;
**function** "and" (L, R: unsigned) **return** unsigned;
**function** "or" (L, R: unsigned) **return** unsigned;
**function** "nand" (L, R: unsigned) **return** unsigned;
**function** "nor" (L, R: unsigned) **return** unsigned;
**function** "xor" (L, R: unsigned) **return** unsigned;
**function** "xnor" (L, R: unsigned) **return** unsigned; -- 1076-1993 only

**function** "not" (L:signed) **return** signed;
**function** "and" (L, R: signed) **return** signed;
**function** "or" (L, R: signed) **return** signed;
**function** "nand" (L, R: signed) **return** signed;
**function** "nor" (L, R: signed) **return** signed;
**function** "xor" (L, R: signed) **return** signed;
**function** "xnor" (L, R: signed) **return** signed; -- 1076-1993 only

*Relational Operators*

**function** “>” (L, R: unsigned) **return** boolean;
**function** “>” (L, R: signed) **return** boolean;
**function** “>” (L: natural; R: unsigned) **return** boolean;
**function** “>” (L: integer; R: signed) **return** boolean;
**function** “>” (L: unsigned; R: natural) **return** boolean;
**function** “>” (L: signed; R: integer) **return** boolean;
**function** “<” (L, R: unsigned) **return** boolean;
**function** “<” (L, R: signed) **return** boolean;
**function** “<” (L: natural; R: unsigned) **return** boolean;
**function** “<” (L: integer; R: signed) **return** boolean;
**function** “<” (L: unsigned; R: natural) **return** boolean;
**function** “<” (L: signed; R: integer) **return** boolean;
**function** “<=” (L, R: unsigned) **return** boolean;
**function** “<=” (L, R: signed) **return** boolean;
**function** “<=” (L: natural; R: unsigned) **return** boolean;
**function** “<=” (L: integer; R: signed) **return** boolean;
**function** “<=” (L: unsigned; R: natural) **return** boolean;
**function** “<=” (L: signed; R: integer) **return** boolean;
**function** “>=” (L, R: unsigned) **return** boolean;
**function** “>=” (L, R: signed) **return** boolean;
**function** “>=” (L: natural; R: unsigned) **return** boolean;
**function** “>=” (L: integer; R: signed) **return** boolean;
**function** “>=” (L: unsigned; R: natural) **return** boolean;
**function** “>=” (L: signed; R: integer) **return** boolean;
**function** “=” (L, R: unsigned) **return** boolean;
**function** “=” (L, R: signed) **return** boolean;
**function** “=” (L: natural; R: unsigned) **return** boolean;
**function** “=” (L: integer; R: signed) **return** boolean;
**function** “=” (L: unsigned; R: natural) **return** boolean;
**function** “=” (L: signed; R: integer) **return** boolean;
**function** “/=” (L, R: unsigned) **return** boolean;
**function** “/=” (L, R: signed) **return** boolean;
**function** “/=” (L: natural; R: unsigned) **return** boolean;
**function** “/=” (L: integer; R: signed) **return** boolean;
**function** “/=” (L: unsigned; R: natural) **return** boolean;
**function** “/=” (L: signed; R: integer) **return** boolean;

*Shift and Rotate Functions*

**function** shift_left (ARG: unsigned, COUNT: natural) **return** unsigned;
**function** shift_right (ARG: unsigned, COUNT: natural) **return** unsigned;
**function** shift_left (ARG: signed, COUNT: natural) **return** signed;
**function** shift_right (ARG: signed, COUNT: natural) **return** signed;
**function** rotate_left (ARG: unsigned, COUNT: natural) **return** unsigned;
**function** rotate_right (ARG: unsigned, COUNT: natural) **return** unsigned;
**function** rotate_left (ARG: signed, COUNT: natural) **return** signed;
**function** rotate_right (ARG: signed, COUNT: natural) **return** signed;

The following shift and rotate operators are only supported in IEEE 1076-1993:

**function** "sll" (ARG: unsigned, COUNT: natural) **return** unsigned;
**function** "srl" (ARG: unsigned, COUNT: natural) **return** unsigned;
**function** "sll" (ARG: signed, COUNT: natural) **return** signed;
**function** "srl" (ARG: signed, COUNT: natural) **return** signed;
**function** "rol" (ARG: unsigned, COUNT: natural) **return** unsigned;
**function** "ror" (ARG: unsigned, COUNT: natural) **return** unsigned;
**function** "rol" (ARG: signed, COUNT: natural) **return** signed;
**function** "ror" (ARG: signed, COUNT: natural) **return** signed;

## 3.7 Type Conversions

As VHDL is a strongly typed language, it does not allow us to assign a literal value or object of one type to an object of another type. If transfers of data between objects of different types are needed, VHDL requires to use type conversion features, for the types that are closely related, or to write conversion functions for types that are not closely related.

Explicit type conversions are allowed between closely related types. Two types are said to be closely related when they are either abstract numeric types (integers or floating point), or if they are array types of the same dimensions and share the same types for all elements in the array. If two subtypes share the same base type, then no explicit type conversion is required.

To convert data from one type to an unrelated type (for example from integer to an array type), conversion functions must be used. Type conversion functions are often found in standard libraries and vendor supplied libraries, but the designer can also write his/her own type conversion functions.

A type conversion function is a function that accepts an argument of a specified type and returns the equivalent value in another type. Two conversion functions needed to convert between integer and std_ulogic_vector types are presented below:

```
-- Convert an integer to std_logic_vector

        function int_to_std_ulogic_vector( size: integer; value: integer) return
        std_ulogic_vector is
                variable vector: std_ulogic_vector (1 to size);
                variable q: integer;
        begin
                q:= value;
                for i in size downto 1 loop
                   if((q mod 2)=1) then
                        vector(i)='1';
                   else
                        vector(i)='0';
                   end if;
                   q:= q/2;
                end loop;
                return vector;
        end int_to_std_ulogic_vector;



-- Convert a std_ulogic_vector to an unsigned integer

        function std_ulogic_vector_to_uint (q: std_ulogic_vector) return integer
                is
                alias av: std_ulogic_vector (1 to a'length) is a;
                variable value: integer:= 0;
                variable b: integer:= 1;
        begin
                for i in a'length downto 1 loop
                   if (av(I) = '1') then
                        value:= value+b;
                   end if;
                   b:= b*2;
                end loop;
```

```
            return value;
      end std_ulogic_vector_to_uint;
```

Some type conversion functions are provided in IEEE std_logic_1164 package. They help to convert data between 1076 standard data types (bit and bit_vector) and IEEE 1164 standard logic data types:

```
      function To_bit (s: std_ulogic; xmap: bit:= '0') return bit;
      function To_bitvector (s: std_logic_vector; xmap: bit:= '0') return
            bit_vector;
      function To_bitvector (s: std_ulogic_vector; xmap: bit:= '0') return
            bit_vector;
      function To_StdUlogic (b: bit) return std_ulogic;
      function To_StdLogicVector (b: bit_vector) return std_logic_vector;
      function To_StdLogicVector (s: std_ulogic_vector) return
            std_logic_vector;
      function To_StdULogicVector (b: bit_vector) return std_ulogic_vector;
      function To_StdULogicVector (s: std_logic_vector) return
            std_ulogic_vector;
```

Other conversion functions found in IEEE std_logic_1164 package are used to convert between integer data types and signed and unsigned data types:

```
      function to_integer (ARG: unsigned) return natural;
      function to_integer (ARG: signed) return natural;
      function to_unsigned (ARG, SIZE: unsigned) return unsigned;
      function to_integer (ARG, SIZE: natural) return signed;
```

The matching functions (std_match) are used to determine if two values of type std_logic are logically equivalent, taking into consideration the semantic values of the 'X' (uninitialized) and '-' (don't-care) literal values. Table 3.3 defines the matching of all possible combinations of the std_logic values.

Table 3.3 Matching std_logic values

| | 'U' | 'X' | '0' | '1' | 'Z' | 'W' | 'L' | 'H' | '-' |
|---|---|---|---|---|---|---|---|---|---|
| 'U' | F | F | F | F | F | F | F | F | T |
| 'X' | F | F | F | F | F | F | F | F | T |
| '0' | F | F | T | F | F | F | T | F | T |
| '1' | F | F | F | T | F | F | F | T | T |
| 'Z' | F | F | F | F | F | F | F | F | T |
| 'W' | F | F | F | F | F | F | F | F | T |
| 'L' | F | F | T | F | F | F | T | F | T |
| 'H' | F | F | F | T | F | F | F | T | T |
| '-' | F | T | T | T | T | T | T | T | T |

**function** std_match (L, R: std_ulogic) **return** boolean;
**function** std_match (L, R: unsigned) **return** boolean;
**function** std_match (L, R: signed) **return** boolean;
**function** std_match (L, R: std_logic_vector) **return** boolean;
**function** std_match (L, R: std_ulogic_vector) **return** boolean;

However, they do not convert between standard logic data types and numeric data types such as integers or unsigned and signed types. Conversion between these types is usually provided by vendors of design tools, or the designer must provide own conversion functions.

## 3.8 Questions and Problems

3.1. What the enumerated types are useful for? Give a few examples of using the enumerated types.

3.2. Is the enumerated type synthesizable?

3.3. Given a bus of a computer system that contains 16-bit address lines, 8-bit data lines, and two control lines to read from and write to memory. Declare the single composite type that describes the bus and its constituent components. Use two approaches: a) declare first the bus and then use aliases to describe its components, and b) declare its components and then integrate them into the bus.

3.4. What is the difference between the following tests:

**if** (clk = '1') **then**

**if** (clk = '1') **and** (clk'event)

**if** (clk = '0') **and** (clk'event)

3.5. Describe the difference between bit and std_logic type.

3.6. What is the IEEE library package std_logic_1164? What the overloaded language operators defined in this package?

3.7. What is the IEEE Standard 1076.3 (The Numeric Standard)? Why is it introduced?

3.8. What are the type conversions? What are closely related types? Explain why some type conversions are synthesizable, while the other are not.

# 4 PROCESSES AND SEQUENTIAL STATEMENTS

The primary concurrent statement in VHDL is a **process** statement. A number of processes may run at the same simulated time. Within a process, sequential statements specify the step-by-step behavior of the process, or, essentially, the behavior of an architecture. Sequential statements define algorithms for the execution within a process or a subprogram. They belong to the conventional notions of sequential flow, control, conditionals, and iterations in the high level programming languages such as Pascal, C, or Ada. They execute in the order in which they appear in the process. In an architecture for an entity, all statements are concurrent. The process statement is itself concurrent statement. It can exist in an architecture and define regions in the architecture where all statements are sequential.

## 4.1 Process Statement

The process statement defines the scope of each process. It determines the part of an architecture, where sequential statements are executed (components are not permitted in a process). The process statement provides a behavioral style description of design. The syntax is:

```
[process_label :]
process [(sensitivity-list)]
        subprogram_declaration or subprogram_body
        type_declaration
        subtype_declaration
        constant_declaration
        variable_declaration
        file_declaration
        alias_declaration
        attribute_declaration
        attribute_specification
        use_clause
begin
        sequential_statements
```

```
end process [process_label];
```

The process statement can have an explicit sensitivity list. This list defines the signals that will cause the statements inside the process statement to execute whenever one or more elements of the list change value. Changes in these values, sometimes called the events, will cause the process to be invoked. The process has either sensitivity list or a wait statement, as we will see later. Sequential statements within process or subprogram body are logical, arithmetic, procedure calls, case statements, if statements, loops, and variable assignments.

Processes are usually used to describe the behavior of circuits that respond to external events. These circuits may be combinational or sequential, and are connected with other circuits via signals to form more complex systems. In a typical circuit specification, a process will include in its sensitivity list all inputs that have asynchronous behavior (such as clocks, reset signals, functional inputs to a circuit, etc.).

An example of a process statement in an architecture is shown below. The circuit counts the number of bits with the value 1 in 3-bit input signal inp_sig.

```
entity bit_count is
        port ( inp_sig: in bit_vector (2 downto 1);
         q: out integer range 0 to 3);
end entity bit_count;

architecture count of bit_count is
begin
        process (inp_sig)
                variable n: integer;
        begin
                n := 0;
                for i in inp_sig'range loop
                        if inp_sig(i) = '1' then
                                n := n + 1;
                        end if;
                end loop;

                q <= n;
        end process;
end architecture count;
```

The entity declares 3-bit input ports for the circuit that form an inp_sig array and one 2-bit output port q. The architecture contains only one statement, a concurrent process statement. The process declaration section declares one local variable called n. The process is sensitive to the signal inp_sig. Whenever the value of any bit in input signal changes, the statements inside the process will be executed. The variable n is assigned to the signal q. After all statements have been executed once, the process will wait for another change in a signal or port in its sensitivity list.

## 4.2 Basic Sequential Statements

VHDL contains a number of facilities for modifying the state of objects and controlling the flow of execution of models. These facilities are introduced in the following sections.

### *4.2.1 Variable Assignment Statement*

A variable assignment statement replaces the current value of a variable with a new value specified by an expression. The syntax is:

```
target :=expression;
```

In the simplest case, the target of an assignment is a variable name, and the value of the expression is given to the named variable. The variable on the left side of the assignment statement must be previously declared. The right side is an expression using variables, signals, and literals. The variable and the value must have the same base type. This statement executes in zero simulation time. Variable assignment happens immediately when the statement is executed. Examples of variable assignment statements are:

```
a := 2.0;
c := a + b;
```

It is important to remember that variables cannot pass values outside of process.

The target of the assignment can be an aggregate. In that case the elements listed must be object names, and the value of the expression must be a composite value of the same type as the aggregate. In this case variable assignment becomes effectively a parallel assignment.

### 4.2.2 If Statement

If statements represent hardware decoders in both abstract and detailed hardware models. The if statement selects for execution one or more of the enclosed sequences of statements, depending on the value of one or more corresponding conditions. The conditions are expressions resulting in Boolean values. The conditions are evaluated successively until one found that yield the value true. In that case the corresponding sequence of statements is executed. Otherwise, if the else clause is present, its statement is executed. The syntax of if statement is:

```
if condition then

        sequence_of_statements

[elseif condition then

        sequence_of_statements]

[else

        sequence_of_statements]

end if;
```

The if statement can appear in three forms as **if...then**, **if...then...else**, and **if...then...elseif**. Examples of these statements are given below:

```
if (x) then
        t:= a;
end if;

if (y) then
        t:= b;
else
        t:=0;
end if;

if (x) then
        t:=a;
elseif (y)
        then t:= b;
else
```

```
            t:=0;
    end if;
```

### 4.2.3 Case Statement

Case statements are useful to describe decoding of buses and other codes. The case statement selects for execution one of a number of alternative sequences of statements. The chosen alternative is defined by the value of an expression. The expression must result either in a discrete type, or a one-dimensional array of characters. The syntax of the case statement is:

```
    case expression is

    case_statement_alternative

    [case_statement_alternative]

    end case;
```

where case_statement_alternative is:

```
    when choices =>

            sequence_of_statements
```

All choices must be distinct. Case statement contains multiple when clauses. When clauses allow designer to decode particular values and enable actions following the right arrow (=>). Choices can be in different forms. Examples are given below:

```
    case (expression) is

            when 1 => statements;
            when 3 | 4 => ....  --| means "or"
            when 7 to 10 => ......
            when others =>.....

    end case;
```

Important rule is that case statement must enumerate all possible values of expression or have an **others** clause. The **others** clause must be the last choice of all the choices. If the expression results in an array, then the choices may be strings or bit strings. Example below documents behaviour of a bcd to seven-segment decoder circuit:

```
case bcd is
        when "0000" => led <= "1111110";
        when "0001" => led <= "1100000";
        when "0010" => led <= "1011011";
        when "0011" => led <= "1110011";
        when "0100" => led <= "1100101";
        when "0101" => led <= "0110111";
        when "0110" => led <= "0111111";
        when "0111" => led <= "1100010";
        when "1000" => led <= "1111111";
        when "1001" => led <= "1110111";
        when others => led <= "1111110";
end case;
```

### *4.2.4 Loop Statement*

Loop statements provide a convenient way to describc bit-sliced logic or iterative-circuit behavior. A loop statement contains a sequence of statements that are to be executed repeatedly, zero or more times. The syntax of loop statement is:

```
[loop_label:]
[iteration_scheme] loop
        sequence_of_statements
end loop [loop_label];
```

Iteration scheme is

```
while condition
for loop_parameter_specification
```

Loop_parameter_specification is

```
identifier in discrete_range
```

There are two different styles of the loop statement: the **for loop** and **while loop**. Examples of the use of these statements are shown below:

```
for k in 1 to 200 loop
```

```
        k_new:=k*k;
end loop;

k:=1;

while (k<201) loop
        k_new := k*k;
        k := k+1;
end loop;
```

In the second example, if a while condition evaluates to true it continues to iterate.

The index value in a for loop statement is locally declared by the for statement. This variable does not have to be declared explicitly in the process, function, or procedure. If another variable of the same name exists in the process, function, or procedure, then these two variables are treated as separate variables and are accessed by context. The index value is treated as an object within the statements enclosed into loop statement, and so may not be assigned to. The object does not exist beyond execution of the loop statement.

### *4.2.5 Next Statement*

The next statement is used to skip execution to the next iteration of an enclosing loop statement. The statement can be conditional if it contains condition. The syntax is:

**next** [loop_label] [**when** condition];

Next statement stops execution of the current iteration in the loop statement and skips to successive iterations. Execution of the next statement causes iteration to skip to the next value of the loop index. The loop_label can be used to indicate where the next iteration starts. If the iteration limit has been reached, processing will stop. In the case that execution of the loop has to stop completely, exit statement is used.

### *4.2.6 Exit Statement*

The exit statement completes the execution of an enclosing loop statement. This completion can be conditional. The syntax is:

```
exit [loop_label] [when condition];
```

Exit stops execution of the iteration of the loop statement. For example:

```
for i in 0 to max loop
        if (p(i) < 0) then exit;
        end if;
                p(i) <= (a * i);
end loop;
```

If p(i)<=0, then exit causes execution to exit the loop entirely. The loop_label is useful to be used in the case of nested loops to indicate the particular loop to be exited. If the exit statement contains loop_label, then it will complete execution of the loop specified by loop_label. The exit statement provides a quick and easy method of exiting a loop statement when all processing is finished or an error or warning condition occurs.

### *4.2.7 Null statement*

The null statement has no effect. It may be used to show that no action is required in specific situation. It is most often used in case statements, where all possible values of the selection expression must be listed as choices, but for some choices no action is required. An example is given below:

```
case op_code is
        when aba => a:=a+b;
        when lda => a:=data;
        ..................
        when nop => null;
end case;
```

### *4.2.8 Assert Statement*

During simulation, it is convenient to output a string message as a warning or error message. The assert statement allows for testing a condition and issuing a message. It checks to determine if a specified condition is true, and displays a message if condition is false. The syntax is:

```
assert condition
        [report expression]
```

```
            [severity expression];
```

Assert writes out text messages during simulation. The assert statement is useful for timing checks, out-of-range conditions, etc. If the severity clause is present, the expression must be of the type severity_level. There are four levels of severity: failure, error, warning, note. If it is omitted the default is error. If the report clause is present, the result of the expression must be a string. This is a message that will be reported if the condition is false. If it is omitted, the default message is "Assertion violation". A simulator may terminate execution if an assertion violation occurs and the severity value is greater than some implementation dependent threshold.

Example of the use of the assert statement is given below:

```
process (clk,din)
        variable x: integer;
        .............
begin
        .............
        assert (x > 3)
        report "setup violation"
        severity warning;
end process;
```

The message "setup violation" will be printed if condition is false.

## 4.3 Wait Statement

Wait statement belongs to sequential statements. It is used in processes for modeling signal-dependent activation. It models a logic block that is activated by one or more signals. It also causes the simulator to suspend execution of a process statement or a procedure, until some conditions are satisfied. The syntax is:

```
wait
        [on signal_name { ,signal_name}]
        [until conditional_expression]
        [for time_expression]
```

A wait statement can appear more than once within a process. Essentially, it can be used in one of three forms: **wait...on**, **wait...until**, and **wait...for**. In the case of

**wait...on** statement, the specified signal(s) must have a change of value event that causes process to resume execution.

The following example represents a process used to generate a basic sequential circuit, in this case a D flip-flop:

```
process
begin
        wait until clock = '1' and clock'event;
        q <= d;
end process;
```

The value of d is clocked to q when the clock input has the rising edge. The attribute 'event attached to input clock will be true whenever the clock input has had an event during the current delta time point.

Another example of D flip-flop with asynchronous Clear signal is given below:

```
process (clock, clear)
begin
        if clear = '1' then
                q <= d;
        elsif clock'event and clock = '1' then
                q <= d;
        end if;
end process;
```

Instead of listing input signals in the process sensitivity list wait statement can be used:

```
process
begin
        if clear = '1' then
                q <= d;
        elsif clock'event and clock = '1' then
                q <= d;
        end if;

        wait on (clear, clock);
end process;
```

The wait statement can be used with different operations together. A single statement can include an **on** signal, **until** expression, and **for** time_expression clauses. However, one must ensure that the statement contains expressions in which at least one signal appears. This is necessary to ensure that wait statement does not wait forever. Only signals have events on them, and only they can cause a wait statement or concurrent signal assignment to reevaluate. Some further properties of signals, concurrent assignment statements, and the use of wait statement will be discussed in the subsequent chapters.

The process that does not include a sensitivity list executes from the beginning of the process body to the first occurrence of a wait statement, then suspends until the condition specified in the wait statement is satisfied. If the process includes only single wait statement, the process resumes when the condition is satisfied and continues to the end of the process body, then begins executing again from the beginning until encounters the wait statement. If there are multiple wait statements in the process, the process executes only until the next wait statement is encountered. In this way very complex behaviors can be described, including multiple-clock circuits and systems.

## 4.4 Subprograms

Subprograms are used to document frequently used functions in behavioral design descriptions. There are two different types of subprograms:

- a procedure, that returns multiple values, and
- a function, that returns single value.

A subprogram contains sequential statements, just like a process. Subprograms can declare local variables that exist only during execution of subprogram. They are declared using the syntax:

**procedure** designator [formal_parameter_list]

or

**function** designator [formal_parameter_list] **return** type_mark

A subprogram declaration in this form names the subprogram and specifies parameters required. The body of statements defining the behavior of the subprogram is deferred. For functions, the declaration also specifies the type of the result returned when function is called. This type of subprograms is typically used in package specifications, where the subprogram body is given in the package body.

The formal_parameter_list contains declaration of interface elements which includes constants, variables and signals. If constants are used they must be in **in** mode.

When the body of a subprogram is specified, the syntax used is as follows:

**procedure** designator [formal_parameter_list] **is**

subprogram_declarative_part

**begin**
subprogram_statement_part
**end** [designator];

or

**function** designator [formal_parameter_list] **return** type_mark **is**

subprogram_declarativc_part

**begin**
subprogram_statement_part
**end** [designator];

The subprogram_declarative_part can contain any number of following:

subprogram declaration
subprogram body
type declaration
subtype declaration
constant declaration
variable declaration
alias declaration

The declarative items listed after the subprogram specification declare things which are to be used locally within the subprogram body. The names of these items are not visible outside of the subprogram, but are visible within locally declared subprograms. They also shadow all things with the same names declared outside of the subprogram.

The subprogram_statement_part contains sequential statements. When the subprogram is called, the statements in the subprogram body are executed until the end statement is encountered or a return statement is executed. The syntax of return statement is:

**return** [expression];

The return statement in the procedure body must not contain an expression. However, in the case of function, there must be at least one return statement with expression, and a function must complete by executing a return statement. The value of the expression is the value returned to the function call.

### *4.4.1 Functions*

User-defined function must be declared in advance, before it is used (called). The function accepts the values of input parameters, but returns only one value. It actually executes and evaluates like an expression.

Consider an example of function declaration:

```
function byte_to_int(alpha: byte) return integer;
```

which converts a variable of the type byte into integer. For functions, the parameter mode must be **in**, and this is assumed. The only parameter alpha is of type byte. If its class is not specified, it is assumed to be **constant**. The value returned by the body of this function must be an integer.

The body of this example function and the cal to the function is given below:

```
function byte_to_int(alpha: byte) return integer is
        variable result: integer:=0;
begin
        for n in 0 to 7 loop
                result:= result*2 + bit'pos(alpha(n));
        end loop;
        return result;
end byte_to_int;

process
        variable data: byte;
begin
        ......
```

```
        byte_to_int(data);
        ......
    end process;
```

The function byte_to_int brings in one argument of type byte, named alpha, and returns a single value of the type integer. The body of the function is enclosed in **begin**...**end** statements. The function declaration goes in the architecture declaration section, between the word architecture and the begin statement. Calls to the function are sequential statements inside a process after begin statement.

```
    entity ... is

    end entity

    architecture ... of ... is

        ... function declaration

    begin
        process
        begin

        ... function calls

        end process;
    end architecture...;
```

Similarly, functions can be declared in an entity or in a package. Vendors usually provide utility functions in a VHDL package. These are source code design units that are compiled and used from VHDL library.

### *4.4.2 Procedures*

A procedure is also a type of subprogram. With the procedure, more than one value can be returned using parameters. The parameters are of type **in**, **out**, and **inout**. If not specified, the default value is **in**. If mode **in** is used, it brings a value in, **out** sends a value back through an argument list, and **inout** brings a value in and sends it back. Parameters can be variables or signals. Signals must be declared. Procedures can contain wait statements, and signal parameters can pass signals to be waited on. Local variables can be declared in a procedure. A procedure call is a statement. The procedure must be declared in a package, in a process header or in architecture declaration prior to its call. Parameters can be assigned a default value that is used when no actual parameter is specified in a procedure call.

Procedure shown below converts a vector of bits into an integer. The procedure body assigns the value to q and converts bits in z to an integer q. The procedure also returns a flag to indicate whether the result was 0.

```
procedure vector_to_int (z: in bit_vector; zero_flag: out boolean;
                                        q: inout integer) is
begin
        q := 0;
        zero_flag := true;
        for i in 1 to 8 loop
                q := q*2;
                if (z(i) = '1') then
                q := q + 1;
        zero_flag := false;
        end if;
        end loop;
end vector_to_unt;
```

In addition to giving back q, an integer, it also returns a zero_flag that tells if the result was a 0 or not; the result is true or false.

The next example is a procedure that is an 8-bit parity generator:

```
procedure parity           (a: in bit_vector (0 to 7),
                result1, result2: out bit) is
variable temp: bit;
begin
        temp := '0';
        for i in 0 to 7 loop
        temp := temp xor a(i);
        end loop;
        result1 := temp;
        result2 := not (temp);
end;
```

This procedure brings in an 8-bit vector a and sends back two values, the even and odd parity. When this procedure is called, variables, but not signals, must be used because the procedure does variable assignments:

```
variable x, y: bit;
parity (vector, x, y);
```

The following example shows two procedure calls to a previously declared procedure parity. It is called twice to check the 16 bits of y: first the upper 8 bits of y, returning top; then the lower 8 bits of y, returning the value back through the variable bottom. The example shows the procedure calls inside the main body of the process; the procedure declaration is not shown.

```
architecture behavior of receiver is

process
        variable top, bottom, odd, dummy: bit;
        variable y: bit_vector (15 downto 0);
begin
        parity (y(15 downto 8), top, dummy);
        parity (y(7 downto 0), bottom, dummy);
        odd := top xor bottom;
end process;
end behavior;
```

## 4.5 Questions and Problems

4.1. What is the difference between concurrent and sequential statements in VHDL?

4.2. Describe the role of processes in VHDL. What is the sensitivity list?

4.3. You have to design a modulo-n counter. It has to be described using two processes: the first one is a free-running counter, and the second one is checking the state of the free-running counter to reset it when the final state has been reached. Describe the role of variables and signals in the description of the counter.

4.4. Describe the role of wait statement within the processes. Compare the use of wait statements and sensitivity lists.

4.5. Describe the use of functions and procedures in VHDL.

# 5 CONCURRENCY AND BEHAVIORAL MODELING

In Chapter 1 we briefly discussed the main design styles supported by VHDL and introduced concepts of concurrent and sequential statements and their basic use in VHDL descriptions. The concept of concurrency is the essential one to VHDL and it differentiates VHDL from high-level programming languages. This concept is useful for both VHDL synthesis and simulation. Also, it helps to describe both combinational and registered logic. Concurrent statements are specified exclusively in one place in VHDL, between the **begin** and **end** statements of an architecture declaration. All statements described in this part of architecture declaration are considered to be parallel in the execution and of equal priority. Besides signal assignment statements which are used to describe parallel execution, processes can be used with the same purpose. There is no order dependency of concurrent statements, and, therefore, we can specify them in any order. The key element in description of concurrent executions is the concept of signals, as the mechanism to communicate between components and between processes. Various assignment statements can be considered as connections between different types of objects, just as it is described by netlists in schematic diagrams.

## 5.1 Behavioral Modeling

In Section 1.6 we discussed all three major types of modeling using VHDL, with emphasis on structural modeling as something more similar to the traditional way of modeling digital systems. Behavioral modeling represents another way of describing digital systems, especially in initial stages of system design, and is closely related to the simulation and synthesis of VHDL models.

The most basic form of behavioral modeling in VHDL is the signal assignments statement, an examples of which is given below:

```
a <= b;
```

The effect of this statement is that the current value of signal b will be assigned to signal a. This statement will be executed whenever signal b changes its value. If the result of execution is a new value that is different than the current value of the signal, then an event is scheduled for the target signal. If the result of the execution is the same value, then no event will be scheduled, but transaction will be generated. Similarly, using **after** keyword, an event of change of the signal can be scheduled for some other time. Signal assignment statement also represents the most basic concurrent statement, designating change of the value of the target signal whenever the change of the source signal takes place. The source signal is in the sensitivity list of the statement.

In order to illustrate the concept of concurrency, consider the more complicated signal assignment statements in the following example of four-input multiplexer with the symbol as in Figure 5.1.

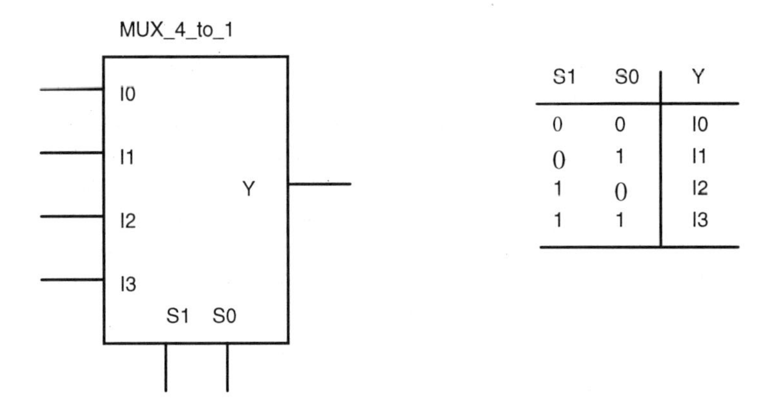


| S1 | S0 | Y |
|---|---|---|
| 0 | 0 | I0 |
| 0 | 1 | I1 |
| 1 | 0 | I2 |
| 1 | 1 | I3 |

Figure 5.1 Four-input multiplexer

The behavioral model for the multiplexer is:

```
library ieee;
use work.std_logic_1164.all;

entity mux_4_to_1 is
        port (i0, i1, i2, i3, s0, s1: in std_logic;
                y: out std_logic);
end entity mux_4_to_1;
```

```
architecture mux4 of mux_4_to_1 is
        signal sel: integer;
begin
        with sel select
        y <=    i0 after 5 ns when 0,
                i1 after 5 ns when 1,
                i2 after 5 ns when 2,
                i3 after 5 ns when 3,
                'X' after 5 ns when others;

        sel <=  0 when s0 = '0' and s1 = '0' else
                1 when s0 = '1' and s1 = '0' else
                2 when s0 = '0' and s1 = '1' else
                3 when s0 = '1' and s1 = '1' else
                4;
end architecture mux_4_to_1;
```

This model contains six input ports and one output port. Based on the value of selection variables s0 and s1, one of the inputs is assigned to the output signal y. Architecture statement part contains two statements. The first statement is a selected signal assignment and will select among a number of options to assign the correct value to a target signal y. The expression, sel , is evaluated and the statement that matches the value of the expression will assign the value to the target signal. All of the possible values of the expression must have a matching choice. If not, the **others** clause must be used. If the values of input signals are unknown values, then the last value, 'X', is assigned to output y.

The second statement in the architecture statement part is called a conditional signal assignment statement. This statement will assign a value to the target signal, in this case local signal sel, based on conditions that are evaluated for each statement. In this statement **when** conditions are executed one at a time in sequential order until the conditions of a statement are met. The first statement that matches the conditions required will assign the value to the target signal.

It is very important to note that both statements will be executed concurrently. The first statement is sensitive to the change of the signal sel, and the second statement is sensitive to the change of signals s0 and s1.

### *5.1.1 Modeling Delays in VHDL*

VHDL provides two types of delays to model behaviour: inertial delay, which is most commonly used, and transport delay, which is used to model wire delays.

Inertial delay is default in VHDL. It is used if no delay type is specified. It provides a behaviour similar to the actual device. That means, the output signal of the device has inertia with the value equal to the delay through the device. Any glitches, pulses, etc. that have periods less than the delay through the device, will have no effect on the output of the device. The value of the output will change only if the input signal values are maintained at a particular value for longer than the delay through the device. Most of the currently available simulators use inertial delay model. One of the reasons for this is that it prevents propagation of spikes throughout the circuit, and it is behaviour wanted by the designers.

The following model shows how to write an inertial delay model. Because the default delay type is inertial, it is not necessary to specify the type of the delay:

```
library ieee;
use work.std_logic_1164.all;

entity and2 is
        port ( a, b: in std_logic;
         y: out std_logic);
end and2;

architecture and2 of and2 is

begin
        y <= a and b after 20 ns;
end architecture and2;
```

Transport delay, on the other hand, must be specified. It is equivalent to wire delay in which any pulse is propagated to the output signal delayed by the delay value specified.

The following is an example of a transport delay model. It is similar in every respect to the inertial delay model except for the keyword **transport** in the signal assignment statement:

```
library ieee;
use work.std_logic_1164.all;

entity inv_with_delay is
        port (a: in std_logic;
        b: out std_logic);
```

```
end inv_with_delay;

architecture inv_with_delay of inv_with_delay is

begin
        b <= transport not (a) after 20 ns;
end inv_with_delay;
```

During simulation, concept of so called simulation deltas is used to order some types of events to produce consistent results. Specifically, if zero-delay events are not properly ordered, results can be different in different simulation runs. Simulation deltas are an infinitesimal amount of time used as a synchronization mechanism when zero-delay events are present. VHDL simulators use delta delay whenever zero delay is specified explicitly or implicitly. This means that the values of the signals will not be propagated immediately, but will be scheduled for the next delta time point which belongs to the same simulation time. All necessary signal values are then reevaluated until they get their final stable value. Simulation time can be incremented only after the all signal reevaluations.

### *5.1.2 Drivers*

Multiply-driven signals are very useful in modeling a data bus, a bi-directional bus, etc. VHDL enables modeling of these kinds of circuits using the concept of signal drivers. A VHDL driver is one contributor to the overall value of a signal. A multiply-driven signal has many drivers. The values of all drivers are resolved together to create a single value for the signal. The method of resolving all of the contributors into a single value is through a resolution function. resolution function is a designer-written function that will be called whenever a driver of a signal changes value. Resolution functions are discussed in Chapter 7.

Drivers are created by signal assignment statements. A concurrent signal assignment statement inside of an architecture produces one driver for each signal assignment. Multiple signal assignments for a same signal will produce multiple drivers for that signal. How these drivers are resolved is left to the designer. In the case that the designer does not want to constrain signal behaviour, the source signals that drive the same output signal will short together.

Consider the following example in which a circuit description does not meet the requirement that a signal is driven by only one driver. The circuit represents actually an attempt to describe a 2-to-1 multiplexer:

```
library ieee;
use work.std_logic_1164.all;

entity mux2_to_1 is
        port (i0, i1, s: in std_logic;
        y: out std_logic);
end mux2_to_1;

architecture mux1 of mux2_to_1 is
begin
        y <= i0 and s;
        y <= i1 and not s;
end mux1;
```

Based on the value of s signal, output signal y should get the value of either i0 or i1 input signal. However, each of the two signal assignments results in a driver being created, resulting in two drivers driving signal y. A possible solution to this problem is to specify output y using a single signal assignment, as in the following:

```
library ieee;
use work.std_logic_1164.all;

entity mux2_to_1 is
        port (i0, i1, s: in std_logic;
         y: out std_logic);
end mux2_to_1;

architecture mux2 of mux2_to_1 is
        signal y1, y2: std_logic;
begin
        y1 <= i0 and s;
        y2 <= i1 and not s;
        y <= y1 or y2;
end mux2;
```

Two intermediate signals y1 and y2 have been used to describe output as a function of these two values.

Another method to provide a single driver might be to use a conditional signal assignment, as in the following:

```
library ieee;
use work.std_logic_1164.all;

entity mux2 is
        port (i0, i1, s: in std_logic;
         y: out std_logic);
end entity mux2;

architecture mux2right of mux2 is

begin
        y <= i0 when s = '0' else
         i1 when s = "1" else
        'X'; -- unknown
end architecture mux2;
```

The use of conditional assignment statement provides only one driver to be created for output signal y. If we used selected conditional statement, at least two drivers would be created, requiring resolution function to resolve the value of the output signal.

Some further topics of behavioral modeling including generics and blocks will be discussed in subsequent chapters.

## 5.2 Signals and Signal Assignments

Signals provide communication between processes and between components. They provide global communication within a design. Signals are used to wire structural designs. Signal assignments are scheduled in simulated time and are not treated like variable assignments.

### *5.2.1 Structural Signals in Netlisting*

The top-level design is frequently done in a structural style with library units instanced as components "netlisted" together using signals.

Signals declared in the port declaration of an entity describe a component's input/output connectors. The example below shows a comparator circuit with signals A, B, and C of type std_logic declared in the entity. Two of the signals are inputs and one of them is output.

```
library ieee;
use.std_logic_1164.all;

entity comparator is
        port (a, b: in std_logic; c: out std_logic);
end entity comparator;

architecture structural of comparator is
signal i: std_logic;

component xor2 port (x, y: in std_logic; z: out std_logic);
end component;

component inv port (x: in std_logic; z: out std_logic);
end component;

begin
        u0: xor2 port map (a, b, i);
        u1: inv port map (i, c);
end architecture structural;
```

Local signal i is used to connect two components, an xor2 gate and an inverter inv. The external signals are A, B, and C.

### *5.2.2 Process Communication*

Signals can also be used for activating processes inside an architecture: one process can be waiting for a signal from another process. An example of a process A waiting until process B is done is shown in Figure 5.2. Process A includes wait ...until B_done, which suspends the execution of process A. The signal B_done comes from process B, and when set to '1', activates process A. This illustrates process synchronization using a single signal. Signal B_done is declared within the architecture and is available and "visible" to all its processes.

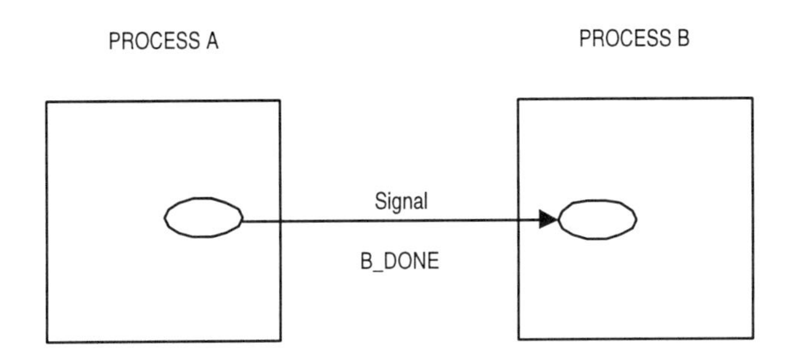


Figure 5.2 An example of process communication and synchronization

```
architecture one of example is
signal b_done: bit := '0';

begin
b: process
begin
        if s1 then b_done <= '1';
..................
end process b;

a: process
begin
        wait until b_done = '1';
...................
end process a;
end architecture one;
```

Process statements contain sequential statements, but are themselves concurrent statements within an architecture.

### *5.2.3 First Look to the Testbench*

One usual way of exercising compiled VHDL programs is to create a process which is used as a testbench of the unit under test (UUT). This is a very convenient way to validate design written in VHDL. Tests are run on the simulated design in a VHDL simulator. The situation is illustrated in Figure 5.3.

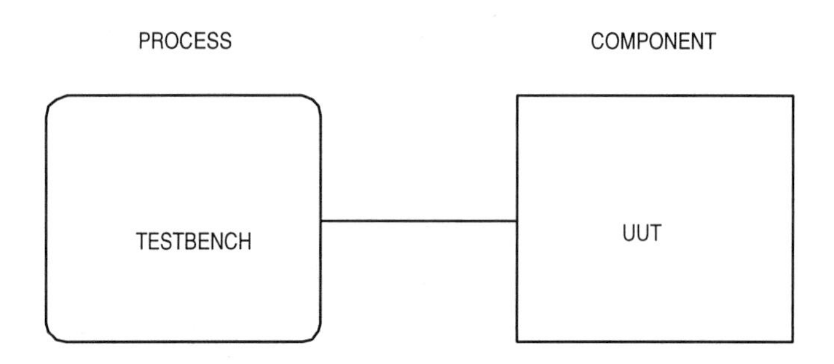


Figure 5.3 Illustration of testbench

Approaches to testing a design can be different, but the most structured one is to use a process to read a file of test vectors from a data file. User formats the vectors as appropriate for the design. Each vector is read and used as a stimulus signal on UUT. The expected response is also in the file; the process monitors whether or not the behaviour of the UUT matches the expected one. More on the testbenches can be found in Chapter 8.

### 5.3 Concurrent Signal Assignments

VHDL provides concurrent statements that allow documenting systems that perform parallel operations. During simulation, components in these systems are "run" at the same simulated time. Concurrent statements are used to describe this kind of parallel behavior. Concurrent descriptions can be structural or behavioral. The key concurrent statement is a process.

A process, as described earlier:

- Runs concurrently with the other processes
- Contains only sequential statements
- Defines regions in architectures where all statements execute sequentially
- Must contain either an explicit sensitivity list or `WAIT` statement
- Provides "programming language"-like capability

- Can access signals defined in architecture and entity.

A detailed description of processes and their use in description of digital designs is covered in the later chapters.

### *5.3.1 Signal Assignments*

A signal assignment schedules one or more transactions to a signal (or port). The syntax of a signal assignment statement is:

target <= [**transport**] waveform;

where target is a signal name or aggregate of signals, and the waveform is:

value_expression [**after** time_expression]

or

**null** [**after** time_expression]

The value_expression is any expression that evaluates to a value of the same type as the target signal or aggregate, and the time_expression specifies the delay specifies the delay with which signal assignment will be scheduled. If the time_expression for the delay is omitted, it defaults to 0 fs. This means that the transaction will be scheduled for the same time as the assignment will be executed, but during the next simulation cycle.

We can consider that a list of transactions giving future values for the signal is associated to each signal. A signal assignment adds transactions to this list. The list is a way of representing the waveform of the signal. For example, if the assignment

a <= '1' **after** 15 ns;

is executed at time 10 ns, the event specified with signal assignment statement will be executed at the time equal 25 ns. The signal a will be up-dated to value '1 at time 25 ns. Similarly, the assignment

a <= '0' **after** 5 ns; '1' **after** 20 ns;
executed at the time 30 ns will add two transactions to the list:

- change of the value of the signal a to '0' after 35 ns and

- change of the value of the signal a to ‘1’ after 50 ns.

This is illustrated with the following list:

| Time | 35 ns | 50 ns |
|---|---|---|
| Signal value | ‘0’ | ‘1’ |

If a signal assignment is executed, and there are already old transactions from the previous assignments in the transaction list, then some from the previous assignments may be deleted. This can happen if the word **transport** is included in the new assignment. The word transport is used to model transport delay. In this case, all old transactions scheduled to occur after the first new transaction are deleted before the new transactions are added. For example, if the assignment

```
a <= transport 'Z' after 15 ns;
```

is executed at time 23 ns, then the transaction scheduled for 50 ns will be deleted, and the new list of scheduled events will become:

| Timc | 35 ns | 38 ns |
|---|---|---|
| Signal value | 0 | Z |

If word transport is omitted, then VHDL assumes the second kind of delay, inertial delay, which is used to model devices which do not respond to input pulses shorter than their output delay. When an inertial delay transaction is added to the transaction list, all old transactions scheduled to occur after the new transaction are deleted, and the new transaction is added. Next, all old transactions scheduled to occur before the new transaction are examined. If there are any with a different value from the new transaction, then all transactions up to the last one with different value are deleted. The remaining transactions with the same values are left.

A concurrent signal assignment is essentially a shorthand way to write an process and is equivalent to a process containing one statement. It is illustrated in the following example:

```
architecture x of mux1 is
begin
        output <= a(index);
end x;
```

is equivalent with:

```
architecture x of mux is
begin
process (a, index);
begin
        output <= a (index);
end process;
end architecture x;
```

Any signal on the right side of a signal assignment is in the process sensitivity list. Anytime a signal changes, the assignment statement is evaluated and a signal value is assigned.

Concurrent signal assignment statements execute at the simulated time. There is no order associated with them. For example two concurrent assignment statements are:

```
a <= b(input1);
c <= d(input2);
```

The order in which statements will be executed depends only on the change of signals on the right side of assignments. They execute synchronously with no relative order. Concurrent signal assignments are used to describe either combinational logic using arbitrary complex expressions on the right side of assignment statement, or they can be used to describe connections between components at the lower level of design hierarchy.

Let us consider the example of an architecture that contains simple concurrent signal assignments:

```
architecture example of logic_gates is
        signal a, b: std_logic;
        signal x, y: std_logic;
begin
        x <= (a and b) or (a and not(b));
        y <= not (a xor b);
end architecture example;
```

In this example, signals are declared locally within the architecture and can be used in the same way as the ports of the corresponding entity. The only difference between ports and locally declared signals is that ports have a direction, or mode (**in**, **out**, **inout**, **buffer**), and the locally declared signals can be used in either way. For example, if the port is declared as mode out, it can only be assigned the value, and cannot be read. Concurrent signal assignments can also include delay specifications, as described earlier. However, delays are useful only for simulation purposes. They are not supported by synthesis tools.

### *5.3.2 Conditional Signal Assignment*

A conditional concurrent signal assignment has one target, but can have more than one expression. Only one of the expressions can be used at a time. The syntax is:

```
target <= {expression when condition else} expression;
```

A conditional signal assignment consists of an assignment to one output (or a collection of outputs) and a series of conditional statements. To ensure that all conditions are covered, a terminating **when others** clause can be used.

An example of conditional signal assignment is shown below:

```
architecture arch1 of decision is
begin
        x <=  a when (inp > 5) else
              b when (inp = 6) else
              c when others;
end arch1;
```

The conditional signal assignment also provides a method to imply (and impose) priority to a list of conditions, giving higher priority to the conditions as they are ordered in **when** clauses.

### *5.3.3 Selective Signal Assignment*

A selective signal assignment statement can have only one target and only one **with** expression. It is similar to conditional signal assignment, but does not imply priority to the input conditions. The input value is tested similarly to the **case** statement. The syntax is:

```
with expression select
        target <= {expression when choices,};
```

An example is:

```
architecture arch2 of selector is
begin
        with x select
                out <=  a when 10,
                        b when 20,
                        c when 30;
                        d when others;
end arch2;
```

The selective signal assignment is outside of the process. Since it is a concurrent signal assignment, it runs whenever any change occurs to selected signal (in our example, x signal). This is effectively a shorthand notation for an equivalent process with a **case** statement.

Another example of a selected signal assignment is given below:

```
architecture behavioral of alu is
begin
        with operation_select select
                alu_result <= operand1+operand2 when op_add;
                        operand1-operand2 when op_sub;
                        operand1+1 when op_inc;
                        operand1-1 when op_dec;
                        operand1 and operand2 when op_and;
                        operand1 or operand2 when op_or;
end architecture behavioral;
```

In this example, the value of the signal operation_select is used to select which signal assignment to alu_result to execute. The statement is sensitive to operation_select, operand1 and operand2. Whenever any of these signals change value, the selected signal assignment is executed again.

### *5.3.4 Concurrent Procedure Call*

A concurrent procedure call is a procedure call that is executed outside of a process; it stands alone in an architecture. Concurrent procedure calls may be considered as independent sequential statements that execute whenever there is a change (event) on any of their inputs. The advantage of a procedure over a process is that the body of the procedure can be specified elsewhere (for example, in a package), and used repeatedly throughout the design. The concurrent procedure call:

- has **in**, **out**, and **inout** parameters modes.
- has more than one return value
- is considered a statement
- is equivalent to a process containing a single procedure call.

In the following example, the procedure dff is called within the concurrent area of the architecture:

```
architecture counter1 of counter is
        signal D, Qreg: std_logic_vector(7 downto 0);
begin
        D <=    Data when (load='1') else
                Qreg(7 downto 0)+'00000001' when (ena='1') else
                Qreg(7 downto  0);
        dff(rst, clk, D, Qreg);
        Q <= Qreg;
end architecture counter1;
```

### 5.4 Block Statement

Blocks are partitioning mechanisms within VHDL that allow the designer to logically group areas of the model. The statement area in an architecture can be broken into a number of separate logical entities. For example, a CPU can be subdivided into blocks so that one block might be an ALU, another a register bank, etc. Each block can declare logical signals, types, constants, etc. Any object that can be declared in the architecture declaration section can be declared in the block declaration section.

A block statement contains a set of concurrent statements. The syntax is:

```
[label:] block
        [block_declaration_item]
begin
        [concurrent_statement]
end block [label];
```

The optional label names the block. A block_declaration_item declares objects local to the block and can be any of the following:

- Use clause
- Subprogram declaration and body
- Type declaration
- Constant declaration
- Signal declaration
- Component declaration

The order of concurrent statements in a block is not significant, since all of these statements are always executing.

Objects declared in a block are visible to that block and all blocks nested within. When a child block inside a parent block declares an object with the same name as one in the parent block, the child's declaration overrides that of the parent. Blocks are only useful for hierarchically organizing concurrent statements. Guarded blocks are equivalent to processes with sensitivity lists and appropriate statements in the process body. An example of the nested blocks is given below:

```
blck1: block
signal s: bit;
begin
        s <= a and b;
blck2: block
        signal s: bit;
begin
        s <= c and d;
        blck3: block
        begin
                z <= s;
        end block;
end block;
        y <= s;
end block;
```

An example of the use of the block statements in the model of a CPU follows:

```
use work.std_logic_1164.all;
```

```
package bit16 is
        type tw16 is array (15 downto 0) of std_logic;
end bit16;

use work.std_logic_1164.all;
use work.bit16.all;
entity cpu is
        port (clk, interrupt: in std_logic;
        address: out tw16; data: inout tw16);
end entity cpu;

architecture cpu_blk of cpu is
        signal ibus, dbus : tw16;
begin
        alu: block
        signal qbus : tw16;
        begin
        -- alu bchavioural statements
        end block alu;

reg: block
        signal zbus : tw16;
        begin
        reg1: block
                signal qbus : tw16;
        begin
                -- reg1 behavioral statements
        end block reg1;

        reg2: block
                signal qbus : tw16;
                -- reg2 behavioral statements
        end block reg2;

        -- other registers statements

end block reg;
end cpu_blk;
```

Entity CPU declares four ports that are used as the model interface. All of these ports are visible to any block declared in an architecture for this entity. The input ports can be read from, and the output ports can be assigned values.

Signals declared inside the architecture cannot be referenced outside of the architecture. However, any block inside the architecture can reference these signals. Any lower-level block can reference signals from a level above, but upper-level blocks cannot reference lower-level local signals. As an example, all of the statements inside of the block alu can reference qbus, but statements outside of block alu cannot use qbus. Another interesting feature of the language can be seen from the example of the use of the same name for two different signals. Signal qbus declared in the block REG2 has the same name as the signal qbus declared in the block alu. Since they are declared in two separate regions, the compiler will consider them as two separate signals with the same name. Each of them can be referenced only in the block that has the declaration of the signal.

Blocks have another interesting feature of containing ports and generics. This allows the designers to reuse blocks written for another purpose in a new design. Let's assume that there is another alu design for example above that performs the functionality needed in a different way. In order to use it in new design we only have to map the signal names and the generic parameters in the design being upgraded to ports and generics created for the new alu block. Illustration of this example follows:

```
use work.std_logic_1164.all;
package math is
        type tw16 is array (15 downto 0) of std_logic;
        function tw_add(a, b: tw16) return tw16;
        function tw_sub(a, b: tw16) return tw16;
end math;

use work.math.all;
use work.std_logic_1164.all;
entity cpu is
        port (clk, interrupt: in std_logic;
                address: out tw16; cont: in integer;
                data: inout tw16);
end entity cpu;
architecture cpu_blk of cpu is
        signal ibus, dbus: tw16;
begin
        alu: block
                port (abus, bbus: in tw16;
                d_out: out tw16;
                ctbus: in integer);
```

```
        port map (abus => ibus, bbus => dbus,
                  d_out => data, ctbus => cont);
        signal qbus: tw16;
        begin
         d_out <= tw_add(abus,bbus) when ctbus=0 else
            tw_sub(abus,bbus) when ctbus=1 else
                  abus;
        end block alu;
end architecture cpu_blk;
```

The port statement declares the number of ports used for the block, their direction, and the type of ports. The port map statement maps the new ports with signals or ports that exist outside of the block. Port abus is mapped to architecture cpu_blk local signal ibus, port bbus is mapped to dbus. Ports d_out and ctbus are mapped to external ports of the entity. Mapping implies a connection between the port and the external signal such that whenever there is a change in value in the signal connected to a port, the port value changes to the new value.

Block statements can be used in the form of so called guarded blocks. A guarded block contains a guard expression which can enable and disable drivers inside the block. The guard expression is a Boolean expression; when true, drivers contained in the block are enabled and, when false, the drivers are disabled. A guarded signal assignment statement is recognised by the keyword **guarded** between the "<=" and expression part of the statement. An example follows:

```
b1: block (cnt ='1')
begin

        a <= guarded b after 5 ns;

end block b1;
```

## 5.5 Some Further Properties of Signals

Although a signal can be used in process, it cannot be declared in a process. Global signals can be declared in packages. If declared in an entity or port declaration, they can be used in any architecture of that entity. Declared signals within a particular architecture are local to that architecture. Signals are initialized by an ":=" just like variables; signal assignments are written with a symbol "<=".

Different signal declarations are given in the following example:

```
package sigdecl is
        signal vcc: std_logic :='1';
end sigdecl

entity board is
port (data_in: in std_logic; data_out: out std_logic);
        signal sys_clk: std_logic :='1';
end board;

architecture data_flow of board is
        signal int_bus: std_logic;
begin
................

end data_flow;
```

Regardless of type of signal declaration, all must occur before a signal is used within an architecture.

The signal can be declared to be input or output port of an entity. In that case it has a direction. The default is input. Signals can be declared to be **in**, **out**, **inout**, and **buffer**. A buffer is the same as inout except there can be only one driver or source. It is important to remember that type declarations must occur before port declarations. They should be declared in a package.

Signal assignments schedule a value at some simulated time. Such an assignment defines a driver of a signal. Signal assignments are sequential within a process or concurrent outside of a process. Within a process signal assignments are delayed until a simulation cycle is run, triggered by the execution of a wait statement.

An example that illustrates a series of sequential signal assignments is given below:

```
process
begin
        sys_clk <= not(sys_clk) after 50 ns;
        int_bus <= data_in after 0 ns;
        data_out <= magic_function (int_bus) after 0 ns;
wait for 0 ns;
end;
```

The assignments are valid only after the wait statement is executed by the simulator.

Signal assignments can schedule a number of values at different time points. This capability is useful for describing repetitive signal forms:

```
s <= '1' after 5 ns; '0' after 10 ns;
t <= not t after 25 ns;
```

Within one process a signal should have only a single driver.

When a signal assignment is simulated, a signal change occurs at a precise simulated time. However, the real hardware may not be able to do exactly what the VHDL signal assignment specifies. Therefore, it is possible to simulate it, but it may not be possible to build it exactly. Zero delay can be specified by not providing an after clause. But it always has to be remembered that actual assignment, even with the zero delay, will not be performed until the wait statement within a process is executed.

## 5.6 Simulation Cycle

VHDL simulators use a two-list technique, which tracks previous and new values of signals. In this method, expressions are first evaluated, then signals are assigned new values in an artificial unit called delta time.

The key points of simulation and delta time are:

- The simulator models zero-delay events using delta time.
- Events scheduled at the same time are simulated in specific order during a delta time step.
- Related logic is then resimulated to propagate the effects for another delta time step.
- Delta time steps continue until there is no activity for the same instant of simulated time.

The execution of a wait statement triggers a simulation cycle to be entered. VHDL simulators do all evaluations before any assignments. Simulated time is not advanced as long as there are expressions unevaluated, or assignments which are not done.

A process contains only sequential statements and normally has one or more wait statements in it. The variable assignments within process occur immediately, while signal assignments do not take effect until a wait statement occurs. A wait statement suspends sequential statements execution, but permits signal assignments to proceed.

Rather then having one or more wait statements inside a process, it is possible to have a sensitivity list. A sensitivity list is a list of names for which a process is going to wait. It follows a process name in parentheses. A sensitivity list is equivalent to a wait statements at the end of a process. If wait statements are used, a sensitivity list is not allowed.

## 5.7 Questions and Problems

5.1. Describe the difference between inertial and transport delay concepts in VHDL.

5.2. Describe the methods that can be used to solve the problem of multiple-driven signals. Show it in an example of a bus which is driven by four different sources of signals.

5.3. List all major features of signals as a means to implement concurrency concepts in VHDL.

5.4. What are the similarities and differences between signals and ports? Do the signals have direction?

5.5. What is the purpose of blocks in VHDL? What is their equivalent in schematic diagrams?

# 6 STRUCTURAL VHDL

The VHDL structural style describes the interconnection of components within an architecture. It is similar to a netlisting language in other CAD systems. In a structural architecture, components that will be used are declared, then instances of components created with particular mappings of signal wires to the various pins of components. Each component instantiation is a concurrent statement, similar to those already described.

Figure 6.1 describes a flip-flop using two-input *nand* gates. There are two

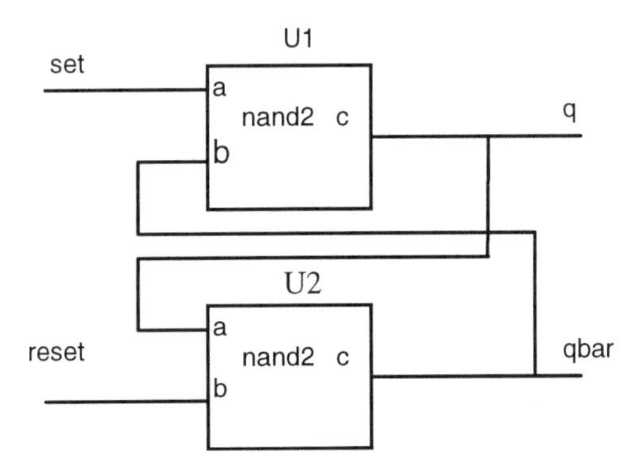


Figure 6.1 SR flip-flop

components in the schematic labeled by U1 and U2. The input signals coming into the flip-flop entity are set and reset signals; q and qbar are the output signals. These signals connect to the nand2 components, whose pins are designated with the inputs a nd b and output c. VHDL description of flip-flop architecture is given below:

```
entity rsflipflop is
        port (set, reset: in bit; q, qbar: inout bit)
end rsflipflop;

architecture structural of rsflipflop is
        component nand2
```

```
            port (a, b: in bit; c: out bit);
            end component;
    begin
            u1: nand2 port map (set, qbar, q);
            u2: nand2 port map (q, reset, qbar);

    end structural;
```

In this example q and qbar must be of the mode **inout** in the entity description. The statement part of the architecture designates the signals connected for u1 and u2 nand2 gates.

Component instances use signals that are previously declared:

- as entity ports or
- signals declared in an entity or
- signals declared in a package or
- signals declared in an architecture

Function calls are allowed in place of pin names. This can be useful for doing type conversion when required in component hookups. If no connection is to be specified, a VHDL reserved word **open** is to be used.

## 6.1 Component Instantiation

In addition to positional association of signals, illustrated in the port map, VHDL supports named association as shown in the next example:

```
    u1: nand2
    port  map (a => set, c =>q, b =>qbar);
```

A right arrow "=>" reads as "gets" (pin c gets signal q). This makes VHDL more readable. Pins may be specified in any order.

If a description of an array of components is necessary, the generate statement can be used. It iterates components similarly to the loop statement. In this way a replication of structure outside a process can be done.

The generate statement is convenient for expressing an array of components with similar connections. An example of the 4-bit shift register is shown in Figure 6.2.

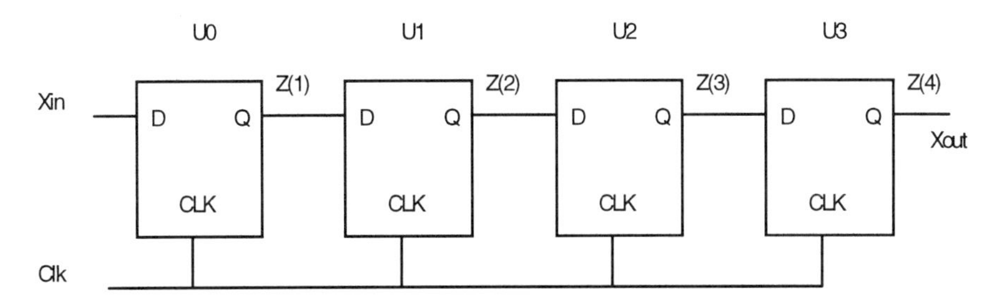


Figure 6.2 4-bit shift register

Complete VHDL description of the shift register from Figure 6.2 is shown below:

```
library ieee;
use ieee.std_logic_1164.all;

entity shiftreg is
        port (xin, clk: in std_logic; xout: out  std_logic);
end entity shift;

library gates;
use gates.all;

architecture gen_shift of shiftreg is
        component dff
                port (d, clk: in std_logic; q: out std_logic);
        end component;
        signal z: bit_vector (0 to 4);
begin
        z(0) <= xin;
        for i in 0 to 3 generate
                u: dff port map (z(i), clk, z(i+1));
        end generate;
        xout <= z(4);
end architecture gen_shift;
```

This example illustrates the use of the **generate** statement that generates four identical dff flip-flops. Each instance will have a unique name that is based on the instance label provided (in this case u) and the index value. For-generate statement can be nested making possible instantiation of multi-dimensional arrays of component instances or other concurrent statements.

Usually, all slices in the design are not identical. In our example first and last flip-flop are distinguished in that they do not interconnect to other flip-flops at their ends. Therefore, a special condition is needed for the first and last element of this array.

The following extension illustrates the use of the conditional **if-generate** statement with some special testing conditions which separates the generation of the pieces designated to the left and the right side:

```
library ieee;
use ieee.std_logic_1164.all;

entity shiftreg is
        port (xin, clk: in std_logic; xout: out  std_logic);
        constant width: integer:=8;
end shift;

library gates;
use gates.all;

architecture gen_shift of shiftreg is
component dff
        port (d, clk: in std_logic; q: out std_logic);
end component;
begin
        signal z: bit_vector (0 to width);

for i in 0 to width generate
if (i=0) generate
        u: dff port map (xin, clk, z(i+1));
end generate;

if (i > 0) and (i < width-1)) generate
        u: dff port map (z(i), clk, z(i+1));
end generate;

if (i=width) generate
        u: dff port map (z(i), clk, xout);
```

```
end generate;
end gen_shift;
```

## 6.2 Hierarchy

A VHDL design consists of a hierarchy of components compiled from behavioral, structural, and dataflow level architectures. Once design is compiled, it goes into the library and can be used as a component for subsequent designs. A component must be declared before its instance from library is taken. A design containing components needs a configuration. Configurations are described in the following section.

In the following example two architectures of comparator are shown. In the first architecture, called first-arch, components xor2 and inv are declared before they are used. In the second architecture second_arch, these component declarations occur in a package gates_pack residing in library work. Architecture second_arch uses use statement to access component declarations. Therefore, the second design is shorter. Package from the library is used to hold frequently used declarations.

```
architecture first_arch of comparator is
signal i: bit;

component xor2 port (x,y: in: bit; z: out bit);
end component;

component inv port (x: in bit; z: out bit);
end component;

begin
        u0: xor2 port map (a, b, i);
        u1: inv port map (i, c);
end first_arch;


use work.gates.all;
architecture second_arch of comparator is
signal i: bit;

begin
        u0: xor2 port map (a, b, i);
```

```
        u1: inv port map (i, c);
end second_arch;
```

Package gates as the design unit is shown below:

```
package gates is

component xor2
        port (x, y: in bit; z: out bit);
end component;

component inv
        port (x: in bit, z: out bit);
end component;

end gates;
```

Two components are declared in the package. Components themselves must have an entity and an architecture and must be compiled in a library. The package contains only component declarations. Description of design entity xor2 is shown in following example. There are two architectures of xor2, called slow_arch and fast_arch, with the different gate delays:

```
entity xor2 is
        port (x, y: in bit; z: out bit);
end xor2;

architecture slow_arch of xor2 is
begin
        z <= x xor y after 1.0 ns;
end slow_arch;

architecture fast_arch of xor2 is
begin
        z <= x xor y after 0.4 ns;
end fast_arch;
```

This means that for an entity more than one architecture can be compiled, and the architectures must have different names. When component entity xor2 is used, that means, by default, the most recently compiled architecture of that entity in the library is included. If a particular architecture of an entity is to be selected, the

**configuration** statement must be used. A configuration, or selection of an architecture, can be coded in an architecture or as a separately compiled design unit. In this way, the entire design need not to be recompiled.

In the following example, a particular (configured) architecture of xor2 is used

```
architecture structural of comparator is
begin
        signal i: bit;
        for u0: xor2 use entity work.xor2(fast_arch);
        begin
        u0: xor2 port map (a, b, i);
        u1: inv port map (i, c);
end structural;
```

The xor2 entity is selected from library WORK and uses the architecture fast_arch. A configuration can be inside of an architecture, as in this axample, or can stand alone and be a separate set of statements that can be separately compiled. In the example below, configuration of xor2 is a separately compiled design unit:

```
configuration xor_1 of xor2 is
        for fast_arch
        end for;
end xor_1;

configuration xor_2 of xor2 is
        for slow_arch
        end for;
end xor_2;
```

Each of configurations simply names the architecture to be used. Using configuration xor_1 is the same as selecting architecture fast_arch of the entity xor2 as shown in the following example:

```
entity xor2 is
        port (x, y: in bit; z: out bit);
end entity xor2;

architecture slow_arch of xor2 is
begin
        z <= x xor y after 1.0 ns;
end architecture slow_arch;
```

```
architecture fast_arch of xor2 is
begin
        z <= x xor y after 0.4 ns;
end architecture fast_arch;
```

Configuration can be compiled and stored in a library. Another feature of configurations is that you can provide remapping of component ports. The example below shows an interesting application of configurations. A nand2 gate is used as substitute for inverter:

```
configuration comp1 of comparator is
for structural
        for all: inv use entity work.nand2(behave)
                port map (a => a, b => vcc, z => z);
        end for;
end for;
end comp1;
```

All inverters are implemented by nand gates from work.nand2 (the behavioral model). There is however an extra pin. The code says that port b of the nand gate is connected to Vcc. This example illustrates the substitution of a part, and remapping of the pins. The configuration allows not only to use another component, but also to use a different component with a similar function, and to do necessary port mapping.

The simplest configuration needs only to identify the entity and architecture name. For example

```
configuration cfg_compare of comparator is
        for structural
        end for;
end cfg_compare;
```

When an architecture does not contain components, it does not need an explicit configuration. The most recently compiled architecture of an entity is the configuration used. This is sometimes referred to as the null configuration.

## 6.3 Generics

In addition to input/output ports, an entity can specify parameters (generic declarations) that allow us to alter a design's behavior when a component is instanced. This is accomplished by using a generic map. An example of a component declaration with a generic parameter with a default value is given below:

```
entity xor2 is
        generic (m: time := 1.0 ns);
        port (x, y: in bit; z: out bit);
end xor2;

architecture general of xor2 is
begin
        z <= x xor y after m;
end general;
```

Generic declaration must come before port declaration. The architecture general uses the generic parameter m to establish a delay value for z. The generic parameter m affects the behavior of the xor2 model when the xor2 is used. An example of using the component xor2 is given below:

```
architecture structural of comparator is
        signal i: bit;
component xor2
        generic (m: time);
        port (x, y: in bit; z: out bit);
end component;

begin
        u0: xor2 generic map (m => 1.5 ns)
        port map (a, b, i);
        ..............
end structural;
```

Generic statement is specified before port statement. These two statements must be separated by a semicolon. The code takes an instance of the xor2 gate previously compiled as a component in a library, and provides a parameter (delay time). If delay time and ports are not specified, the default values are used. Generic parameters can be of any type.

A generic is a general mechanism for passing instance-specific data into a component. This is similar to passing parameters to VHDL functions. The data passed in is a constant and cannot be modified when it is used. Generics can be also used in component declaration as in the following example:

```
package gates is

component xor2
        generic (m: time:= 1.0 ns);
        port (x, y: in bit; z: out bit);
end component;

component inv
..............
end component;

end gates;
```

The package provides a default value for generic when the component is instanced. A generic can only have one default value.

Generic parameters value information can be provided as a part of a configuration:

```
configuration new of comparator is
        for structural
          for u0: xor2 use entity work.xor2 (xor2_gen)
                generic map (m => 1.5 ns);
          end for;
        .....................
        end for;
end new;
```

Generics can be used also in behavioral architectures, usually as the parameters passed to some functions.

## 6.4 Configurations

Configurations are a primary design unit used to bind component instances to entities. As we have illustrated in previous chapters, design entities can have more than one architecture. The configuration can be used to specify which architecture to use for a specific instance if many architectures for an entity are available. For example, one architecture might be a behavioral model for entity, while another architecture might be a structural model for entity. The architecture used in the current design can be selected by specifying which architecture to use in the configuration, and recompiling only the configuration. In this way it also documents different versions of a design during the design process. A configuration, or selection of architectures, can be coded in an architecture or as a separately compiled design unit.

The simplest way of configuring is not to specify configuration at all. In that case the last architecture compiled will be used for an entity.

### *6.4.1 Configuration Outside of an Architecture - Default Configuration*

The default configuration is the simplest form of explicit configuration. It can be used for models that do not contain any blocks or components to configure. The default configuration specifies the configuration name, the entity being configured, and the architecture to be used for the entity. Consider the following example of two configurations called decode_2 and decode_3 of the entity *decoder*, that have two architectures decode_2_to_4 and decode_3_to_8:

```
use work.std_logic_1164.all;
entity decoder is
        port(en: in bit;
                a: in integer;
                q: out integer);
end decoder;

architecture decode_2_to_4 of decoder is
begin
        q <= 1 when (en = '1' and a = 0) else
                2 when (en = '1' and a = 1) else
                4 when (en = '1' and a = 2) else
                8 when (en = '1' and a = 3) else
                0;
end decode_2_to_4;

architecture decode_3_to_8 of decoder is
begin
        q <= 1 when (en = '1' and a = 0) else
                2 when (en = '1' and a = 1) else
                4 when (en = '1' and a = 2) else
                8 when (en = '1' and a = 3) else
                15 when (en = '1' and a = 4) else
                31 when (en = '1' and a = 5) else
                63 when (en = '1' and a = 6) else
```

```
                127 when (en = '1' and a = 7) else
                0;
end decode_3_to_8;

configuration decode_2 of decoder is
        for decode_2_to_4
        end for;
end decode_2;

configuration decode_3 of decoder is
        for decode_3_to_8
        end for;
end decode_3;
```

Configurations above have identified only the entity and architecture names. Two different architectures for a decoder entity are configured using two default configurations. The decoder entity does not specify number of bits on the input or number of bits on the output of the decoder. The data type of both input and output is integer. This enables multiple types of decoder to be supported, depending on the limitations of VHDL compiler. This type of configuration does not contain blocks or components to configure. The block configuration area from **for** clause to **end**...**for** clause is empty, and default will be used. The first configuration called decode_2 binds architecture decode_2_to_4 with the entity decoder, and the second configuration binds architecture decode_3_to_8 with the entity decoder and forms a simulatable (and synthesizable) object called decode_3.

Although both architectures from this example are behavioral, they can be synthesized by most of VHDL compilers.

### *6.4.2 Component Configurations*

Architectures that contain instances of components can also be configured. If the architecture contains other components then it is of a structural type. The components are configured in that case through component configuration statements. Consider as an example a multiplexer of the type 4-to-1. This multiplexer can be represented by a schematic diagram as in Figure 6.3. The components used in its design are defined using the following VHDL description:

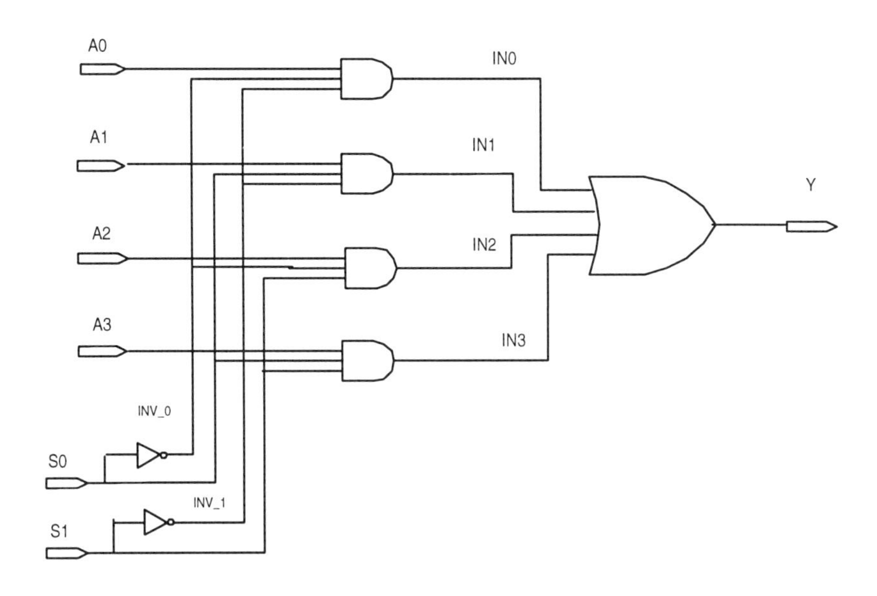


Figure 6. 3 Gate-level schematic for 4-to-1 multiplexer

```
use work.std_logic_1164.all;

entity inv is
        port( a: in std_logic;
              b: out std_logic);
end inv;

architecture beh of inv is
begin
        b <= not(a);
end beh;

configuration inv_1 of inv is
        for beh
        end for;
end inv_1;

use work.std_logic_1164.all;
```

```
entity and3 is
        port( a1, a2, a3: in std_logic;
                b1: out std_logic);
end and3;

architecture beh of and3 is
begin
        b1 <= a1 and a2 and a3;
end beh;

configuration and3_1 of and3 is
        for beh
        end for;
end and3_1;

use work.std_logic_1164.all;

entity or4 is
        port( a1, a2, a3, a4: in std_logic;
                b1: out std_logic);
end entity or4;

architecture beh of ior4 is
begin
        b1 <= a1 or a2 or a3 or a4;
end architecture or4;

configuration or4_1 of or4 is
        for beh
        end for;
end or4_1;
```

Once all components and their configurations are defined, entity and architecture for multiplexer are as follows:

```
use work.std_logic_1164.all;
entity multiplexer is
        port( a0, a1, a2, a3: in std_logic;
                s0, s1: in std_logic;
                y: out std_logic);
end multiplexer;
```

```
architecture struct of multiplexer is
        component inv
        port( a: in std_logic;
                b: out std_logic);
        end component;

        component and3
        port( a1, a2, a3: in std_logic;
                b1: out std_logic);
        end component;

        component or4
        port( a1, a2, a3, a4: in std_logic;
                b1: out std_logic);
        end component;

        signal inv_0,inv_1, in0,in1,in2,in3: std_logic;

begin
        i0: inv  port map (s0, inv_0);
        i1: inv  port map (s1, inv_1);
        a0: and3 port map (inv_0, inv_1, a0, i0);
        a1: and3 port map (s0, inv_1, a1, i1);
        a2: and3 port map (inv_0, s1, a2, i2);
        a3: and3 port map (s0, s1, a3, i3);
        o0: or4 port map (i0, i1, i2, i3, y);
end struct;
```

The simulator will use the last compiled architecture to build the executable design. This will work fine until more than one architecture exists for an entity. In that case a better method is to specify exactly which architecture to use for each entity. One possible approach is to specify configurations for each component at the lower level. An example of such a configuration for the multiplexer entity is as follows:

```
configuration muliplex_1 of multiplexer is
        for struct
                for i0:inv use configuration work.inv_1;
                end for;

                for i1:inv use configuration work.inv_1;
                end for;
```

```
            for all: and3 use configuration work.and3_1;
            end for;

            for o0: or4 use configuration work.or4_1;
            end for;
        end for;
end muliplex_1;
```

The other approach to component configuration is to use entity-architecture pair style. The same example of the multiplexer with this style of configuration is presented below:

```
configuration multiplex_2 of multiplexer is
        for struct
            for all: inv use entity work.inv(beh);
            end for;
            for a0: and3 use entity work.and3(beh);
            end for;
            for others: and3 use entity work.and3(beh);
            end for;
            for o0: or4 use entity work.or4(beh);
            end for;
        end for;
end multiplex_2;
```

The clause **others** is used for all components that have not yet been configured. In this way component a0 can use an architecture that is different from the other components to describe its behavior. This enables mixed-level modeling. One component can be at the gate level, and the other can be modeled at the behavioral level.

Another useful feature of configuration is that it can provide remapping of component ports. For this purpose the configuration port map clause can be used. It looks exactly the same as the component instantiation port map clause used in an architecture. The configuration port map specifies which of the component ports map to the actual ports of the entity. If the port names are different, then the port map clause will specify the mapping. The port map clause will map the port names of the component declarations, called the formal ports, to the port names of the entities from the library, called actuals.

VHDL further allows not only remapping of the ports, but also remapping of entities from libraries to components. This capability allows the names of components to differ from actual entities being mapped to them. The designer can choose a different kind of component with a similar function, and also do some port mapping at the same time.

### *6.4.3 Architecture Configurations*

Configuration can be used in the architecture declarative region. It specifies the configuration of parts used in the architecture. A separate configuration declaration is not needed to configure the components used in the architecture. This allows the designer to specify either the configuration or the entity-architecture pair to use for a particular component type. An example of configuration in the architecture is as follows:

```
architecture struct of multiplexer is
        signal inv_0,inv_1, in0,in1,in2,in3: std_logic;

        for i0: inv use entity work.inv(beh);
        for all: and3 use entity work.and3(beh);

begin
        i0: inv port map (s0, inv_0);
        i1: inv port map (s1, inv_1);
        a0: and3 port map (inv_0, inv_1, a0, i0);
        a1: and3 port map (s0, inv_1, a1, i1);
        a2: and3 port map (inv_0, s1, a2, i2);
        a3: and3 port map (s0, s1, a3, i3);
        o0: or4 port map (i0, i1, i2, i3, y);
end struct;
```

This example shows configuring of the gates i0, a0, a1, a2, and a3 in the architecture struct. It is selected from the library work and uses the architecture beh. Other gates in this example have not yet been configured. Obviously, there are some advantages in having a configuration clause in an architecture and there are disadvantages in compiling configurations separately. Configurations are useful for managing large designs. With the proper use of configurations, a top-down design approach can be implemented. This allows different descriptions of the design to be used for the different models needed at any point in the design process.

## 6.5 Questions and Problems

6.1. Draw a schematic diagram representing a 3-to-8 decoder using single and 2-input standard logic gates. Describe this decoder in VHDL as a pure structural design.

6.2. Draw a schematic diagram representing a 4-bit binary counter implemented using T-type flip-flops. Describe this counter in VHDL as a pure structural design.

6.3. Extend the design from the preceding example to 12 bits. Use a generate statement to describe the counter.

6.4. Given VHDL description:

```
library ieee;
use ieee.std_logic_1164.all;
use ieee.numeric_std.all;

entity example1 is
        port( a: in unsigned(3 downto 0);
                b: in unsigned(3 downto 0);
                y: out unsigned(3 downto 0));

architecture arch1 of example1 is
begin
        process(a,b)
        begin
                y(0) <= a(0) xor b(0);
                y(1) <= a(1) and b(1);
                y(2) <= a(2) or b(2);
                y(3) <= a(3) nand b(3);
        end process;
end arch1;
```

Draw a schematic diagram that represents a synthesized circuit from this description.

6.5. A small 1-bit arithmetic logic unit (ALU) performs the following operations:

$$R \leftarrow A+1$$
$$R \leftarrow A+B+C_{in}$$
$$R \leftarrow A$$
$$R \leftarrow A-1$$

R $\leftarrow$ not (A)
R $\leftarrow$ A and B
R $\leftarrow$ A or B
R $\leftarrow$ A xor B

Design the 1-bit ALU using VHDL and any type of modeling. Using the designed 1-bit ALU and structural-type modeling describe 16-bit ALU. Show the hierarchy of designs. Assume that the arithmetic addition generates the output carry $C_{out}$. Use a single selection signal to select the operation of the ALU.

6.6. Repeat the preceding problem assuming that the ALU consists of two components called arithmetic unit (AU) and logic unit (LU). Assume that a separate signal selects between arithmetic and logic operations.

6.7. Repeat the preceding problem under the assumption that you are allowed to use only your own components with simple gate components (and, or, not) as the starting point.

# 7 ADVANCED TOPICS IN VHDL

In this Chapter some of the advanced topics of VHDL are considered. They belong to the areas such as user packages and libraries, overloading, resolution functions and multiple drivers. They demonstrate how the user can create his/her own packages and libraries, reusability of previous designs, and some of the major advantages of VHDL to create one's own data types and operations, as was partially shown in the discussion on enumerated types and subtypes.

## 7.1 Packages and Libraries

A package is a compilable source design unit, written by a vendor or a user, which contains frequently used declarations. Once in a library, the group of declarations can be called with a use statement. Packages usually contain type declarations, constant declarations, and subprogram declarations. A package allows us to share information in different designs and isolate changes.

A package consists of two compilable parts: a package header declaration, and a package body declaration. A header declares the names of things that are declared in the package body declaration. The package header must be declared before the corresponding package body. A package contains:

1) Package header with

- subprogram declarations
- type declarations
- component declarations
- deferred constant declarations

2) Package Body with

- subprogram body
- deferred constant value

The separation of header and body in separate files allows isolating changes and minimizing the number of other recompiled design units.

Most vendors distribute packages in VHDL source form. The packages contain VHDL source-design units: entities, architectures, and packages. These can be compiled and put into a library. Such library is referred to as resource library. The compiled version of these packages can then be accessed with a use statement or configuration statement.

### 7.1.1 Constants and Deferred Constant Declarations

In the package my_units below a constant called unit_delay is declared. The package header gives the value of 2 ns to unit_delay, and there is no package body. After my_units is compiled into a library, unit_delay can be used in any subsequent design.

```
package my_units is
constant unit_delay: time := 2 ns;
end package my_units;

entity comparator is
port (a,b: in bit; c: out bit);
end entity comparator;

library my_library;
use my_library.my_units.all;

architecture new of comparator is
begin
        c <= not(a xor b) after unit_delay:
end architecture new;
```

The architecture new is separately compiled and refers to the previously compiled package my_units to access the value unit_delay for the signal assignment statement. One disadvantage of this type of constant declaration is that if the constant value (2 ns) is changed, the package and any design that refers to it must be recompiled.

Instead of recompiling a deferred constant can be used. The deferred constant is declared in the package header. Its actual value is deferred to the package body. When the simulation runs, the actual value is obtained from the package body. The package body can be recompiled without the recompilation of the package header. In this way a header description (interface) has not to be recompiled. The only condition is that the package header and package body are in separate files. An example of deferred constant declaration is given below:

```
package first is
        alpha: integer;
end package first;

package body first is
        constant alpha: integer :=15;
end body;
```

### *7.1.2 Subprograms in Packages*

Functions can be declared in packages so that the interface to the function is declared in the package header, and the actual body of the function in the package body. Header and body are then separately compiled. If the algorithm in the package body is changed, the package header does not need to be changed. Any architecture using the function does not have to be recompiled. As long as the arguments are the same, only the package body has to be recompiled. If we recompile the package header, we must recompile the body along with any design that uses the package function.

### *7.1.3 Component Declarations*

Rather than having component declarations in line in a structural design, it is more convenient to have it in a package. An example of component declaration without using package is given below:

```
entity comparator is
        port (a, b: in bit; c: out bit);
end comparator;

architecture structural of comparator is
signal i: bit;
component xor2
        port (x,y: in bit; z: out bit);
end component;
```

```
component inv
        port (x: in bit; z: out bit);
end component;
begin
        u0: xor2 port map (a, b, i);
        u1: inv port map (i, c);
end structural;
```

Another option is to use a library, called my_lib, which contains a compiled package named my_gates. The use statement identifies the library name and the package name. Component declarations are in the package. They do not have to be declared in the architecture. They occur in the package header. It is necessary to compile the package components prior to compiling the entity comparator since the use statement refers to something that already exists in the library. A package header is shown below:

```
package my_gates is

component xor2
        port (x, y: in bit; z: out bit);
end component;

component inv
        port (x: in bit; z: out bit);
end component;

end my_gates;
```

This package is used in the following example:

```
library my_lib;
use my_lib.my_gates.all;

entity comparator is
        port (a, b: in bit; c: out bit);
end comparator;

architecture structural is
signal i: bit;
begin
        u0: xor2 port map (a, b, i);
        u1: inv port map (i, c);
```

**end** structural;

It is important to remember that components defined by entity and architecture are not in the package, but separately in library WORK or my_lib.

### *7.1.4 Use Statement*

The use statement is used to access a compiled package from a library and make its contents visible in a subsequent design. This statement must be placed before the entity declaration or architecture declaration. There is the choice of selecting all items or a particular (named) item. The syntax is:

**use** library_name.package_name. {**all**, item}

If the all option is used, everything in the package can be accessed. The alternative is to identify particular functions or definitions that are to be used. For example, if there are two packages, pack1 and pack2, declarations that are to be used from these packages can be accessed as in the example below:

**use** my_lib.pack1.adder;
**use** my_lib.pack2.subtractor;
**entity** ......

The scope of the use statement refers only to the design unit it precedes.

Selected name can be used to get an object from a package or a library. For example unit_delay in a package my_units can be referred as to

my_units.unit_delay

To access a procedure in a package TEXTIO we specify:

textio.readline (1,a);

The general formats for a selected name are:

library_name.library_unit_name
package_name.package_element

However, before we can refer to a library, we first must declare it in a library declaration statement.

## 7.2 Resolution Functions and Multiple Drivers

VHDL provides a technology-independent and user-definable extensions to support hardware concepts, such as bussing, wiring, and three-state operation. The language allows user-written or vendor-written resolution functions, which resolve the meaning of shorting or "dotting" signals together. These functions are called automatically, which introduces the implicit function call. This call is made whenever a statement assigns a value using a signal assignment statement to a "resolved bus". Resolution function is associated with a particular data type, and any assignment to a signal of that type invokes the particular resolution function. The types like wired-and, wired-or, or three-state bus are possible.

Any signal in VHDL has its driver that determines the values on the signal. In the case of multiple signal drivers, resolution function must determine meaning (wired-or, wired-and, three-state, etc.). It is illegal in VHDL to have a signal with multiple drivers without a resolution function attached to that signal. A resolution function consists of a function that is called whenever one of the drivers for the signal has an event occur on it. The resolution function will be executed and will return a single value from all of the driver values. A resolution function has a single-argument input and returns a single value. The single-input argument consists of an unconstrained array of driver values for the signal that the resolution function is attached to. The resolution function will examine the values of all of the drivers and return a single value called the resolved value of the signal.

In VHDL, only one active driver for a signal per process should exist. In the following example, there is a signal y driven by two processes, process p1, and process p2.

```
architecture wrong of design is
signal y: bit3;
p1: process....
begin
        y <= '1';
end process;

p2: process
begin
        y <= 'z';
end process;
end wrong;
```

Although the signal drivers are in separate processes, this example requires resolution function for bit3 to indicate the meaning when 1 and z are driven. The resolution function should be automatically called when s is driven. The resolution functions needs to return a value that sets y to some value ('0', '1', or 'Z'). In this case a particular type should be associated to a resolution function, for example wired-or or wired-and type. A resolution function is associated with a subtype. Any signal that needs to be resolved is declared of that subtype, which is how a resolution function is tied to a subtype. The syntax is:

```
subtype resolve_type is resolution_function_base_type
```

For example:

```
type bit3 is ('0', '1', 'Z');
type bit3_vector is array (natural range <>) of bit3;

function resolve (s: bit3_vector) return bit3;

subtype res3 is resolve bit3;

signal data: res3;
```

Typically, a VHDL user does not write resolution functions, but uses prewritten functions. For simulation purposes, an arbitrary function can be written to resolve bus conflicts. Generally, as we have already shown in preceding example, a resolved signal is created in four steps:

- The signal's base type is declared.
- The resolved signal subtype is declared, which defines the name of the subtype and the resolution function.
- The resolution function is declared and defined.
- Resolved signals of the resolved type are declared.

Sometimes, it is convenient to declare resolved types in a separate package, and use them subsequently throughout designs. The following example illustrates this situation:

```
package resolved_types is
function my_function (data: in std_logic_vector) return bit;
```

```
        subtype resolved_bit is my_function bit;
end;
package body resolved_types is
  function my_function (data: in std_logic_vector) return bit is
        for i in data'range loop
        if data(i) = '0' then
        return '0';
        end if;
        end loop;
        return '1';
  end;
end;

use work.resolved_typesall;
entity new_design is
port(xin, yin: in bit; qout: out resolved_bit);
end entity new_design;

architecture arch1 of new_design is
begin
        qout <= xin;
        qout <= yin;
end architecture arch1;
```

While simulation tools support practically any user defined resolution function, synthesis tools usually support only wired-and, wired-or, and three-state functions.

## 7.3 Overloading

VHDL permits substituting already defined operators, such as +, -, and, or ,etc. with new definitions associated with the user data types. The operators can use existing symbolics to invoke vendor-written functions. This feature makes VHDL object-oriented, where an object is the data type and the operator, defined for that data type, works appropriately. When an existing operator or function is given a new or additional meaning, it is called overloading. There are a number of ways to overload objects in VHDL: by overloading subprogram names, overloading a number of parameters, or by overloading the operators.

Two functions can have the same name as long as they have different argument types. The functions are declared with a function declaration statement, and each function has the same name. VHDL discriminates the function based upon the type of argument. The following example shows how a subprogram can be overloaded by the argument type:

```
use.work.std_logic_1164.all;
package p_shift is
        type s_int is range 0 to 255;
        type s_array is array(0 to 7) of std_logic;
        function shifter(a: s_array) return s_array;
        function shifter(a: s_int) return s_int;
end package p_shift;

package body p_shift is
function shifter(a: s_array) return s_array is
         variable result: s_array;
begin
        for i in a'range loop
        if i = a'high then
                result(i):='0';
         else
                result(i):= a(i+1);
        end if;
        end loop;
        return result;
end shifter;

function shifter(a: s_int) return s_int is
begin
        return (a/2);
end shiftr;
end body p_shift;
```

The package p_shift contains two functions, both named shifter. Both functions provide a right-shift capability, but each function operates on a specific type. The compiler will pick up the appropriate function, based on the calling argument(s) and return argument. If, for example, return type does not match one given in definition of a function, the compiler will produce an error.

Another way of overloading is with a varying number of arguments. The functions can have the same name, but in their declarations they have a different number of arguments. The functions are discriminated at the time of call based upon the number of arguments that are listed in the argument list.

In VHDL, the name of a function can be in the form of an operator. For example, we call a function with the symbols '+', '-', '>', etc. Any of these operators can be overloaded by defining in a package a new meaning of the operator for a different data type. For example, a '+' operator (function) can be declared for arguments of bit_vector data type.

## 7.4 Questions and Problems

7.1. Specify when it is appropriate to use packages in VHDL and the advantages of using VHDL packages?

7.2. What are multiple drivers? Give a few examples of situations in which a signal has multiple drivers. What is the problem in simulation and design if the signals have multiple drivers?

7.3. What are the resolution functions? Give an example of the resolution function that resolves the problem of driving the bus signal from three different sources.

7.4. What is the overloading? Give examples of overloaded operators in the IEEE std_logic_1164 and numeric_std packages.

# 8 VHDL IN SIMULATION - TEST BENCHES

Before processing a design by synthesis tool, the designer usually wants to verify that the design performs according to the specification. This is almost always done by running a simulation. To simulate the circuit we need to provide not only a design description itself, but we need also to write a test bench which will provide verification of the operation of the circuit over time. Writing test benches consumes a considerable amount of time, sometimes even half of the cycle needed to finalize a new design.

The test bench may be written also in VHDL, and can be considered as a virtual circuit tester which generates and applies stimulus to our design description, and optionally verifies that the circuit is performing properly. Another way of specifying stimulus to the circuit is to use a waveform editor which enables visual entry of information which will be provided to the circuit as stimulus. The waveforms are in the form of timing diagrams. The simulator takes these waveforms as its inputs, together with the circuit description, and produces output waveforms.

In this Chapter we will concentrate on the more general way of specifying a test bench which assumes the use of one or more VHDL processes to specify stimulus to the circuit under test, or sometimes called unit under test. Writing test benches is a very tedious task because it requires us to specify the time for each signal transition. This process is error-prone due to the inconvenient visualization of time relationships between different waveforms. The only mechanisms available in VHDL to describe these relationships are wait statements or the use of transport clause in signal assignment statements. Still, these mechanisms are very powerful for describing test-vectors.

Test benches can be very simple, for example, just providing a sequence of test inputs to the circuit over time. Sometimes this sequence represents a simple periodic waveform used to describe a system clock. However, sometimes it can be a very complex one and stored in a file. The test bench has to provide a reading of test vectors from the file, and also the writing of results to a report file. Even more, it can be required to provide reference vectors and compare them with the output of the unit under test during simulation to provide a pass or fail indication.

There is an important advantage in writing a test bench in the same language as the unit under test: the designer does not need any specialized simulation tool or simulation language. The unit under test becomes a part of the test bench, and can be considered as an instance of the actual design.

VHDL provides another facility to communicate information on the status of the simulation process to the designer. It is an assert and report statement, which eases reporting on the status of simulation, and is used to provide conditional debugging messages on the status of the unit under test.

## 8.1 Writing Simple Test Benches

The simplest way to understand how test benches work is to write one for a simple example design. We will consider two example circuits in this case: a 4-to-1 multiplexer as a combinational circuit, and a timer, as a typical sequential circuit.

### *8.1.1 Multiplexer Test Bench*

First of all, lets look to a VHDL description of a 4-to-1 8-bit multiplexer:

```
library ieee;
use ieee.std_logic_1164.all;
entity mux41 is
          port ( a, b, c, d: in std_logic_vector(7 downto 0);
                    sel: in std_logic_vector(1 downto 0);
                    y: out std_logic_vector(7 downto 0);
end mux41:

architecture arch1 of mux41 is
begin

          y <=      a when sel = "00" else,
                    b when sel = "01" else,
                    c when sel = "10" else,
                    d when sel = "11";
end arch1;
```

In order to test it, this model requires generation of multiplexer inputs a, b, c, and d, as well as generation of select inputs. Depending on the value of the select input, output from the unit under test will be equal to the value of the selected input. The relationship between the waveform generator, which also provides test-vectors, and the unit under test is illustrated in Figure 8.1 The waveform generator appears as a component at the same level of design hierarchy as the unit under test itself. However, the whole test bench appears as an entity without ports, on the hierarchical level just above the waveform generator and unit under test. The waveform generator has to provide test-vectors for input and select signals, but also can observe the values of the generated output from the multiplexer and report on the operation of the unit under test. The mechanism readily available in VHDL to describe the waveform generator is the process. In this case a single process is

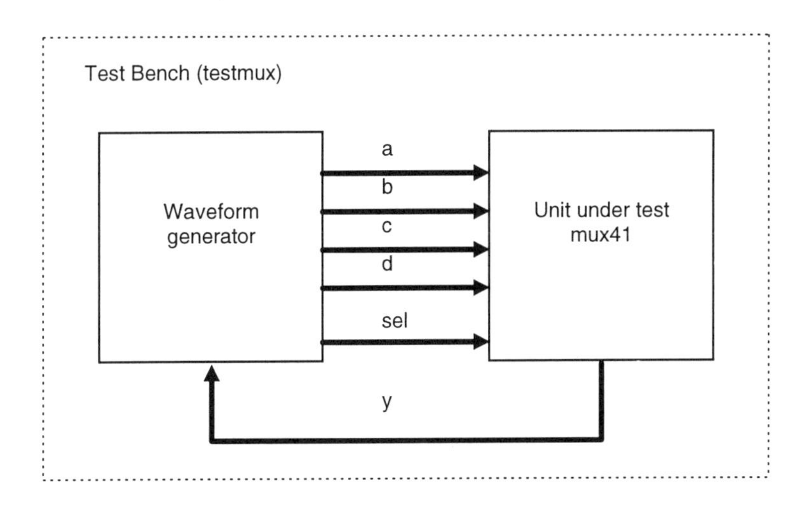


Figure 8.1 Test bench for 4-to-1 multiplexer

sufficient to describe possible scenarios. In more complex situations, multiple processes will be a more appropriate way to describe generation of the input waveforms to the unit under test. The topmost entity which represents test bench, called testmux, has no ports, but still must be declared. The unit under test is declared in the architecture declarative section, and instantiated within the architecture body. The new component, waveform generator, is described as a process. Within this process, wait statements are used to control times at which signals generated by the waveform generator change their values. Inputs to the multiplexer are held permanently at constant values, and select lines are changing their values every 50 ns (specified by constant interval). All waveforms, including output y from the multiplexer, are obtained running simulation and shown in Figure 8.2. The VHDL description of the testbench for the 4-to-1 multiplexer is given below:

```
library ieee;              -- Load the ieee 1164 library
use ieee.std_logic_1164.all; -- Make the package 'visible'

use work.all;              -- Make the contents of 'work' available

-- The top level entity of the test bench has no ports...

entity testmux is
end testmux;

architecture test of testmux is

  -- First, declare the lower-level entity (mux41)...

component mux41
 port  (a, b, c, d: in std_logic_vector(7 downto 0);
         sel: in std_logic_vector(1 downto 0);
         y: out std_logic_vector(7 downto 0);
end component;

--Next, declare internal signals to connect internal components

signal a, b, c, d: std_logic_vector(7 downto 0);
signal intsel: std_logic_vector(1 downto 0);
signal inty: std_logic_vector(7 downto 0);


begin
-- Create the instance of the multiplexer unit under test...

uut: mux41 port map(a , b, c, d, sel, y);

-- Now run a process to apply stimulus over time...
        waveforrmgen: process
                constant interval: time := 50 ns;
        begin
                a <= "00010001";  -- hex 11
                b <= "00100010";  -- hex 22
```

```
                    c <= "00110011";  -- hex 33
                    d <= "01000100";  -- hex 44
                    sel <= 0;
            wait for interval;
                    sel <= 1;
            wait for interval;
                    sel <= 2;
            wait for interval;
                    sel <= 3;
            wait for interval;
                    sel <= 0;
            wait; -- introduced to stop simulation; otherwise it continues
                    -- indefinitely
            end process waveformgen;
    end test;
```

The last wait statement has no condition. It is used to stop simulation after the first execution of the process body. Otherwise, the process would , after reachir g the last statement, repeat execution from its beginning. The simulation run was performed using Accolade's PeakVHDL system.

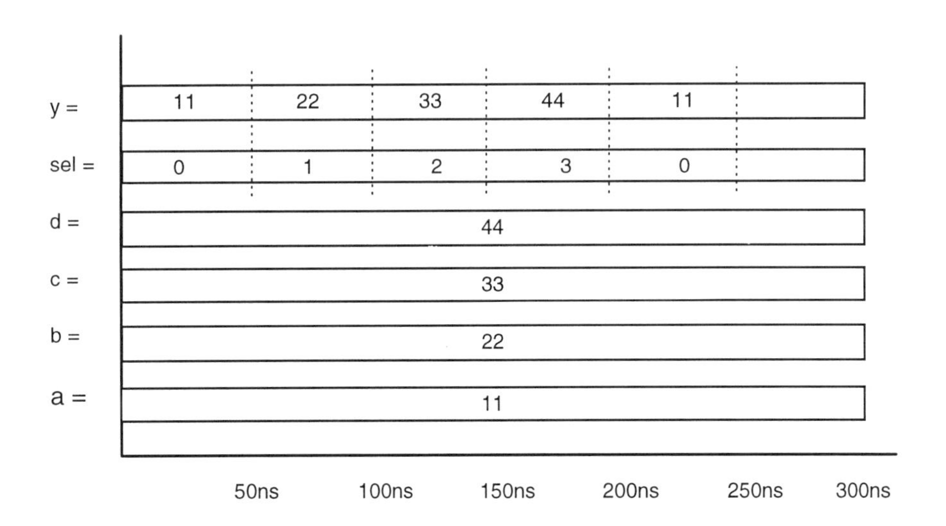


Figure 8.2 Waveforms produced by the Simulator running test bench

### *8.1.2 Timer Test Bench*

The second example is a 16-bit timer which allows to be initialized to an initial value (count) supplied through input data lines. The initial count is loaded into the timer upon activation of inittimer control signal. The timer can be started by activating starttimer signal to count down until reaches value 0, when it generates time-out signal to indicate that the desired time interval has expired. The external circuitry has to deactivate starttimer signal upon detection of the timeout active output.

The VHDL description of the timer is given below:

```
library ieee;
use ieee.std_logic_1164.all;

entity timer is
port(clk, reset, inittimer, starttimer: in std_logic;
        data: in integer range 0 to 2**16-1;
        timeout: out std_logic);
end timer;

architecture beh of timer is

begin

        tim: process(clk)
                variable cnt: integer range 0 to 2**16-1;
        begin
        if reset = '1' then
                cnt:=1;
        elsif (clk'event and clk ='1') then
                if inittimer = '1' then
                        cnt:= data;
                elsif starttimer = '1' then
                        cnt:= cnt-1;
                else
                        cnt:=cnt;
                end if;
                if cnt = 0 then
                        timeout <= '1';
                else
                        timeout <= '0';
                end if;
```

```
        end if;

        end process tim;

    end beh;
```

The test bench for this example is illustrated in Figure 8.3.

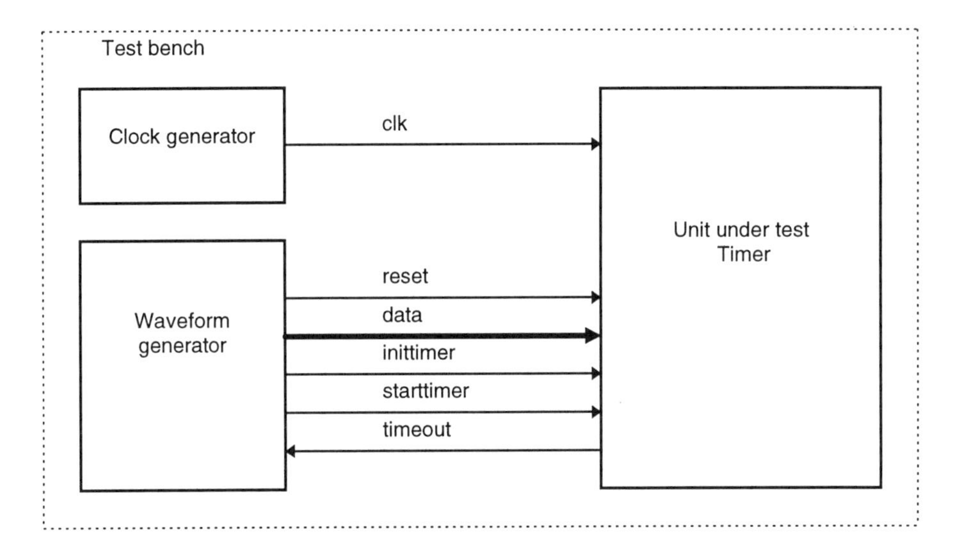


Figure 8.3 Timer circuit and its test bench

```
library ieee;
use ieee.std_logic_1164.all;

entity testtimer is
end testtimer;

use work.all;

architecture test of testtimer is
        component timer
        port (clk, reset, inittimer, starttimer: in std_logic;
        data: in integer range 0 to 2**16-1;
         timeout: out std_logic
        );
```

```
        end component;

        signal clk, reset, inittimer, starttimer : std_logic; -- top level
                                                        -- signals
        signal data: integer range 0 to 2**16-1;
        signal timeout: std_logic;

        signal clockcycle: natural := 0; -- used for debugging purposes to
                -- keep track of number of clock cycles in simulation run
begin
        uut: timer port map(clk, reset, inittimer, starttimer, data,
                timeout);

        -- first process sets up a 10MHz background clock
        clock: process
                begin
                clockcycle <= clockcycle + 1;
                        clk <= '1';
                wait for 50 ns;
                        clk <= '0';
                wait for 50 ns;
        end process clock;

-- this process applies stimulus to reset, initialize and start the timer...
        stimulus: process
        begin
                reset <= '1';
                wait for 140 ns;
                reset <= '0'; -- reset the timer
                inittimer <= '1';
                data <= 10;    -- initialize timer to 10
                wait for 120 ns;
                inittimer <= '0'; -- finish initialization
                starttimer <= '1'; -- start counting down
                wait until timeout ='1'; -- wait
                  assert (timeout = '1')
                  report "Timing violation" severity error; -- if timeout
                                                -- signal does not appear
                starttimer <='0';
                inittimer <= '1';
                data <=5;
                wait for 80 ns;
```

```
                    inittimer<='0';
                    starttimer <= '1';
                    wait until timeout ='1';
                       assert (timeout = '1')
                       report "Timing violation" severity error;
                    starttimer <='0';
                    inittimer <= '1';
                    data <=7;
                    wait for 120 ns;
                    inittimer<='0';
                    starttimer <= '1';
                    wait until timeout ='1';
                       assert (timeout = '1')
                       report "Timing violation" severity error;
                    wait;
            end process stimulus;
    end test;
```

Two processes are provided to generate proper waveforms to the unit under test. The clock process generates periodic waveform with the cycle time of 100 ns with 50% duty cycle. The other process, called stimulus, generates the waveforms which provide initial reset, initialization of the timer with the constant initial values, and start of the count-down process. It also senses the timeout line. The assert statements, followed by report statements, are used to report of certain events, in this case the absence of the timeout signal. If the timeout signal does not change the value to ‘1’, message “Timing violation” appears, and the simulation process stops. The final wait statement is used to suspend simulation indefinitely after desired inputs have been applied.

Both presented examples use the method of generating test vectors within the test bench itself. This is achieved using:

- signal assignment statements to specify values of the signals which have relatively low number of changes, such as resets, inittimer, etc.
- relative time generated signals (if absolute time generated signals are used they must not be mixed with relative time assignments) with wait-for statements providing desired timing conditions
- wait-until statements to achieve test-vector generation while waiting on the occurrence of specified events (in this case timeout =’1’).

## 8.2 More on Test Benches

Even from the simple examples, we can see the main principles used in test bench design. First, we have to specify VHDL design of the unit under test. Full power of VHDL can be used in this case specifying the entity of the unit under test and one or more its architectures. For testing purposes, it is often desirable to have more than one architecture, because it enables faster exploration of various alternatives, but also faster simulation if using more abstract architecture(s) first. The test bench allows easy change of architecture by using VHDL configuration which binds the entity with a specific architecture at simulation time.

Another advantage of writing a test bench is that we can extend it with the unit which will automatically verify behavior of the unit under test. This can be accomplished by adding to the test bench a unit which compares results produced by the unit under test and expected (or desired) values for specific states of the unit under test. This requires not only generation of the stimulus to the unit under test, but also generation of expected response to the stimulus.

Further, mechanism for test vectors generation must be selected. In the examples of the preceding section, all test vectors were generated within the test bench, being part of the processes which generate waveforms. This method is suitable for situations in which there are not too many transitions of stimulus signals, or if those transitions can be described by repetitive patterns, such as clocks, that fit nicely into the VHDL constructs. However, more complex test-vectors require separation and a higher degree of independence from the processes that use them. To achieve this, generally two methods are available:

- to specify test vectors in the form of arrays of constants that are separated from the test bench processes, possibly in test-vector packages, or
- by specifying test-vector in the VHDL files, being more suitable for situation of very large sequences of test vectors

Also, results of simulation usually have to be stored and made available for further analysis. Some simulators provide graphical presentation of both the stimulus and the results in the form of timing diagrams. However, the generic VHDL mechanisms are in the form of reporting using assert-report mechanism, and external files which can be used to save results of simulation. The assert-report mechanism is useful for generation of simulation results messages to the screen, and is usually combined with debugging mechanism of the simulator (such as step-by-step simulation). In the case that results have to be analyzed off-line, all messages can be stored to the files as they are generated.

The framework of the test bench, presented in the examples in the preceding section, can now be extended to a more general one as it is illustrated in Figure 8.4.

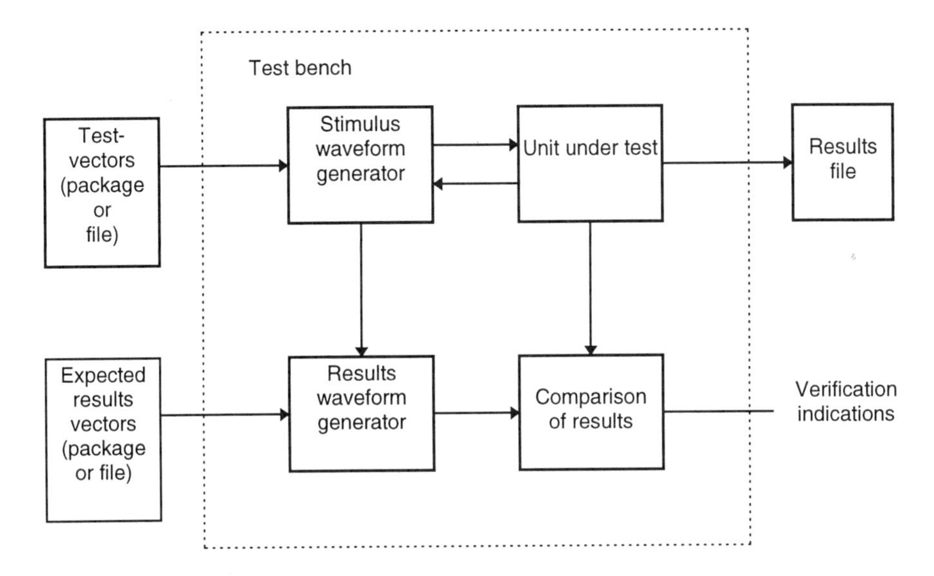


Figure 8.4 Test bench framework

The most important parts of a test bench are stimulus and result waveform generators. They can be just VHDL processes that generate waveforms based on pre-defined test patterns sequentially, or conditionally based on responses from the unit under test. Sometimes it is desirable to include the parts of the modeled hardware system within the test bench to make it easier to generate the required test and expected results vectors.

### *8.2.1 Generation of Repetitive Signals*

Most of the digital systems are driven by clocks. Depending on the design approach, and one or more system clocks (or phases of the clock) may be used. Also, in the simulation some events that even do not have a periodic nature can be represented by repetitive patterns without losing generality, to simplify the description of stimulus vectors. VHDL provides a number of ways to generate repetitive signals, and they are explored in this subsection.

It is convenient to specify generation of the repetitive waveforms in separate processes, because that sort of behavior is reusable in many designs. Examples of models that generate repetitive (clock) signals are shown below.

```
entity clocks is
        port(singlephase1, singlephase2: inout bit;
                twophase0, twophase1: inout bit);
end clocks;

architecture various of clocks is
        constant periodshort: time:=10ns;
        constant periodmedium: time := 25ns;
        constant periodlong: time:=50ns;
begin
        -- clock with 50% duty cycle and frequency 100MHz
        singlephase1 <= not singlephase1 after periodshort/2;

        -- clock with 50% duty cycle and frequency 20MHz
clk1: process
begin
        wait for (periodlong/2);
                singlephase2 <= '1';
        wait for (periodlong/2);
                singlephase2 <= '0';
end process clk1;

twophase0 <= not twophase0 after periodmedium/2;
twophase1 <= twophase0 after (periodmedium/2);

end various;
```

Clocks generated by the above VHDL description are shown in Figure 8.5.

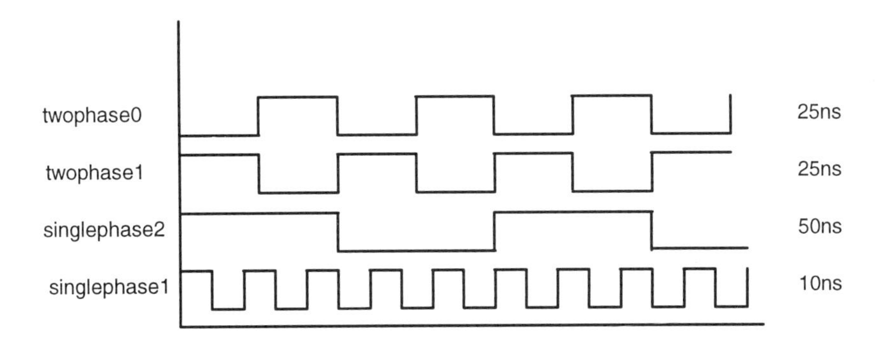


Figure 8.5 Various clocks generated by preceding Example

It is important to note that the clock output port signals are declared in inout mode because they are both read and written in order to generate correct values.

Another example is of multiphase clock often required in different sorts of digital systems. The example of four-phase clock is given below:

```
entity multphase is
        port(phase0, phase1, phase2, phase3: inout bit);
end multphase;

architecture various of multphase is
        constant timeshort: time:=10ns;
begin
clk4: process
begin
        -- period of the clock is 4 * timeshort = 40ns
        wait for (timeshort);
                phase0 <='0';
        wait for (timeshort * 3);
                phase0 <= '1';
end process clk4;
        phase1 <= phase0 after (timeshort);
        phase2 <= phase1 after (timeshort);
        phase3 <= phase2 after (timeshort);
end various;
```

The waveforms generated by the simulator which correspond to this example are shown in Figure 8.6

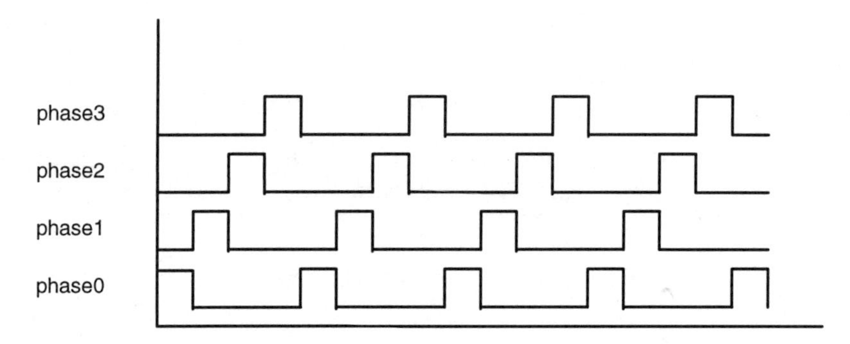


Figure 8.6 Four-phase clock corresponding to preceding Example

In both preceding examples, clocks are generated in separate entities that have only output ports. Obviously, when used for test benches these components can be instantiated as any other components and connected to the rest of the test bench. Another possibility, which is used throughout this Chapter, is to use a clock's architecture body description to implement a clock of desired features as the process within a test bench architecture body.

### *8.2.2 Using Other VHDL Features*

VHDL provides a number of mechanisms that help simulation and writing test benches. The assert statement is used to conditionally display a text message to the screen. Its format is

```
assert condition_expression
        report text_string
        severity severity_level;
```

When executing an assert statement, the condition_expression is evaluated to a Boolean value (true or false). If the condition expression evaluates to false, the text specified in the optional report statement is displayed in the simulator transcript window. The effect of an assert statement is similar to an if statement, but the action is undertaken if the condition evaluates to false, not to true value. The severity statement clause indicates the severity level to both designer and the simulator. Depending on the severity level and the simulator implementation, appropriate action will be undertaken by the simulator. Severity levels are presented in Table 8.1.

Table 8.1 Severity levels

| Severity level | Meaning and use |
|---|---|
| note | General information on the status of model simulation |
| warning | Existence of potential problems |
| error | Conditions occurred that will cause the error |
| failure | Conditions with disastrous effects occurred |

The most common use of assert and report statements is to display information about the status of the model during a simulation run. The limitation of VHDL is that report supports only a single string as its argument, and different work arounds are needed to display multi-line strings. Usually, we have to use concatenation of single-line strings using the special character constants CR (carriage return) and LF (line feed). The more serious limitation is that there is no support to display other data types. In order to display data types such as standard logic, integers, arrays o other types, the designer must provide conversion functions that will convert from a specific data type to a formatted string.

Another way to simplify test benches is to use more advanced features of VHDL, especially multiple processes with loops. The timer test bench shown in the preceding section demonstrated how multiple processes can be used to make an efficient test bench for a simple circuit. However, the power of processes and loops can be seen from the following example of a more complicated test bench for the same timer circuit.

```
library ieee;
use ieee.std_logic_1164.all;

-- package to store constants and test vectors
package test_vector is
          constant novectors: integer:=10;
          subtype word is integer range 0 to 2**16-1;
          type tvector is array(0 to novectors-1) of word; -- new type
          constant tv: tvector := (10, 15, 8, 11, 35, 4, 23, 17, 2, 12); -- test
                                                                          -- vectors
end test_vector;
```

```
entity testtimer is -- test bench entity
end testtimer;

library ieee;
use ieee.std_logic_1164.all;
use work.test_vector.all; -- use package with test vectors
use work.all;

architecture testnew of testtimer is
        component timer
        port (clk, reset, inittimer, starttimer: in std_logic;
        data: in integer range 0 to 2**16-1;
         timeout: out std_logic
        );
        end component;
        constant clockperiod: time:=100ns;
        constant datahold: time:=20ns;
        signal clk, reset, inittimer, starttimer : std_logic; -- top level

                                                        -- signals
        signal data: integer range 0 to 2**16-1;
        signal timeout: std_logic;
        signal finished: boolean := false; -- new signal which
                -- indicates that all test vectors have been applied

        begin
        uut: timer port map(clk, reset, inittimer, starttimer, data,
                        timeout);

        -- The first process sets up a 10MHz background clock
        clock: process
                variable temp: std_logic:= '1';
                begin
                while finished /= true loop -- loop performed until
                                                        -- finished is true
                wait for clockperiod/2;
                        temp := not temp;
                        clk <= temp;
                end loop;
                wait;           -- stop simulation
        end process clock;
```

```
-- this process applies stimulus to reset, initialize and start the timer...
        stimulus: process
        begin
                reset <= '1';
                wait for clockperiod;
                reset <= '0'; -- reset the timer

                for i in 0 to novectors -1 loop
                inittimer <= '1';
                data <= tv(i);   -- initialize timer with next test vector
                wait for clockperiod + datahold;
                inittimer <= '0'; -- finish initialization
                starttimer <= '1'; -- start counting down
                wait until timeout ='1'; -- wait
                        assert (timeout = '1')
                        report "Timing violation" severity error; -- if
                                -- timeout signal does not appear

                end loop;
                finished <= true;
                wait;
        end process stimulus;
end testnew;
```

A single loop is used to perform the sequence of operations of initializing the timer by data from the array of constants stored in package test_vector, starting it and waiting for the timeout signal. The constants are of the new type, word, and obviously can be any type that allows arrays. It becomes easy to change the values of test vectors without respecifying the stimulus process, and perform another simulation run. Other circuit parameters, such as clock period and data hold time are kept as constants providing easy change without the change of body of the respective process.

### *8.2.3 Using Files with Text I/O*

VHDL makes it possible to store test vectors in data files, open one or more data files, read lines from those files, and parse the lines to form individual data elements that are applied as stimulus to the signals of the unit under test. The main advantage of this approach to store test vectors is that only one simple test bench has to be written, and the change of the test actually is done by changing the file from which the test vectors are read.

To support the use of the files, VHDL provides a file data type, and a number of built-in functions for opening, reading from, and writing to file data types. The standard TEXTIO package expands on built-in features for manipulation of files by adding text parsing and formatting functions, and some other extensions. While VHDL 1076-1987 supports file data type, VHDL 1076-1993 adds features needed for better control of opening and closing data files, as well as more consistent syntax constructs. When declaring an object to be of file data type, it is implicitly assumed that file is opened, and there is no mechanism to explicitly close files. Objects of file type can be read from and written to using functions and procedures that are provided in the standard library (such as read, write, and endfile), and additional functions and procedures for formatting of data read from files is provided in the textio library. Files in VHDL are sequential with no provision to read from a specific location, or to write to a specific location in the file. In order to use files, the type of its contents must first be declared, such as

**type** file_of_integers **is** file **of** integer;

The new type, file_of_integers, consists of a sequence of integer values. If we want to create a new object of this file type we have to declare it as in the following example:

**file** newfile: file_of_integers **is in** "filename.asc";

This statement creates the object newfile and opens the indicated disk file. From this point on, we can use built-in procedures to access data in the file (read them, or write them).

The 1993 VHDL has improved provisions to manipulate files, especially to control opening and closing of files, which help to resolve situations such as enabling two parts of the same design description to open the same file at different points. These new provisions are given in the form of built-in functions:

file_open(f, fname, fmode) - opens the file with a given file object f, file name fname, and file mode fmode (read_mode, write_mode, or append_mode).

file_open(status, f, fname, fmode) - opens the file, but also returns the status of file open request (open_ok, status_error, name_error, or mode_error)

file_close(f) - closes the specified file.

read(f, object) - reads one field of data from the file into the object; after reading data, the file marker is advanced to the start of the next data field in the file.

write(f, object) - write the data contained in the object to the file; after writing, the file marker is advanced to the next position in the file.

endfile(f) - returns a Boolean true value if the current file marker is at the end of the file.

If we want to access data in the file we can write an architecture like the following one:

```
architecture vhdl1987 of accessfile is
begin
read_data: process
        type file_of_integers is file of integer;
        file newfile: file_of_integers is in "testvector.asc"; -- test vectors
                                        -- in testvecor.asc ascii file
        variable vect: integer;         -- input test vector
        variable count: integer:=0; -- keeps track of number of data read
        begin
                while not endfile(newfile) loop
                read(newfile, vect); -- read next test vector
                count:=count + 1;
                end loop;

        end process read_data;
end vhdl1987;
```

The following example is the modified testbench for the timer circuit.

```
library ieee;
use ieee.std_logic_1164.all;
use std.textio.all; -- make visible textio package
use work.all;

entity testtimer is -- test bench entity
end testtimer;
```

```
architecture testnew of testtimer is
        component timer
        port (clk, reset, inittimer, starttimer: in std_logic;
        data: in integer range 0 to 2**16-1;
         timeout: out std_logic
        );
        end component;

        constant clockperiod: time:=100ns;
        constant datahold: time:=20ns;
        signal clk, reset, inittimer, starttimer : std_logic; -- top level
                                                          -- signals
        signal data: integer range 0 to 2**16-1;
        signal timeout: std_logic;
        signal finished: boolean := false; -- new signal which indicates
                             -- that all test-vectors have been applied


begin
        uut: timer port map(clk, reset, inittimer, starttimer, data,
                             timeout);

        -- The first process sets up a 10MHz background clock
        clock: process
                variable temp: std_logic:= '1';
                begin
                while finished /= true loop -- loop performed until
                                                          -- finished is true
                wait for clockperiod/2;
                        temp := not temp;
                        clk <= temp;
                end loop;
                wait;              -- stop simulation
        end process clock;

-- this process applies stimulus to reset, initialize and start the timer...
        stimulus: process
                type file_of_integers is file of integer;
                file testfile: file_of_integers is in "testvector.asc"; -- test
                -- vectors in testvecor.asc ascii file
                variable tv: integer;          -- input test vector
```

```
        begin
                reset <= '1';
                wait for clockperiod;
                reset <= '0'; -- reset the timer

                while not endfile(testfile) loop -- do for all test
                                    -- vectors in file
                read (tv, testfile); -- read next test vector
                inittimer <= '1';
                starttimer <= '0';
                data <= tv;    -- initialize timer with next test
                                                -- vector
                wait for clockperiod + datahold;
                inittimer <= '0'; -- finish initialization
                starttimer <= '1'; -- start counting down
                wait until timeout ='1'; -- wait
                  assert (timeout = '1')
                  report "Timing violation" severity error; -- if
                -- timeout signal does not appear
                end loop;
                finished <= true;
                wait;
        end process stimulus;
end testnew;
```

Obviously, the type of object which is stored in the file can be much more complex than the one shown in the preceding example. For example, objects can be records containing fields representing various input signals.

The TEXTIO package is a part of language and resides in a VHDL library called std. It defines a single file type called text to represent a file consisting of variable length strings. An access type, line, is also provided to point to such strings. A number of overloaded procedures called read and write provide the reading and writing of data to or from an object of type line. The TEXTIO package is shown in Appendix.

## 8.3 Questions and Problems

8.1. How do you see the role of simulation in the verification of a digital system design.

8.2. What is a test bench? How does VHDL facilitate the writing of a test bench?

8.3. What is the role of components and processes in writing test benches?

8.4. What VHDL means are used to implement input stimuli to a unit under test?

8.5. How are results of the simulation transferred to the modeler? What is the role of assert and report statements?

8.6. Describe generation of repetitive signals (single and multi-phase clocks)?

8.7. Describe the use of arrays of constants for the generation of input vectors.

8.8. Describe the use of files as data types that support simulation in VHDL.

# 9 SYNTHESIZING LOGIC FROM VHDL DESCRIPTION

The purpose of this Chapter is to demonstrate the different ways in which all types of logic can be modeled and synthesized. First, we consider purely combinational logic including some of the standard combinational logic functions, then standard sequential logic circuits, and, finally, general sequential logic represented by finite state machines.

## 9.1 Combinational Logic

Combinational logic circuits are commonly used in both the data path and control path of more complex systems. They can be modeled in different ways using signal assignment statements which include expressions with logic, arithmetic and relational operators, and also can be modeled using if, case and for statements. If conditional signal assignment statements are used, then the selected signal assignment statement corresponds to the if statement, and the conditional signal assignment statement corresponds to the case statement. However, all signal assignment statements can also be considered as processes that include single statement. This further means that they are always active (during the simulation process) waiting for events on the signals in the expressions on the right side of the assignment statement. Another way of modeling combinational logic is to use processes and sequential statements within processes.

### *9.1.1 Logic and Arithmetic Expressions*

Both logic and arithmetic expressions may be modeled using logic, relational and arithmetic operators. The expressions take the form of continuous dataflow-type assignments. Some VHDL operators are more expensive to compile because they require more gates to implement ( like programming languages where some operators take more cycles to execute). The designers need to be aware of these factors. If an operand is a constant, less logic will be generated. If both operands are constants, the logic can be collapsed during compilation, and the cost of the operator is zero gates. Using constants wherever possible means that the design description will not contain extra functionality. The design will be compiled faster and produce a smaller implementation.

## Logical Operators

VHDL logical operators were presented in Table 1.2 of Chapter 2. These operators are defined for the types bit, std_logic, boolean and arrays of bit, std_logic or boolean (for example, bit_vector or std_logic_vector). The synthesis of logic is fairly direct from the language construct, to its implementation in gates, as shown in the following examples:

```
library ieee;
use ieee.std_logic_1164.all;

entity logic_operators_1 is
        port (a, b, c, d: in std_logic; y: out std_logic);
end logic_operators_1;

architecture arch1 of logic_operators_1 is
signal e: bit;
begin
        y <= (a and b) or c; --concurrent signal assignments
        e <= c  or d;
end arch1;
```

Schematic diagram corresponding to this example is shown in Figure 9.1.

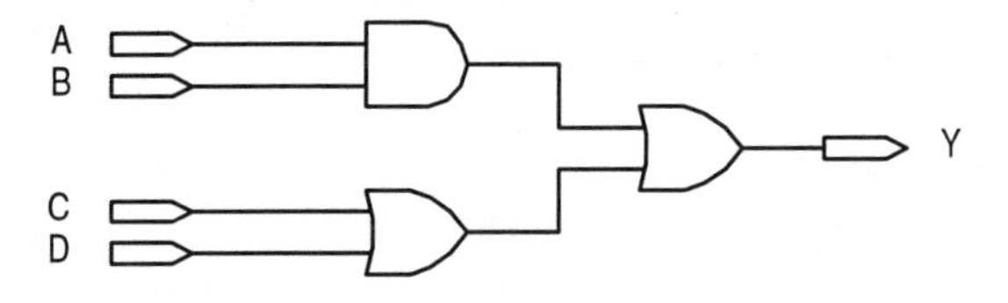


Figure 9.1 Schematic diagram corresponding to example

```
library ieee;
use ieee.std_logic_1164.all;

entity logic_operators_2 is
        port (a, b: in std_logic_vector (0 to 3);
                y: out std_logic_vector (0 to 3));
```

```
end entity logical_ops_2

architecture arch2 of logic_operators_2 is
begin
        y <= a and b;
end architecture arch1;
```

Schematic diagram corresponding to this example is shown in Figure 9.2.

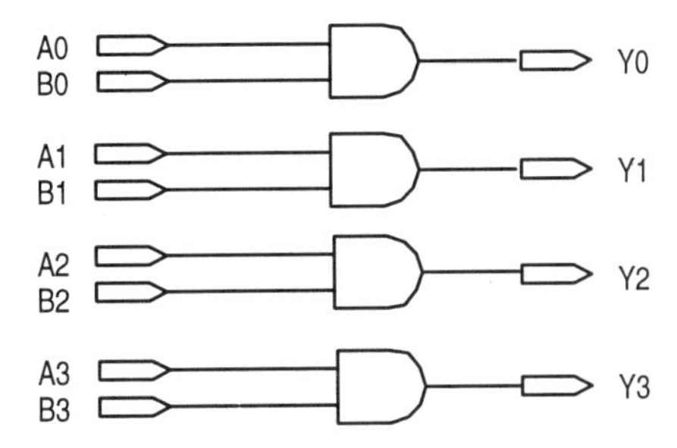


Figure 9.2 Schematic diagram corresponding to 4-bit "and"

## Relational Operators

VHDL relational operators were presented in Table 1.2 of Chapter 2. The simple comparisons operators ( = and /= ) are defined for all types. The ordering operators ( >=, <=, >, < ) are defined for numeric types, enumerated types, and some arrays. The resulting type for all these operators is boolean. The simple comparisons, equal and not equal, are cheaper to implement (in terms of gates) than the ordering operators. To illustrate, the example below uses an equal operator to compare two 4-bit input vectors:

```
library ieee;
use ieee.std_logic_1164.all;

entity relational_operators_1 is
        port (a, b: in std_logic_vector (0 to 3); y: out boolean);
end relational_operators_1;

architecture arch1 of relational_operators_1 is
```

```
begin
        y <= a = b;
end arch1;
```

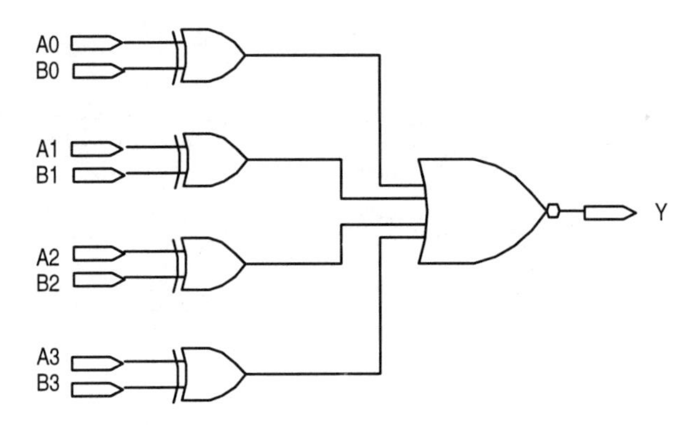


Figure 9.3 Schematic diagram corresponding to 4-bit equality comparator

Schematic diagram corresponding to this example is shown in Figure 9.3. The second example uses a greater-than-or-equal-to operator ('>='):

```
library ieee;
use ieee.std_logic_1164.all;

entity relational_operators_2 is
        port (a, b: in integer range 0 to 15; y: out boolean);
end relational_operators_2;

architecture arch2 of relational_ops_2 is
begin
        y <= a >= b
end arch2;
```

As it can be seen from the schematic corresponding to this example, presented in Figure 9.4, it uses more than twice as many gates as the previous example.

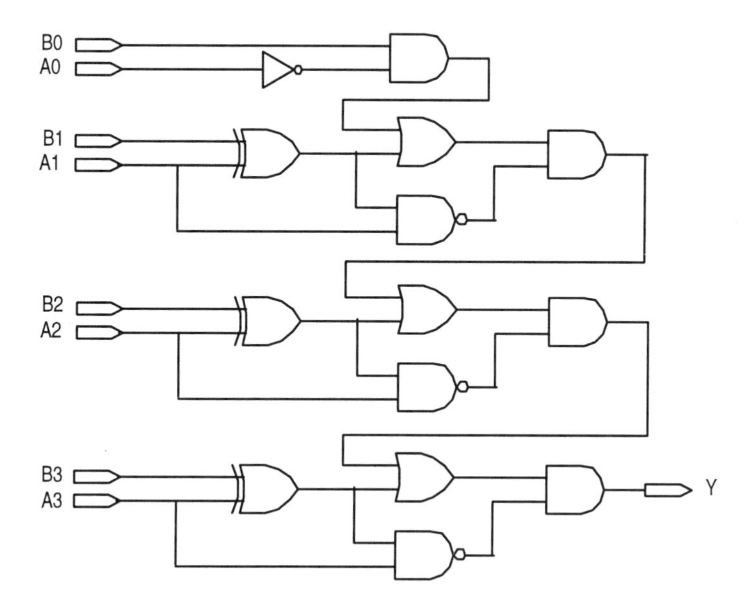


Figure 9.4 Schematic diagram corresponding to "a =>b" comparator

## Arithmetic Operators

The list of VHDL arithmetic operators is given in Table 1.2 of Chapter 2. While the adding operators ( +, - ) are fairly expensive in terms of number of gates to implement, the multiplying operators (*, /, **mod**, **rem**) are very expensive. Implementation of these operators is highly dependent on the target technology. Some implementation aspects for Altera's FPLDs are discussed in chapters of Part 2. Examples of the use of arithmetic operators and parentheses that can be used to control synthesized logic structure are shown below:

```
library ieee;
use ieee.std_logic_1164.all;
use ieee.numeric_std.all;

entity arithmetic_operators is
        port (a, b, c, d: in unsigned(7 downto 0);
        y1, y2: out unsigned(9 downto 0);
```

```
end arithmetic_operators;

architecture arch1 of arithmetic_operators is
begin
        y1 <= a + b + c + d;
        y2 <= (a + b) + (c+d);
end arithmetic_operators;
```

Another possibility is to enclose signal assignment statements into a process with all input signals in the sensitivity list of a process. From the synthesis point of view, there will be no difference. However, simulation can be simpler if a process is used to describe the same circuit.

### *9.1.2 Conditional Logic*

VHDL provides two concurrent statements for creating conditional logic:

- conditional signal assignment, and
- selected signal assignment

and two sequential statements for creating conditional logic:

- if statement, and
- case statement

Examples of the use of these statements for creating conditional logic are given below:

```
library ieee;
use ieee.std_logic_1164.all;

entity condit_stmts_1 is
        port (sel, b, c: boolean; y: out boolean);
end condit_stmts_1;

architecture concurrent of condit_stmts_1 is
begin
        y <= b when sel else c;
```

```
end concurrent;
```

The same function can be implemented using sequential statements and occur inside a process statement. The condition in an if statement must evaluate to true or false (that is, it must be a boolean type).

```
architecture sequential of condit_stmts_1 is
begin
process (s, b, c)
        variable n: boolean;
begin
        if sel then
                n := b;
        else
                n := c;
        end if;
                y <= n;
end process;
end architecture sequential;
```

The schematic diagram of the circuit generated from the above examples is shown in Figure 9.5.

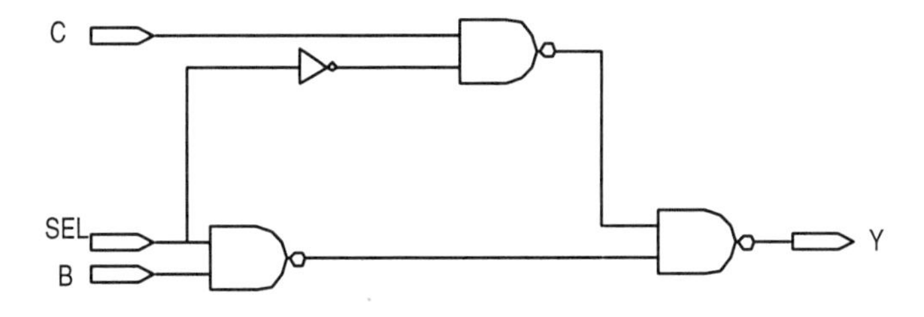


Figure 9.5 Schematic diagram corresponding to example

The following example shows the use of the selected signal assignment for creating of conditional logic that implements a multiplexer. All possible cases must be used for selected signal assignments. The designer can be certain of this by using an **others** case:

```
library ieee;
use ieee.std_logic_1164.all;

entity condit_stmts_2 is
        port (sel: in std_logic_vector (0 to 1);
        a,b,c,d : in std_logic; y: out bit);
end condit_stmts_2;

architecture concurrent of condit_stmts_2 is
begin
with sel select
        y <= a when '00',
        y <= b when '01',
        y <= c when '10',
        y <= d when others;
end example;
```

The same function can be implemented using sequential statements and occur inside a process statement. The following example illustrates the case statement:

```
architecture sequential of condit_stmts_2 is
begin
process (sel,a,b,c,d)
begin
        case sel is
                when '00' => y <= a;
                when '01' => y <= b;
                when '10' => y <= c;
                when others => y <= d;
        end case;
end process;
end sequential;
```

Schematic diagram illustrating this example is shown in Figure 9.6.

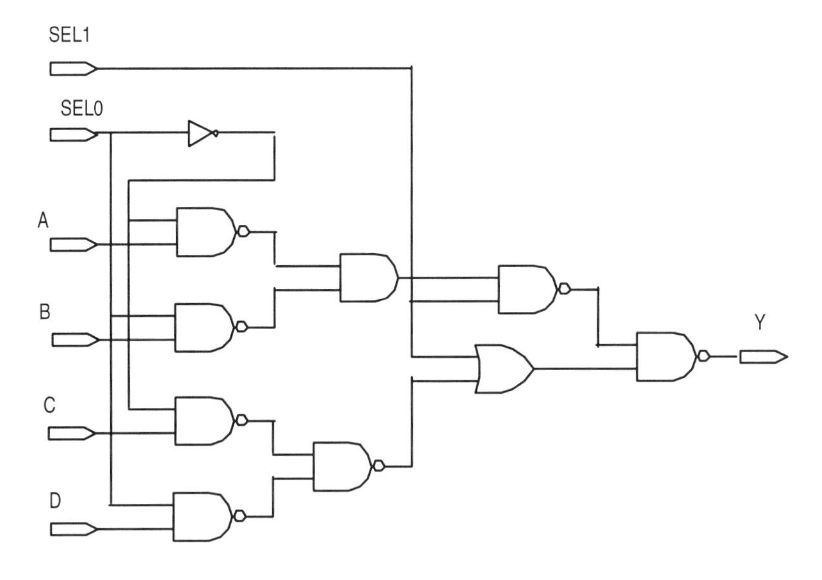


Figure 9.6 Schematic diagram corresponding to example

Using a case statement (or selected signal assignment ) will generally compile faster and produce logic with less propagation delay than using nested if statements (or a large selected signal assignment). VHDL requires that all the possible conditions be represented in the condition of a case statement. To ensure this, use the others clause at the end of a case statement to cover any unspecified conditions.

### *9.1.3 Three-State Logic*

When data from multiple possible sources need to be directed to one or more destinations we usually use either multiplexers or three-state buffers. This section shows the different ways in which three-state buffers may be modeled for inference by synthesis tools. VHDL provides two methods to describe three-state buffers: either by using the 'Z' high-impedance value in standard logic defined in IEEE std_logic_1164, or using an assignment of null to turn off a driver. The first method applies to the type std_logic only, the second method applies to any type. Three-state buffers are modeled then using conditional statements:

- if statements,
- case statements,
- conditional signal assignments

A three-state buffer is inferred by assigning a high-impedance value 'Z' to a data object in the particular branch of the conditional statement. In the case of modeling multiple buffers that are connected to the same output, each of these buffers must be described in separate concurrent statement.

An example of a four bit three-state buffer is given below:

```
library ieee;
use ieee.std_logic_1164.all;
entity tbuf4 is
port (enable : std_logic;
        a : std_logic_vector(0 to 3);
        y : out std_logic_vector(0 to 3));
end tbuf4;

architecture arch1 of tbuf2 is
begin
process (enable, a)
begin
        if enable='1' then
                y <= a;
        else
                m <= 'Z';
        end if;
end process;
end tbuf4;
```

The same function can be achieved by using the equivalent concurrent statement:

```
architecture arch2 of tbuf4 is
begin
        y <= a when enable='1' else 'Z';
end;
```

Schematic diagram of the circuit corresponding to this example is shown in Figure 9.7.

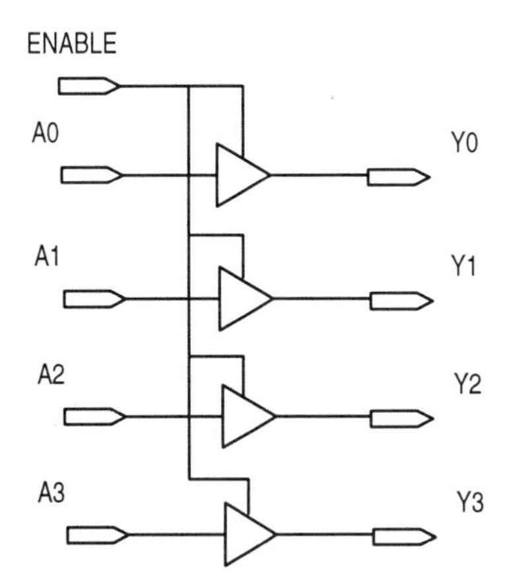


Figure 9.7 Schematic diagram corresponding to example

An internal tristate bus may be described as in the following example.

```
library ieee;
use ieee.std_logic_1164.all;

entity tbus is
port (enable1, enable2, enable3 : std_logic;
          a, b, c : std_logic_vector(0 to 3);
          y : out std_logic_vector(0 to 3));
end
tbus;


architecture arch of tbus is
begin
          y <= a  when enable0 = '1' else 'Z';
          y <= b  when enable1 = '1'  else 'Z';
          y <= c when enable2 = '1' else 'Z';
end arch;
```

Schematic diagram corresponding to this example is shown in Figure 9.8.

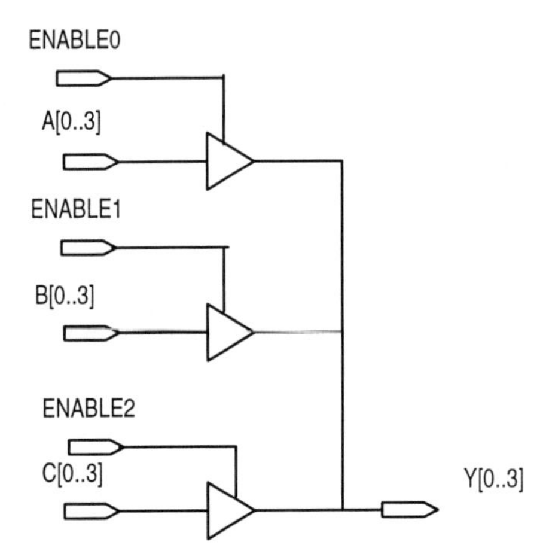


Figure 9.8 Schematic diagram corresponding to example

Three-state buffers can be modeled using case statements. The following example shows the use of case statement:

```
library ieee;
use ieee.std_logic_1164.all;

entity tbuf is
port (a : in std_logic_vector(0 to 2);
          enable: in integer range 0 to 3;
          y : out std_logic;
end tbuf;

architecture arch1 of tbuf is
begin
process(enable, a)
          case enable is
                    when 0   y <= a(0);
                    when 1  y <= a(1);
                    when 2  y <= a(2);
                    when others  y <= 'Z';

          end case;
```

```
end process;
end arch1;
```

The problem with case statement is that others clause cannot be used to assign both three-state and don't-care output value to reduce logic. In that case the solution is to use case statement for minimization of logic by employing don't-care conditions, and to use a separate conditional signal assignment to assign the high-impedance value to infer three-state buffer.

Another way to model three-state buffers is to use the assignment of **null** to a signal of kind **bus** to turn off its driver. When embedded in an if statement, a **null** assignment is synthesized to a three-state buffer:

```
library ieee;
use ieee.std_logic_1164.all;

package pack_bus is
        subtype bus8 is integer range 0 to 255;
end pack_bus;
use work.pack_bus.all;

entity tbuf8 is
        port (enable: in boolean; a: in bus8; y: out bus8 bus);
end tbuf8;

architecture arch1 of tbuf is
begin
process (enable, a)
begin
if enable then
        y <= a;
else
        y <= null;
end if;
end process;
end arch1;
```

### 9.1.4 Combinational Logic Replication

VHDL provides a number of constructs for creating replicated logic. In Chapter 6 we have considered component instantiation and the use of a generate statement as a concurrent loop statement in structural type of VHDL models. However, a number of other constructs are also used to provide replication of logic, namely:

- loop statement,
- function, and
- procedure.

Functions and procedures are referred to as subprograms. These constructs are synthesized to produce logic that is replicated once for each subprogram call, and once for each iteration of a loop. If possible, loop and generate statement ranges should be expressed as constants. Otherwise, the logic inside the loop may be replicated for all the possible values of loop ranges. This can be very expensive in terms of gates.

An example of loop statement used to replicate logic is shown below:

```
library ieee;
use ieee.std_logic_1164.all;

entity loop_stmt is
        port (a: in std_logic_vector (0 to 3);
        y: out std_logic_vector (0 to 3));
end loop_stmt;

architecture arch1 of loop_stmt is
begin
process (a)
        variable temp: std_logic;
begin
        temp := '1';
        for i in 0 to 3 loop
                temp := a(3-i) and temp;
                y(i) <= temp;
        end loop;
end process;
end arch1;
```

Schematic diagram illustrating synthesized circuit from this example is shown in Figure 9.9.

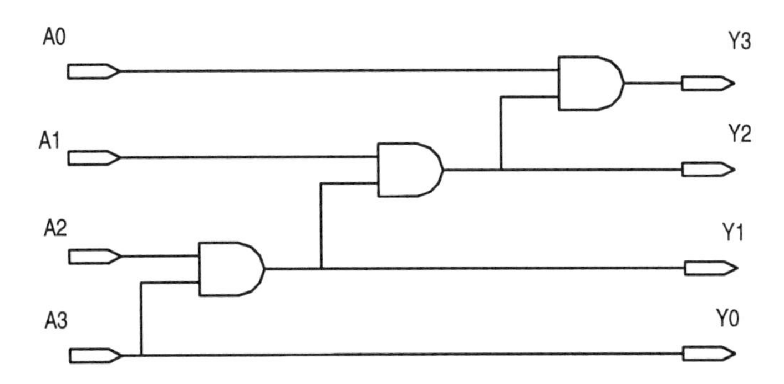


Figure 9.9 Schematic diagram corresponding to example of loop

A loop statement replicates logic, therefore, it must be possible to evaluate the number of iterations of a loop at compile time. Loop statements may be terminated with an exit statement, and specific iterations of the loop statement terminated with a next statement as it was shown in Chapter 4. While exit and next can be useful in simulation, in synthesis they may synthesize logic that gates the following loop logic. This may result in a carry-chain-like structure with a long propagation delay in the resulting hardware.

A function is always terminated by a return statement, which returns a value. A return statement may also be used in a procedure, but it never returns a value. An example of using function to generate replicated logic is shown below:

```
library ieee;
use ieee.std_logic_1164.all;

entity replicate is
        port (a: in std_logic_vector (0 to 3);
        y: out std_logic_vector (0 to 3));
end replicate;

architecture arch1 of replicate is
function replica (b, c, d, e: std_logic) return std_logic is
begin
        return not ( (b xor c) ) and (d xor e);
end;
```

```
begin
process (a)
begin
        y(0) <= replica(a(0), a(1), a(2), a(3));
        y(1) <= replica(a(3), a(0), a(1), a(2));
        y(2) <= replica(a(2), a(3), a(0), a(1));
        y(2) <= replica(a(1), a(2), a(3), a(0));
end process;
end architecture arch1;
```

Schematic diagram which illustrates this example is shown in Figure 9.10.

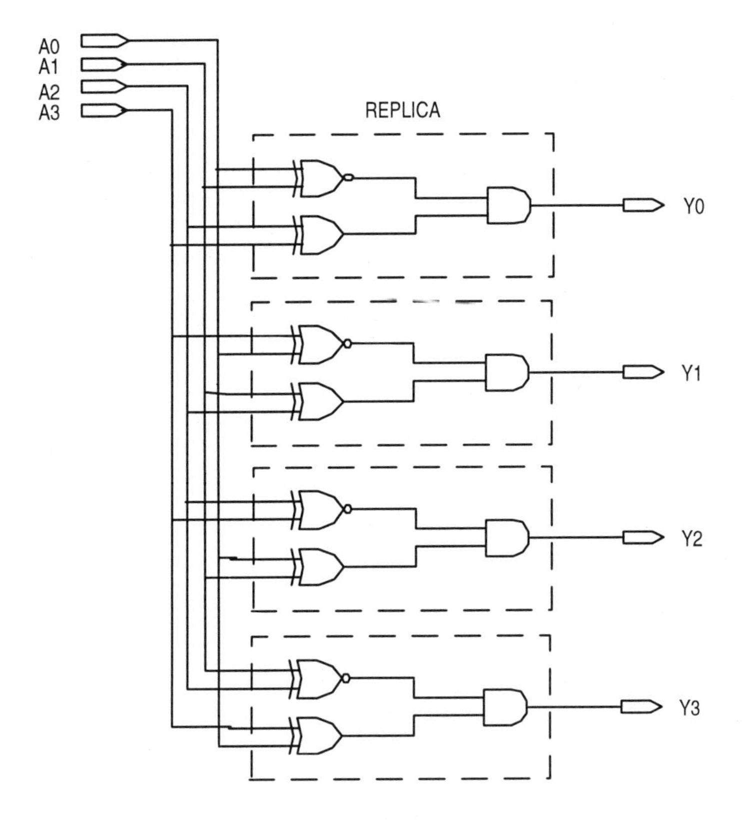


Figure 9.10 Schematic diagram corresponding to example

## 9.2 Sequential Logic

In VHDL we describe the behavior of a sequential logic element, such as a latch or flip-flop, as well as the behavior of more complex sequential machines. This section shows how to model simple sequential elements, such as latches and flip-flops, or more complex standard sequential blocks, such as registers and counters. The behavior of a sequential logic element can be described using a process statement (or the equivalent procedure call, or concurrent signal assignment statement) because the sequential nature of VHDL processes makes them ideal for the description of circuits that have memory and must save their state over time. At this point it is very important to notice that processes are equally suitable to describe combinational circuits, as it was shown in preceding section. If our goal is to create sequential logic (using either latches or flip-flops) the design is to be described using one or more of the following rules:

1. Write the process that does not include all entity inputs in the sensitivity list (otherwise, the combinational circuit will be inferred).

2. Use incompletely specified if-then-elsif logic to imply that one or more signals must hold their values under certain conditions.

3. Use one or more variables in such a way that they must hold a value between iterations of the process.

### *9.2.1 Describing Behavior of Basic Sequential Elements*

The two most basic types of synchronous element, which are found in majority of FPLD libraries, which synthesis tools map to, are:

- the D-type flow-through latch, and
- the D-type flip-flop

Some of the vendor libraries contain other types of flip-flops, but very often they are derived from the basic D-type flip-flop. In this section we consider the ways of creating basic sequential elements using VHDL descriptions.

A D-type flow-through latch, or simply latch, is a level sensitive memory element that is transparent to signal passing from the D input to Q output when enabled (ENA = 1), and holds the value of D on Q at the time when it becomes disabled (ENA = 0). The model of the latch is presented in Figure 9.11.

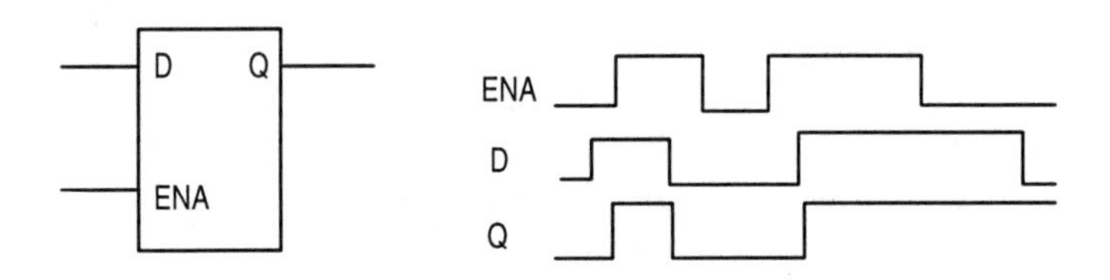


Figure 9.11 Model of the level sensitive latch

The D-type flip-flop is an edge-triggered memory element that transfers a signal's value on its input D to its Q output when an active edge transition occurs on its clock input. The output value is held until the next active clock edge. The active clock edge may be transition of the clock value from '0' to '1' (positive transition) or from '1' to '0' (negative transition). The Qbar signal is always the complement of the Q output signal. The model of the D-type flip-flop with positive active clock edge is presented in Figure 9.12.

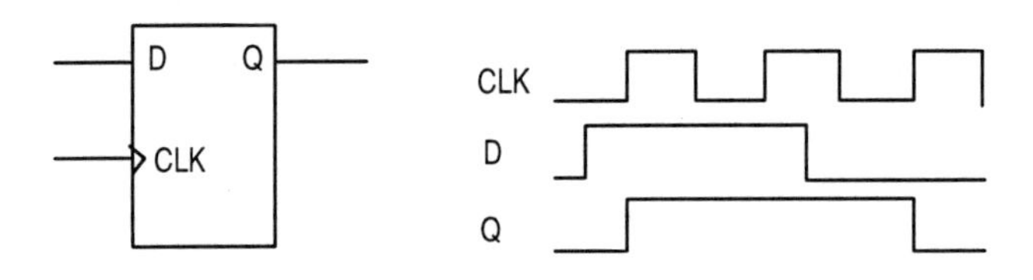


Figure 9.12 Model of the edge-triggered D-type flip-flop

There are three major methods to describe behavior of basic memory elements:

- using conditional specification,
- using a wait statement, or
- using guarded blocks.

Conditional specification is the most common method. This relies on the behavior of an if statement, and assigning in only one condition. The following example describes the behavior of a latch:

```
process (clk)
begin
if clk = '1' then
        y <= a;
        else -- do nothing
end if;
end process;
```

If the clock is high the output (y) gets a new value, if not the output retains its previous value. Note that if we had assigned in both conditions, the behavior would be that of a multiplexer.

The key to specification of a latch or flip-flop is incomplete assignment using the if statement. However, that incomplete assignment is within the context of the whole process statement. A rising edge flip-flop is created by making the latch edge sensitive:

```
if clk and clk'event then
        y <= a;
end if;
```

The second method uses a wait statement to create a flip-flop. The evaluation is suspended by a wait-until statement (over time) until the expression evaluates to true:

```
wait until clk and clk'event
        y <= a;
```

It is not possible to describe latches using a wait statement.

Finally, the guard expression on a block statement can be used to specify a latch.:

```
lab : block (clk)
begin
        q <= guarded d;
end block;
```

It is not possible to describe flip-flops using guarded blocks.

### *9.2.2 Latches*

The following examples describe a level sensitive latch with an and function connected to its input. In all these cases the signal "y" retains it's current value unless the enable signal is '1':

```
library ieee;
use ieee.std_logic_1164.all;

entity latch1 is
        port(enable, a, b: in std_logic; y: out std_logic);
end latch1;

architecture arch1 of latch1 is
begin
process (enable, a, b)
begin
        if enable = '1' then
        y <= a and b;
end if;
end process;
```

Schematic diagram corresponding to this latch is presented in Figure 9.13.

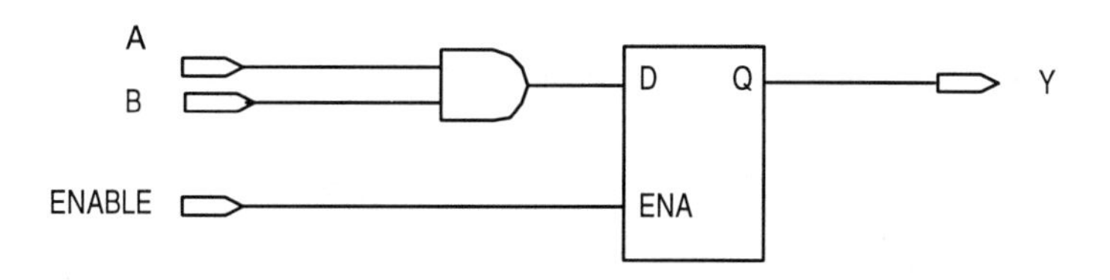


Figure 9.13 Model of the latch with input as a logic function

This example can be easily extended to inputs to the latch implementing any boolean function. Another way to create latches is to use procedure declaration to create latch behavior, and then use a concurrent procedure call to create any number of latches:

```
library ieee;
use ieee.std_logic_1164.all;

package my_latch is
procedure latch2 (signal enable, a, b : std_logic; signal y : out std_logic)
begin
if enable ='1' then
          y <= a and b;
end if;
end;
end my_latch;
use work.my_latch.all;
entity dual_latch is
          port(enable, a, b, c, d: in std_logic; y1, y2: out std_logic)
end dual_latch;

architecture arch1 of dual_latch is
begin
          label_1: latch2 (enable, a, b, y1 );
          label_2: latch2 (enable, c, d, y2 );
end arch1;
```

Latches can be modeled to have additional inputs, such as preset and clear. Preset and clear inputs to the latch are always asynchronous. The following example shows a number of latches modeled within a single process. All latches are enabled by a common input enable.

```
library ieee;
use ieee.std_logic_1164.all;

entity latches is
          port(enable,
                    a1, preset1,
                    a2, clear2,
                    a3, preset3, clear3: in std_logic;
                    y1, y2, y3: out std_logic);
end latches;

architecture arch1 of latches is
begin
process(enable, a1, preset1, a2, clear2, a3, preset3, clear3)
begin
```

```
        -- latch with active high preset
        if (preset1 = '1') then
                y1 <= '1';
        elsif (enable = '1') then
                y1 <= a1;
        end if;

        -- latch with active low clear
        if (clear2 = '0') then
                y2 <= '0';
        elsif (enable = '1') then
                y2 <= a2;
        end if;

        -- latch with active high preset and clear
        if (clear3 = '1') then
                y3 <= '0';
        elsif (preset3 = '1') then
                y3 <= '1';
        elsif (enable = '1') then
                y3 <= a3;
        end if;

    end process;
    end arch1;
```

### *9.2.3 D-Type Flip-Flops*

D-type flip-flops are inferred in VHDL using wait and if statements within a process. They are inferred by detecting a clock signal's edge. The following are some examples of edge-detecting expressions:

```
clock'event and clock = '1' -- rising edge detection using 'event attribute

not clock'stable and clock ='0' -- falling edge detection using 'stable
                                -- attribute
rising_edge(clock)  -- rising edge detection using function call

falling_edge(clock)  -- falling edge detection using function call
```

The functions rising_edge and falling_edge also use VHDL attributes, but represent a more reliable way of detecting clock transitions especially in the case of using multi-valued data types like std_logic. When using std_logic data type, it is necessary to ensure that transitions like "X to 1" are not detected. Functions rising_edge and falling_edge are defined in the IEEE 1164 package std_logic_1164 for clock signals of type std_logic, and in the IEEE 1076.3 synthesis package numeric_std for clocks of type bit.

The following examples describe an edge sensitive flip-flop. In all these cases the output signal retains its current value unless the clock is changing:

```
library ieee;
use ieee.std_logic_1164.all;
entity d_flip_flops is
        port(clk: in std_logic;
                a, a1, a2, a3, a4, a5: in std_logic;
                y, y1, y2, y3, y4, y5: out std_logic);
end entity d_flip_flops;

architecture arch1 of d_flip_flops is
-- A Procedure declaration, this creates a flip-flop
-- when used as a concurrent procedure call
procedure my_ff (signal clk, a: std_logic; signal y : out std_logic)
begin
        if not clk ='1' and clk'event then -- clock falling
        y <= a;
end if;
end;

begin
-- A process with an if statement
p1: process (clk) -- a list of all signals that result in propagation delay
begin
if clk ='1' and clk'event then -- clock with rising edge
        y1 <= a1;
end if;
end process p1;

-- A Process statement containing a wait statement:
p2: process -- No list of all signals used in the process
begin
wait until not clk = '1' ; -- clock falling
```

```
        y2 <= a2;
end process p2;

-- A d-type flip-flop created by concurrent procedure call
lab_1: my_ff(clk, a3, y3);
-- A d-type flip-flop using concurrent conditional signal assignment,
-- note that  y is both used and driven
y 4<= a4 when clk ='1' and clk 'event else y;

-- A d-type flip-flop using edge detection function
y5 <= a5 when rising_edge(clk);

end architecture arch1;
```

In most of examples of various sequential circuits presented in this book we assume the clock is a simple signal. In principle, any complex Boolean expression could be used to specify clocking. However, the use of a complex expression implies a gated clock. As with any kind of hardware design, there is a risk that gated clocks may cause glitches in the register clock, and hence produce unreliable hardware. Generally, only use of simple logic in the if expression is recommended. It is possible to specify a gated clock with a statement such as:

```
if clk and enable then -- clk and enable must be of type Boolean
-- signal assignments
end if;
```

which implies a logical AND in the clock line. To specify a clock enable use nested if statements:

```
if clk then -- clk and enable must be of type Boolean
if enable then
-- signal assignments
end if;
end if;
```

This will connect enable to the flip-flop clock enable if the clock enable input exists. If the clock enable option does not exist then the data path will be gated with enable. In neither case will enable gate the clk line.

Other behaviors of flip-flops may be described in VHDL. To add the behavior of synchronous set or reset we simply add a conditional assignment of a constant immediately inside the clock specification.

```
library ieee;
use ieee.std_logic_1164.all;

entity d_flip_flops is
          port(clk: in std_logic;
                    a1, a2: in std_logic;
                    preset1, reset2: in std_logic;
                    y1, y2: out std_logic);
end entity d_flip_flops;

architecture arch1 of d_flip_flops is

-- Synchronous set
p1: process (clk)
begin
if clk = '1' and clk'event then-- clock rising
if preset1 = '1' then
          y1 <= '1';
else
          y1 <= a1;
end if;
end if;
end process p1;

-- Synchronous reset using process and wait statement
p2: process
begin
wait until rising_edge(clk);
if reset2 = '1' then
          y2 <= '0';
else
          y2 <= a2;
end if;
end process p2;
end architecture arch1;
```

In similar way we can describe the behavior of d-type flip-flop with asynchronous set or/and reset. We simply add a conditional signal assignment of a constant immediately outside the clock specification:

```
library ieee;
use ieee.std_logic_1164.all;

entity d_flip_flops is
        port(clk: in std_logic;
                a1, a2, a3, a4, a5: in std_logic;
                reset1, reset2, reset3,
                reset4, reset5, preset4, preset5: in std_logic;
                y1, y2, y3, y4, y5: out std_logic);
end entity d_flip_flops;

architecture arch1 of d_flip_flops is
signal q3, q5: std_logic;

-- Reset using a concurrent statement statement:
y1 <= '0' when reset1= '1' else a1 when clk ='1' and clk 'event else y1;

-- Resct using the function rising_edge described earlier :
y2 <= '0' when reset2 = '1' else a2 when rising_edge(clk );

-- Reset using sequential statements:
p3: process (clk, reset3)
begin
if reset3='1' then
        q3 <= '0';
else
if clk = '1' and clk'event then-- clock rising
        q3 <= a3;
end if;
end if;
end process p3;
y3 <= q3;

-- asynchronous set and reset using concurrent statement
y4 <= '0' when reset4 = '1' else
'1' when preset4 = '1' else
a4 when clk = '1' and clk 'event;
```

```
-- asynchronous reset and set using a sequential statements
p5; process (clk, reset5, preset5)
begin
if reset ='1' then
        q5 <= '0';
elsif preset ='1' then
        q5 <= '1';
else
if clk ='1' and clk'event then-- clock rising
        q5 <= a5;
end if;
end if;
end process p5;
y5 <= q5;
end architecture arch1;
```

Some other types of d flip-flops, for example those having synchronous or asynchronous load control input, will be described as the part of other more complex sequential circuits. Also, multi-bit registers, will be introduced in later chapters, as a simple extension of single flip-flop type circuits to multiple flip-flops with identical behavior. Most of the above examples can be simply modified to describe multi-bit registers by using std_logic_vector data type instead of std_logic type.

## 9.3 Finite State Machines

Finite State Machines (FSMs) represent an important part of design of almost any more complex digital system. They are used to sequence specific operations, control other logic circuits, and provide synchronization of different parts of more complex circuit. FSM is a circuit that is designed to sequence through a specific patterns of states in a predetermined manner. Sequences of states through which an FSM passes depend on the current state of the FSM, and previous history of the FSM. A state is represented by the binary value held on the current state register. FSM is clocked from a free running clock source. The general FSM model is presented in Figure 9.14.

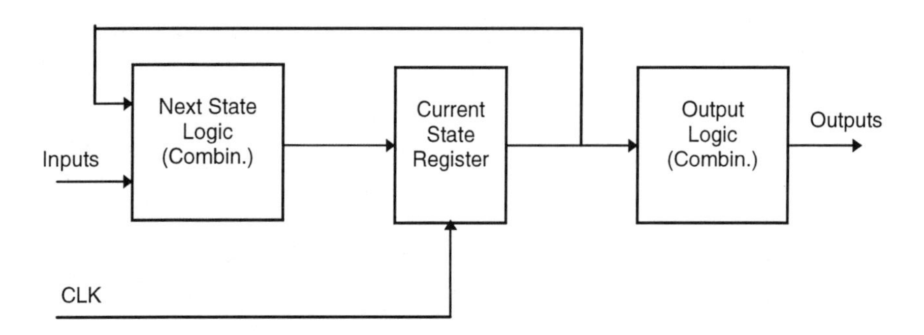


Figure 9.14 General model of FSM

It contains three main parts:

1. Current State Register. It is a register of n flip-flops used to hold the current state of the FSM. The current state is represented by the binary value contained in this register. State changes are specified by sequences of states through which the FSM passes after changes of inputs to the FSM.

2. Next State Logic. It is a combinational logic used to generate the transition to the next state from the current state. The next state is a function of the current state and the external inputs to the FSM. The fact that the current state is used to generate transition to the next state means that feedback mechanism within the FSM must be used to achieve desired behavior.

3. Output Logic. It is a combinational circuit used to generate output signals from the FSM. Outputs are a function of the current state and possibly FSM's inputs. If the output is a function of only the current state, then we classify the FSM as Moore FSM. If the outputs depend also on the inputs to the FSM, then we classify the FSM as Mealy FSM. Both these types of FSMs are discussed in more details in sections 9.3.2 and 9.3.3, respectively. Sometimes, combined Mealy/Moore models are suitable to describe specific behavior.

The behavior of an FSM is usually described either in the form of a state transition table or a state transition diagram. These types of descriptions very often represent a starting point for the FSM modeling and description in VHDL.

An important issue when designing FSMs is state encoding, that is assignment of binary numbers to states. There are a number of ways to perform state encoding. For small designs or those in which there are not too tight constraints in terms of resources, the common way is to let a synthesis tool encode the states automatically. However, for bigger designs, a kind of manual intervention is necessary. The obvious solution is to use explicit state encoding using constant declaration. First, if states are declared in separate enumerated type, then the enumerated declaration can be replaced by encoding using constant declarations, as it is shown in the following example. The declaration:

```
type state is (red, yellow, green);
signal current_state, next_state: state;
```

can be replaced with:

```
type states is std_logic_vector(1 downto 0);
constant red: state:= "00";
constant yellow: state:= "01";
constant green: state:= "10";
signal current_state, next_state: state;
```

Obviously, in this case we are using simple sequential state encoding with the increasing binary numbers. The other possibility is to use some other encoding scheme, such as using Gray code or Johnson state encoding, which has some advantages in terms of more reliable state transitions, but also usually results in a more expensive circuit. One particular way is to use the so-called one-hot encoding scheme in which each state is assigned its own flip-flop, and in each state only one flip-flop can have value '1'. This encoding scheme is not optimal in terms of number of flip-flops, but is still very often used by FPLD synthesis tools. The reason for this is the relatively high number of available flip-flops in FPLDs, as well as the assumption that a large number of flip-flops used for state representation leads to a simpler next state logic.

Some synthesis tools provide a non-standard, but widely used, attribute called enum_encoding, which enables explicit encoding of states represented by strings of binary state variables. Our previous example can be described by using the enum_encoding attribute as:

```
type state is (red, yellow, green);
attribute enum_encoding of state: type is "00 01 10";
signal current_state, next_state: state;
```

The enum_encoding attribute is declared elsewhere in the design as a string:

```
attribute enum_encoding: string;
```

Another important issue is ability to bring an FSM to a known state regardless of its current state. This is usually achieved by using (implementing) a reset signal, which can be synchronous or asynchronous. An asynchronous reset ensures that the FSM is always brought to a known initial state, before the next active clock and normal operation resumes. Another way of bringing an FSM to an initial state is to use synchronous reset. This usually requires the decoding of unused codes in the next state logic, because the FSM can be stuck in an uncoded state.

### *9.3.1 Using Feedback Mechanisms*

VHDL provides two basic ways to create feedback: using signals and using variables. With the addition of feedback, the designer can build FSMs. This will be discussed in the sections that follow. VHDL has the constructs which make it possible to describe both combinational and sequential (registered) feedback systems. A simple example of using feedback on signals is presented below:

```
library ieee;
use ieee.std_logic_1164.all;

entity signal_feedback is
        port(clk, reset, a: in std_logic; y: inout std_logic);
end entity signal_feedback;

architecture arch1 of signal_feedback is
signal b: std_logic;
function rising_edge (signal s : std_logic) return boolean is
begin
return s = '1' and s'last_value ='0' and s'event; -- positive transition from
        -- 0 to 1
end;
begin

p1: process (clk, reset)
begin
if reset = '1' then
        y <= '0';
elsif rising_edge(clk)
        y <= b;
```

```
end if;
end process p1;

p2: process (a, c)-- a combinational process
begin
        b <= a nand y;
end process p2;
end architecture arch1;
```

An internal signal b is used to provide a feedback within the circuit. Schematic diagram of the circuit inferred from this VHDL description is shown in Figure 9.15.

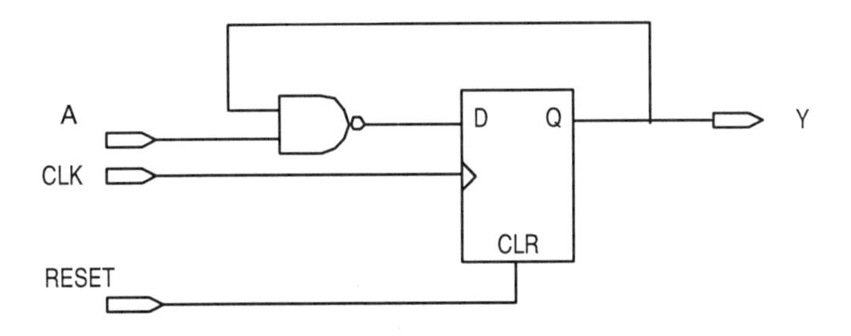


Figure 9.15 Circuit with feedback on signal

The same feedback can be synthesized by the following VHDL description:

```
library ieee;
use ieee.std_logic_1164.all;

package new_functions is
function rising_edge (signal s : std_logic) return boolean is
begin
return s = '1' and s'last_value ='0' and s'event; -- positive transition from
0 to 1
end;
end new_functions;

use ieee.std_logic_1164.all;
use work.new_functions.all;
```

```
entity signal_feedback is
        port(clk, reset, a: in std_logic; y: inout std_logic);
end signal_feedback;

architecture arch1 of signal_feedback is
begin
process(clk,reset)
begin
        if reset = '1' then
                y <= '0';
        elsif rising_edge(clk)
                y <= a nand y;
        end if;
end process;
end arch1;
```

In this case, signal c is both driven and used as a driver.

Another way to implement feedback in VHDL is by using variables. Variables exist within a process and are used to save states from one to another execution of the process. If a variable passes a value from the end of a process back to the beginning, feedback is implied. In other words, feedback is created when variables are used (placed on the right hand side of an expression, for example in an if statement) before they are assigned (placed on the left hand side of an expression). Feedback paths must contain registers, so you need to insert a wait statement to enable the clock to change the value of variable.

An example of the feedback implemented using variables is shown below:

```
library ieee;
use ieee.std_logic_1164.all;

entity variable_feedback is
port(clk, reset, load, a: in std_logic; y: out std_logic)
end variable_feedback;

architecture arch1 of variable_feedback is
begin
process
        variable v: bit;
begin
```

```
            wait until clk = '1';
            if reset = '1' then
                    y <= '0';
            elsif load = '1' then
                    y <= a;
            else
                    v:= not v;  -- v used before it is assigned
                    y <= v;
            end if;
    end process;
    end arch1;
```

A flip-flop is inserted in the feedback path because of the wait statement. This also specifies registered output on signal y.

### *9.3.2 Moore Machines*

A Moore state machine has the outputs that are a function of the current state only. The general structure of Moore-type FSM is presented in Figure 9.16. It contains two functional blocks that can be implemented as combinational circuits:

- next state logic, which can be represented by function next_state_logic, and
- output logic, which can be represented by function output_logic

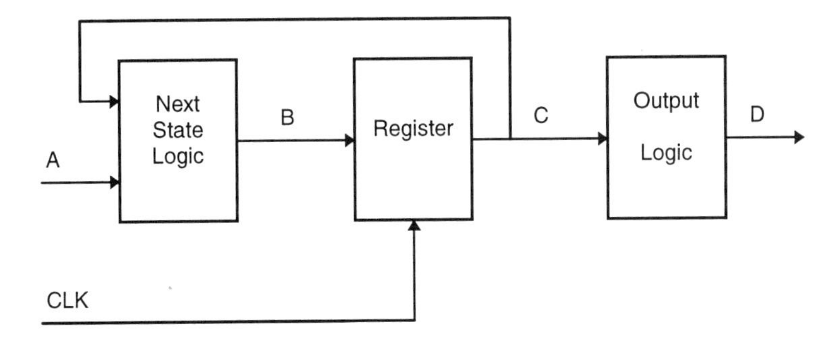


Figure 9.16 Moore-type FSM

Outputs of both of these functions are the functions of their respective current inputs. The third block is a register that holds the current state of the FSM. The Moore FSM can be represented by three processes each corresponding to one of the functional blocks:

```
entity system is
        port (clock: std_logic; a: some_type; d: out some_type);
end system;

architecture moore1 of system is
        signal b, c: some_type;
begin
next_state: process (a, c) – next state logic
begin
        b <= next_state_logic(a, c);
end process next_state;

system_output: process (c)
begin
        d <= output_logic(c);
end process system_output;

state_reg: process
begin
        wait until rising_edge(clock);
c <= b;
end process state_reg;
end moore1;
```

A more compact description of this architecture could be written as follows:

```
architecture moore2 of system is
        signal c: some_type;
begin

system_output: process (c)-- combinational logic
begin
        d <= output_logic(c);
end process system_output;

next_state: process-- sequential logic
begin
        wait until clock;
```

```
c <= next_state_logic(a, c);
end process next_state;
end moore2;
```

In fact, a Moore FSM can often be specified in a single process. Sometimes, the system requires no logic between system inputs and registers, or no logic between registers and system outputs. In both of these cases, a single process is sufficient to describe behaviour of the FSM.

### *9.3.3 Mealy Machines*

A Mealy FSM has outputs that are a function of both the current state and primary system inputs. The general structure of the Mealy-type FSM is presented in Figure 9.17.

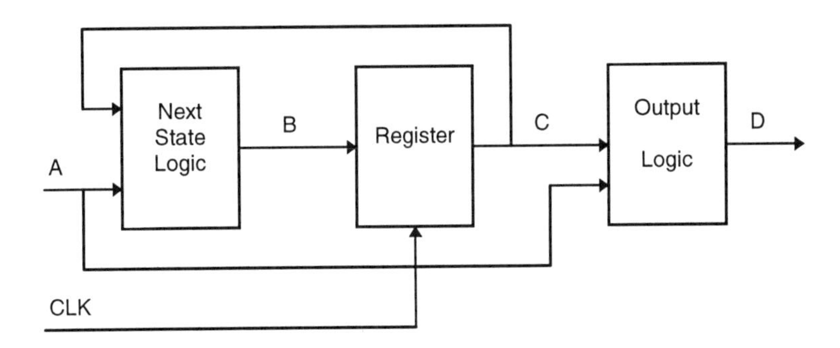


Figure 9.17 Mealy-type FSM

The Mealy FSM can be represented by the following general VHDL model:

```
entity system is
        port (clock: std_logic; a: some_type; d: out some_type);
end system;

architecture mealy of system is
        signal c: some_type;
begin

system_output: process (a, c)-- combinational logic
begin
        d <= output_logic(a, c);
```

```
end process system_output;

next_state: process -- sequential logic
begin
        wait until clock;
                c <= next_state_logic(a, c);
end process next_state;
end mealy;
```

It contains at least two processes, one for generation of the next state, and the other for generation of the FSM output.

Other approaches can be used when designing state machines. Sometimes, a combined Mealy/Moore FSM having both types of output is designed. In some applications it is possible to decompose a more complex state machine into several interacting state machines. Interaction can be unidirectional or bi-directional. All these approaches are not discussed in detail in this Chapter, but are left to be introduced throughout examples and case studies in the remaining chapters.

## 9.4 Questions and Problems

9.1. Under what conditions does a typical synthesizer generates a combinational circuit from the VHDL process?

9.2. Given a VHDL entity with two architectures:

```
library ieee;
use ieee.std_logic_1164.all;
use ieee.numeric_std.all;

entity example1 is
        port (a, b, c, d: in unsigned(7 downto 0);
                y: out unsigned(9 downto 0));
end example1;

architecture arch1 of example1 is
begin
        process(a, b, c, d)
        begin
                y <= a + b + c + d;
        end process;
```

```
end arch1;

architecture arch2 of example1 is
begin
        process(a, b, c, d)
        begin
                y <= (a + b) + (c + d);
        end process;

end arch2;
```

What is the difference between the circuits synthesized for these two architectures? Draw the synthesized circuits.

9.3. How are three-state buffers synthesized in VHDL? What conditional statements can be used to describe three-state logic?

9.4. A 64K memory address space is divided into eight 8K large segments. Using VHDL describe an address decoder that decodes the segment from the 16-bit address.

9.5. The address space from the preceding example is divided into seven segments of equal length (8K) and the topmost segment is divided into four segments of 2K size. Using VHDL describe an address decoder that decodes the segment from the 16-bit address. Describe a decoder using processes and different sequential statements (loop or case).

9.6. Describe the use of sequential statements for the generation of replicated combinational logic.

9.7. Specify conditions under which a VHDL compiler generates sequential logic from the VHDL process.

9.8. Describe a J-K flip-flop using the VHDL process.

9.9. How can feed-back in the FSMs be described in VHDL?

9.10. Write a VHDL template for a description of Mealy and Moore-type FSMs. Apply it to the example of a system that describes driving the car with four states: stop, slow, mid, and high, and two inputs representing acceleration and braking. The output is represented by a separate indicator for each of the states. The states are coded using a one-hot encoding scheme.

# PART2

# FPLDs as Prototyping and Implementation Technology

# 10 FPLDS: ARCHITECTURES AND EXAMPLES OF DEVICES

The purpose of this Chapter is to introduce Field Programmable Logic Devices (FPLDs) as an ideal target technology for VHDL based designs for both prototyping and production volumes. The FPLDs make the same paradigm shift that drove the success of logic synthesis with ASIC design, namely the move from schematics to HDL based design tools and methodologies with a major advantage of much shorter turnaround time, and no need for special processes. This chapter discusses the FPLDs from the point of view of the system designer who uses FPLDs as either prototyping or production technology for complex digital systems and explains them from user's point of view. First, a very short introduction to FPLDs is made, which is followed with description of typical architectural features. Then, main FPLD families from Altera Co. are presented, and finally the design frameworks and environments which support VHDL design simulation and synthesis for FPLDs are introduced. Finally, a bridge between VHDL as an abstract design tool presented in the preceding chapters, and complex examples of designs presented in the following chapters that are often supported only with FPLDs, and not with other semiconductor technologies, is established.

## 10.1 Introduction to FPLDs

FPLDs represent a relatively new development in the field of VLSI circuits. They implement a large number of logic gates in multilevel structures. The architecture of FPLD is similar to that of Mask-Programmable Gate Array (MPGA), consisting of an array of logic cells that can be interconnected by programming to realize different designs. The major difference between FPLDs and MPGAs is that an MPGA is programmed using integrated circuit fabrication to form metal interconnections, while an FPLD is programmed using electrically programmable switches similar to ones in traditional Programmable Logic Devices (PLDs) such as PLAs/PALs/GALs. FPLDs can achieve much higher levels of integration than traditional PLDs due to their more complex routing architectures and logic implementation. FPLD routing architectures provide a more efficient MPGA-like routing, where each connection typically passes through several switches. Logic is implemented in FPLDs using multiple levels of lower fan-in gates, which is often more compact than two-level implementations.

An FPLD manufacturer makes a single, standard device that users program to carry out desired functions. Field programmability comes at a cost in logic density and performance. Programming is done by end users at their site with no IC masking steps. FPLDs are currently available in densities even more than 200,000 gates in a single device with a perspective to go beyond 1 million gates in the next few years. This size is large enough to implement many digital systems on a single chip, and larger systems can be implemented on multiple FPLDs on the standard PCB or in the form of Multi-Chip Modules (MCM). Although unit costs of FPLDs are higher than for MPGAs of the same density, there are no up-front engineering charges to use an FPLD, so they are cost-effective for many applications. The result is a low-risk design style, where the price of logic error is small, both in money and project delay.

FPLDs are useful for rapid product development and prototyping. They provide very fast design cycles, and, in the case that the major value of the product is in algorithms, or fast time-to-market, they prove to be even cost-effective as the final deliverable product. FPLDs are fully tested after manufacture, so user designs do not require test program generation, automatic test pattern generation, and design for testability. Due to the capability of "on the fly" changes in functionality, some FPLDs have found a suitable place in designs that require reconfiguration of the hardware structure during system operation.

### *10.1.1 General Features of FPLDs*

In this section main features of FPLDs are presented, and comparison to the other available technologies made.

#### Speed

FPLDs offer devices which operate at very high speeds approaching 200 MHz in many applications. Obviously, the speeds that can be achieved are higher than in systems implemented by SSI circuits, but also lower than the speeds of MPGAs. The main reason for this comes from FPLD's programmability. Programmable interconnect points add resistance to the internal path, while programming points in the interconnect mechanism add capacitance to the internal path. On another side, despite these disadvantages in comparison to MPGAs, FPLDs speed is adequate for most applications. Also, some dedicated architectural features of FPLDs can eliminate unneeded programmability in speed-critical paths. By moving FPLDs to faster processes, applications can be sped up by simply buying and using a faster device without design modification. The situation with MPGAs is quite different; new processes require new mask-making and increase the overall product cost.

### Density

FPLDs programmability introduces on-chip programming overhead circuitry requiring area that cannot be used by designers. This results in the fact that for the same amount of logic, FPLDs will always be larger and more expensive than MPGAs. However, I/O pad limitation has already resulted with the fact that a large area of die cannot be used for core functions in MPGAs. The use of this wasted area for field programmability does not result in the increase of area of resulting part - FPLD. Thus, for a given number of gates the size of an MPGA and FPLD is dictated by the I/O count, so the FPLD and MPGA capacity will be the same. This has especially become true with the migration of FPLDs to submicron processes. MPGA manufacturers have already shifted to high-density products, leaving designs with less than 20,000 gates to FPLDs.

### Development Time

FPLDs development is followed by the development of tools for system design based on FPLDs. All those tools belong to high-level tools affordable even for very small design houses. The development time primarily includes prototyping and simulation, while the other phases, including time-consuming test pattern generation, mask-making, wafer fabrication, packaging, and testing are completely avoided. This leads to the typical development times for FPLD designs measured in days or weeks, in contrast to MPGA development times in several weeks or months.

### Prototyping and Simulation Time

While the MPGA manufacturing process takes usually weeks or months from completion of design to the delivery of finished parts, FPLDs require practically only design completion. Design modifications to correct a design flaw are quickly and easily done, providing short turn-around time leading to faster product development and shorter time-to-market for new FPLD-based products.

Proper verification requires MPGA users to verify their designs by extensive simulation before manufacture introducing all of the drawbacks of the speed/accuracy trade-off connected with any simulation. In contrast, FPLDs simulations are much simpler due to the fact that timing characteristics and models are known in advance. Also, many designers avoid simulation completely and choose in-circuit verification. The designer implements the design and uses a functioning part as a prototype which operates at full speed and absolute time accuracy. A prototype can be easily changed and reinserted into the system within minutes or hours.

FPLDs provide low-cost prototyping, while MPGAs provide low-cost volume production. This leads to prototyping on the FPLD, and then switching to an MPGA for volume production. Usually there is no need for design modification when retargeting to an MPGA, except sometimes for re-verification of all timing paths. Some FPLD vendors offer mask-programmed versions of their FPLDs, giving users flexibility and the advantages of both implementation methods.

### Manufacturing time

All integrated circuits must be tested to verify proper manufacturing and packaging. The test is different for each design. MPGAs typically require three types of costs associated with testing:

- on-chip logic to enable easier testing
- generation of test programs for each design, and
- testing the parts when manufacturing is complete

Due to the simple and repeatable structure, the test program for one FPLD device is the same for all designs and all users of that part. It further justifies all reasonable efforts and investments to produce extensive and high-quality test programs that will be used during the life-time of the FPLD. FPLD users are not required to write design-specific tests for their designs because the manufacturer's test program verifies that every FPLD will be functional for all possible designs that may be implemented on it. The consequences from the manufacturing of chips from both categories are obvious. Once verified, FPLDs can be manufactured in any quantity, and delivered as fully tested parts ready for design implementation, while MPGAs require separate production preparation for each new design.

### Future modifications

Instead of customizing the part by customizing the manufacturing process as for MPGAs, FPLDs are customized by electrical modifications to the part. The electrical customization takes milliseconds or minutes, and can even be performed without special devices, or with low cost programming devices. Even more, it can usually be performed in-system, meaning that the part can already be on the printed circuit board, reducing the danger of the damage due to careless handling. On the other hand, every modified design to be implemented in an MPGA requires a custom mask that costs several thousands of dollars, and the cost can be amortized only over the total number of units manufactured.

### Development and Support

Generally, FPLDs are connected with very low design risk in both terms of money involved and design delays. Rapid and easy prototyping enables all errors to be corrected with short delays, but also gives designers the chance to try more risky logic designs in the early stages of product development. Development tools used for FPLD designs usually integrate the whole range of design entry, processing and simulation tools which enable easy reusability of all parts of design proved to be correct.

FPLD designs can be made with the same design entry tools used in traditional MPGAs and Application-Specific Integrated Circuits (ASICs). The resulting netlist is further manipulated by FPLD-specific fitting, placement, and routing algorithms that are available either from FPLD manufacturers or CAE vendors. However, FPLDs also allow designing on the very low device-dependent level to provide the best device utilization if it is needed.

### Cost

Finally, the above introduced options reflect in the costs. The major benefit from an MPGA-based design is low cost in large quantities. The actual volume of the products determines which technology is more appropriate to be used. FPLDs introduce much lower costs of design development and modification, including low initial Non-Recurring Engineering (NRE) charges, tooling and testing costs. However, larger die area and lower circuit density result in higher manufacturing costs per unit. The break-even points depend on the application and volume, and usually are at the figure of between ten and twenty thousand units for large capacity FPLDs. This limit goes even higher in the case that an integrated volume production approach is applied using a combination of FPLDs and their corresponding masked-programmed counterparts. This also introduces further flexibility allowing to satisfy short-term needs with FPLDs, while satisfying long-term needs at the certain volume level with the masked-programmed devices.

### *10.1.2 Architectural Features of FPLDs*

The general architecture of an FPLD is illustrated in Figure 10.1. A typical FPLD consists of a number of logic cells that are used to implement logic functions. Logic cells are arranged in some form of matrix. Interconnection resources provide connections between logic cell outputs and inputs, as well as with input/output blocks that are used to connect FPLD with the outer world.

The concrete implementations of FPLDs, despite of the same general structure, differ in terms of all of the major components. There exists a difference in approach to circuit programmability, internal logic cell structure, and routing mechanisms.

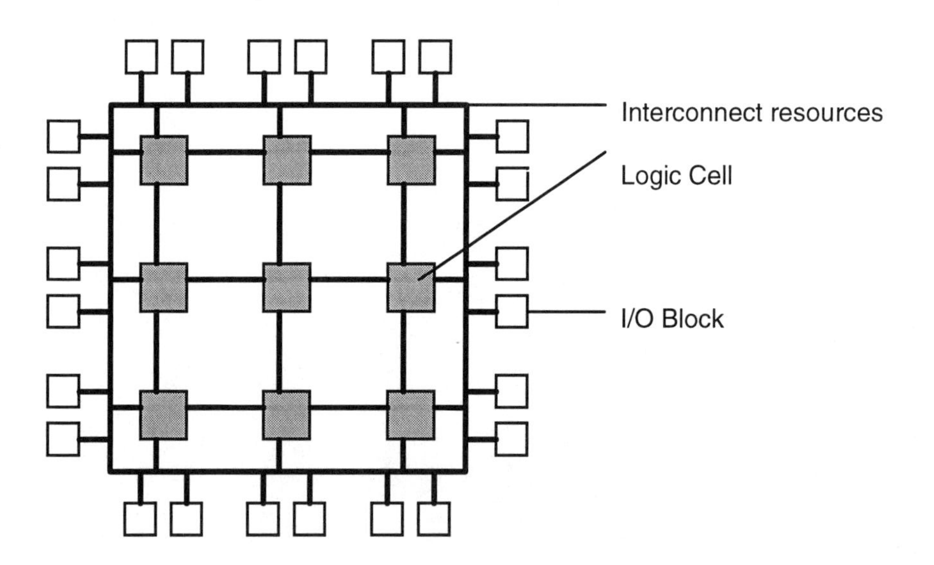


Figure 10.1 FPLD architecture

## Logic Cells

An FPLD logic cell can be as simple as a transistor or as complex as a microprocessor. Typically, it is able to implement combinational and sequential logic functions of different complexity. Current commercial FPLDs employ logic cells that are based on one or more of the following:

- Transistor pairs
- Basic small gates, such as two-input NANDs or XORs
- Multiplexers
- Look-up tables (LUTs)
- Wide-fan-in AND-OR structures

## Programming Technologies

Three major programming technologies, each associated with area and performance costs, are commonly used to implement the programmable switch for FPLDs. These are:

- Static Random Access Memory (SRAM) , where the switch is a pass transistor controlled by the state of a SRAM bit
- EPROM, where the switch is a floating-gate transistor that can be turned off by injecting charge onto its floating gate, and
- Antifuse, which, when electrically programmed, forms a low resistance path.

In all cases, a programmable switch occupies larger area and exhibits much higher parasitic resistance and capacitance than a typical contact or via used in the custom MPGAs. Additional area is also required for programming circuitry, resulting in higher density and lower speed of FPLDs compared to MPGAs.

## Routing Architectures

An FPLD routing architecture incorporates wire segments of varying lengths which can be interconnected via electrically programmable switches. The density achieved by an FPLD depends on the number of wires incorporated. If the number of wire segments is insufficient, only a small fraction of the logic cells can be utilized; on the other hand, an excessive number of wire segments also wastes area. The distribution of the lengths of the wire segments greatly affects both the density and performance achieved by an FPLD. For example, if all segments are chosen to stretch over the entire length of the device (so called long segments), implementing local interconnections becomes too costly in area and delay. On the other hand, employment of only short segments requires long interconnections to be implemented using too many switches in series, resulting in unacceptably long delays.

Both density and performance can be optimized by choosing the appropriate granularity and functionality of logic cell, and by designing the routing architecture to achieve a high degree of routability while minimizing the number of switches. Various combinations of programming technology, logic cell architecture, and routing mechanisms lead to various designs suitable for specific applications. A more detailed presentation of all major components of FPLD architectures is given in the chapters that follow.

### *10.1.3 Design Process*

The complexity of FPLDs has surpassed the point where manual design is either desirable or feasible. The utility of an FPLD architecture becomes more and more dependent on effective automated logic and layout synthesis tools to support it and provide high overall FPLD utilization.

The design process with FPLDs is similar to other gate array design. Input can come from a schematic netlist, a hardware description language, or a logic synthesis system. After defining what has to be designed, the next step will be design implementation. It consists of fitting the logic into the FPLD structures. This step is called "logic partitioning" by some FPLD manufacturers, and "logic fitting" in reference to CPLDs.

After partitioning, the design software assigns the logic, now described in terms of functional units on the FPLD, to particular physical locations on the device and chooses the routing paths. This is much similar to placement and routing in traditional gate arrays.

There exists a variety of design tools used to perform all or some of the above tasks. Part 3 is devoted to the high level design tools with the emphasis on those that enable behavioral level specification and synthesis, primarily high-level hardware description languages.

An application targeted to an FPLD can be designed on any one of a number of logic or ASIC design systems, including schematic capture and hardware description languages. To target FPLD, the design is passed to FPLD-specific implementation software. The interface between design entry and design implementation is a netlist that contains the desired nets, gates, and references to specific vendor-provided macros. Manual and automatic tools can be used interchangeably, or an implementation can be done fully automatically.

Combination of moderate density, reprogrammability and powerful prototyping tools provides a novel capability for system designers: hardware that can be designed with a software-like iterative-implementation methodology. Figure 10.2 is presented in order to compare and contrast a typical ASIC and typical FPLD-targeted design methodology.

In a typical ASIC design cycle, the design is verified by simulation at each stage of refinement. Accurate simulators are slow. ASIC designers use the whole range of simulators across the speed/accuracy spectrum in an attempt to verify the design. Although simulation can be used in designing for FPLDs, an FPLD designer can replace simulation with in-circuit verification, simulating the circuitry in real time with a prototype. The path from design to prototype is short, allowing a designer to verify operation over a wide range of conditions at high speed and high accuracy.

The fast design-place-route-load loop is similar to the software edit-compile-run loop and provides the same benefits. Design can be verified by trial-and-error method. A designer can verify that the design works in a real system, not merely in a potentially-erroneous simulation model of the system.

Design-by-prototype does not verify proper operation with worst-case timing, but rather that the design works on the "typical" prototype part. To verify worst-case timing, designers can check speed margins in actual voltage and temperature corners with a scope and logic analyzer, speeding up marginal signals.. They also may use a software timing analyzer or simulator after debugging to verify worst-case paths; or simply use faster speed- grade parts in production to ensure sufficient speed margins over the complete temperature and voltage range.

As with software development, reprogrammable FPLDs remove the dividing line between prototyping and production. A working prototype may qualify as a production part if it meets performance and cost goals. Rather than re-design, a designer may chose to substitute a faster speed FPLD using the same programming bitstream, or a smaller, cheaper compatible FPLD with more manual work to squeeze the design into a smaller device. A third solution is to substitute a mask-programmed version of the logic array for the field-programmable one. All three options are much simpler than a system redesign, which must be done for traditional MPGAs or ASICs.

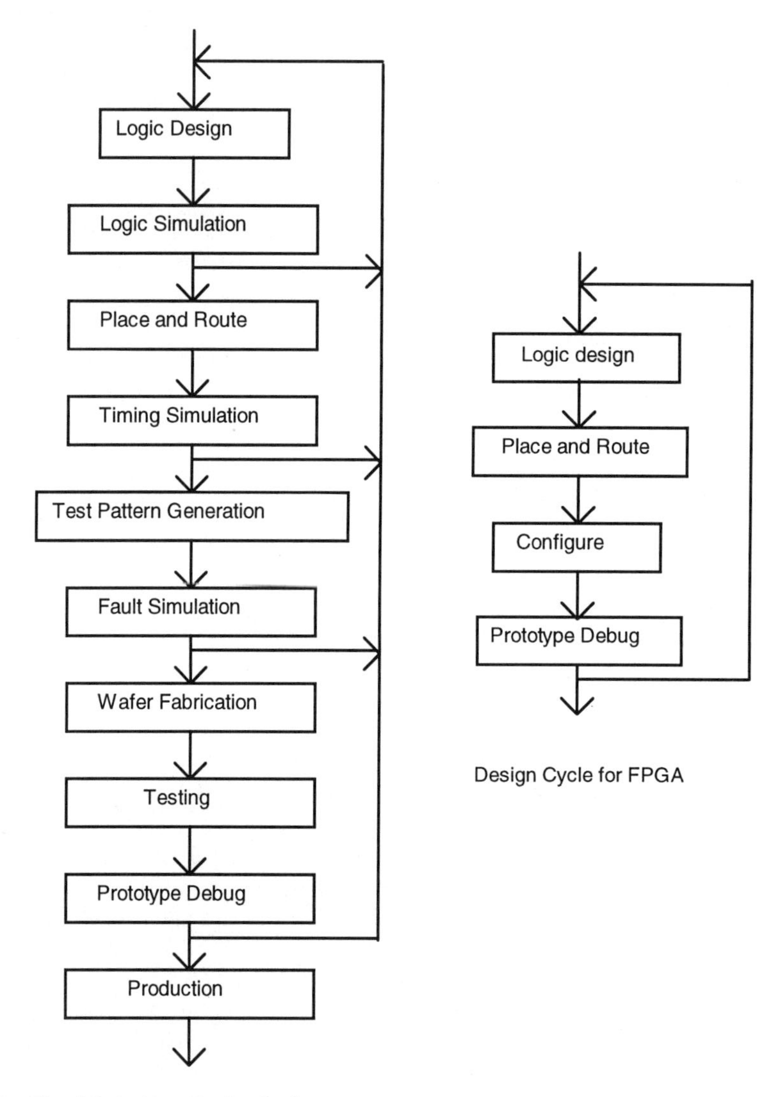


Figure 10.2 Comparing Design Cycles

The design process usually begins with capture of the design in some computer readable format. Most users enter their designs as schematics built of macros from a library. Another alternative is to enter designs in terms of Boolean equations, state machine descriptions, or functional specifications. Different portions of design can be described in different ways, compiled separately, and the results merged at some higher hierarchical level in the form of top-level schematic.

Several guidelines are suggested for reliable design with FPLDs, mostly the same as those for users of MPGAs. The major goal is to make the circuit function properly independent of shifts in timing from one part to the next.

Rapid system prototyping is most effective when it becomes rapid product development. Reprogrammability allows a system designer another option: to modify the design in the FPLD by changing the programming bitstream after the design is in the customer's hands. The bitstream can be stored in a dedicated (E)PROM or elsewhere in the system. In some existing systems, manufacturers send modified versions of hardware as a new bitstream on a floppy disk or as a file sent over modem.

## 10.2 Example 1: Altera EPLD Devices

In this and the following section we present two different types of FPLDs offered by Altera: general-purpose Erasable Programmable Logic Devices (EPLDs) based on floating gate programming technology that belong to MAX 5000, 7000 and 9000 series, and SRAM based Flexible Logic Element matriX (FLEX) 6000, 8000 and 10K series, which belong to Complex Programmable Logic Devices (CPLDs). All Altera device families use CMOS process technology, which provides lower power dissipation and greater reliability than bipolar technology. FPLDs from the MAX series are targeted for combinatorially intensive logic designs. In the following text we will concentrate on the MAX 7000 as an example of EPLDs and as the most widely used series, although the 9000 series devices have larger capacity. The FLEX device family provides logic density from 2,500 to more than 100,000 usable gates and pin counts from 84 to more than 500 pins. Predictable interconnect delays combined with the high register counts, low standby power, and in-circuit reconfigurability of FLEX devices, make these devices suitable for high-density, register-intensive designs. The most complex CPLDs from Altera are FLEX10K devices with embedded static memory blocks making these devices suitable for a wider range of more advanced applications.

### *10.2.1 Altera MAX 7000 Devices General Concepts*

The MAX 7000 family of high-density, high-performance EPLDs provides dedicated input pins, user-configurable I/O pins, and programmable flip-flop and clock options that ensure flexibility for integrating random logic functions. The MAX 7000 architecture supports emulation of standard TTL circuits and integration of SSI, MSI, and LSI logic functions. It also easily integrates multiple programmable logic devices from standard PALs and GALs, to FPLDs. MAX 7000 EPLDs use CMOS EEPROM cells as the programming technology to implement logic functions. MAX 7000 EPLDs contain from 32 to 256 logic cells, called macrocells, combined into groups of 16 macrocells, called Logic Array Blocks (LABs). Each macrocell has a programmable-AND/fixed-OR array and a configurable register with independently programmable Clock, Clock Enable, Clear, and Preset functions.

Each EPLD contains an AND array that provides product terms, which are essentially n-input AND gates. EPLD schematics use a shorthand AND-array notation to represent several large AND gates with common inputs. Product terms can also be used to generate complex control signals for use with programmable register (Clock, Clock Enable, Clear, and Preset) or the Output Enable signal for the I/O pins. These signals are called array control signals.

The Altera EPLDs support programmable inversion allowing software to generate inversions wherever necessary without wasting macrocells for simple functions. Software also automatically applies De Morgan's inversion and other logic synthesis techniques to optimize the use of available resources.

In the remaining parts of this section we will present all functional units of the Altera EPLD with sufficient levels of details necessary for full understanding of their operation and potentials of their applications.

### *10.2.2 Macrocell*

The fundamental building block of an Altera EPLD is the macrocell. The MAX 7000 macrocell can be individually configured for both combinational and sequential operation. Each macrocell consists of three parts:

- The logic array that implements all combinational logic functions
- The product-term select matrix that selects product terms which take part in implementation of logic function, and

- The programmable register that provides D, T, JK, or SR options, but also can be bypassed

One typical macrocell of the high-end members of MAX 7000 series is presented in Figure 10.3. The logic array consists of a programmable-AND/fixed-OR array, known as PLA. Inputs to the AND array come from the true and complement of the dedicated input and clock pins, and from macrocell and I/O feedback paths. A typical logic array contains 5 p-terms that are distributed among the combinational and sequential resources. Connections are opened during the programming process. Any p-term may be connected to the true and complement of any array input signal. The p-term select matrix allocates these p-terms for use as either primary logic inputs (to the OR and XOR gates) to implement logic functions, or as secondary inputs to the macrocell's register Clear, Preset, Clock, and Clock Enable functions. One p-term per macrocell can be inverted and fed back into the logic array. This "shareable" p-term can be connected to any p-term within the LAB.

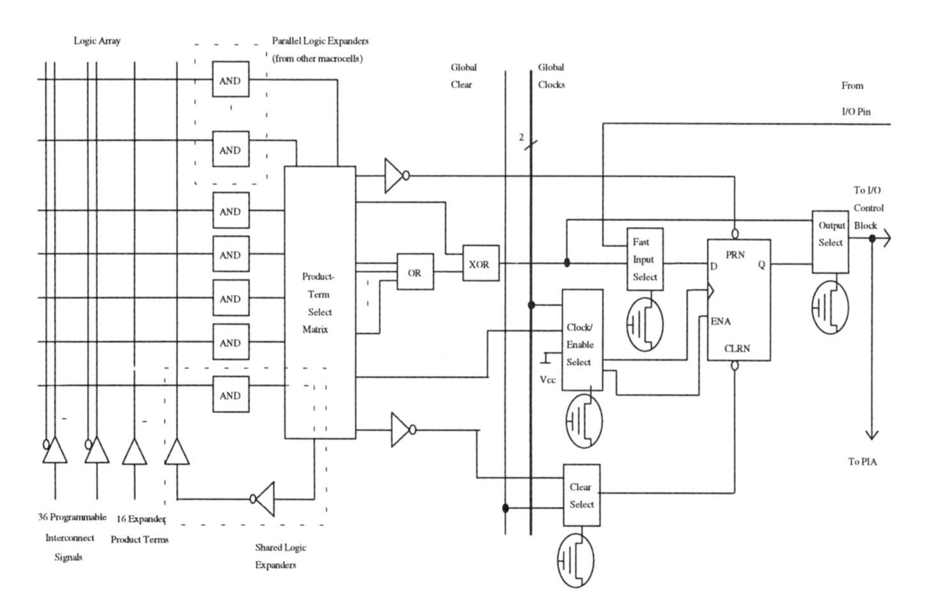


Figure 10.3 Macrocell architecture

Each macrocell flip-flop can be individually programmed to emulate D, T, JK, or SR operations with a programmable Clock control. If necessary, the flip-flop can be bypassed for combinational (non-registered) operation. Each flip-flop can be clocked in three different modes:

- By a global clock signal. This mode results with the fastest Clock-to- output performance.
- By a global Clock signal and enabled by an active-high Clock Enable. This mode provides an Enable on each flip-flop while still resulting in the fast Clock-to-output performance of the global Clock.
- By an array Clock implemented with a p-term. In this mode, the flip-flop can be clocked by signals from buried macrocells or I/O pins.

Each register also supports asynchronous Preset and Clear functions by the p-terms selected by p-term select matrix. Although these signals are active-high, active-low control can be obtained by inverting signals within the logic array. In addition, the Clear function can be driven by the active-low, dedicated global Clear pin.

The flip-flops in macrocells also have a direct input path from the I/O pin, which bypasses PIA and combinational logic. This input path allows the flip-flop to be used as an input register with a fast input set-up time (3 ns).

The more complex logic functions, that require more than five p-terms, can be implemented using shareable and parallel expander p-terms instead of additional macrocells. These expanders provide outputs directly to any macrocell in the same LAB.

Each LAB has up to 16 shareable expanders that can be viewed as a pool of uncommitted single p-terms (one from each macrocell) with inverted outputs that feed back into logic array. Each shareable expander can be used and shared by any macrocell in the LAB to build complex logic functions.

Parallel expanders are unused p-terms from macrocells that can be allocated to a neighboring macrocell to implement fast, complex logic functions. Parallel expanders allow up to 20 p-terms to directly feed the macrocell OR logic, with 5 p-terms provided by the macrocell itself and 15 parallel expanders provided by neighboring macrocells in the LAB.

When both the true and complement of any signal are connected intact, a logic low 0 results on the output of the p-term. If both the true and complement are open, a logical "don't care" results for that input. If all inputs for the p-term are programmed opened, a logic high (1) results on the output of the p-term.

Several p-terms are input to a fixed OR whose output connects to an exclusive-OR (XOR) gate. The second input to the XOR gate is controlled by a programmable resource (usually p-term) that allows the logic array output to be inverted. In this way active-low or active-high logic can be implemented, as well as the number of p-terms can be reduced (by applying De Morgan's inversion).

### *10.2.3 I/O Control Block*

The EPLD I/O control block contains a tri-state buffer controlled by one of the global Output Enable signals or directly connected to GND or VCC as shown in Figure 10.4. When the tri-state buffer control is connected to GND, the output is in high-impedance and the I/O pin can be used as a dedicated input. When the tri-state buffer control is connected to VCC, the output is enabled.

I/O pins may be configured as dedicated outputs, bi-directional lines, or as additional dedicated inputs. Most EPLDs have dual feedback, with macrocell feedback being decoupled from the I/O pin feedback.

In the high-end devices from MAX 7000 family the I/O control block has six global Output Enable signals that are driven by the true or complement of two Output Enable signals, a subset of the I/O pins, or a subset of the I/O macrocells.

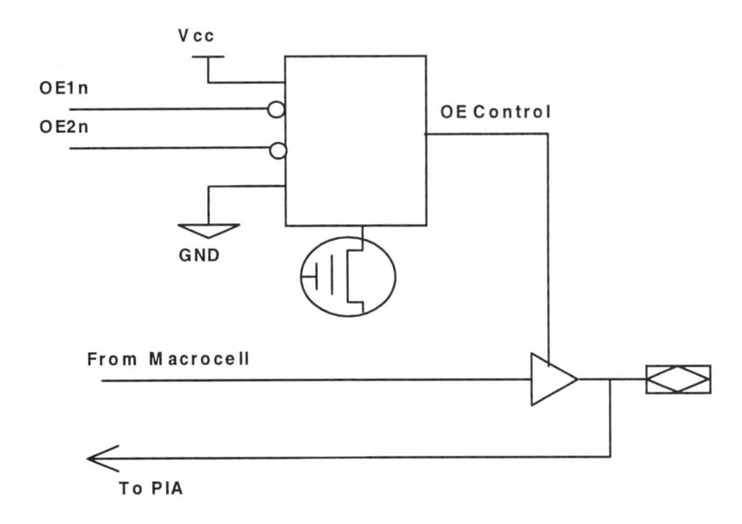


Figure 10.4 I/O control block

Some additional features are found in the MAX 7000 series. Each individual macrocell can be programmed for either high-speed, or low-power operation. The output buffer for each I/O pin has an adjustable output slew rate that can be configured for low-noise or high-speed operation. The fast slew rate should be used for speed-critical outputs in systems that are adequately protected against noise.

### *10.2.4 Logic Array Blocks*

Programmable logic in EPLDs is organized into Logic Array Blocks (LABs). Each LAB contains a macrocell array, an expander product-term array, and an I/O control block. The number of macrocells and expanders varies with each device. The general structure of the LAB is presented in Figure 10.5. Each LAB is accessible through Programmable Interconnect Array (PIA) lines and input lines. Macrocells are the primary resource for logic implementation, but expanders can be used to supplement the capabilities of any macrocell. The outputs of macrocells feed the decoupled I/O block, which consists of a group of programmable 3-state buffers and I/O pins. Macrocells that drive an output pin may use the Output Enable p-term to control the active-high 3-state buffer in the I/O control block. This allows complete and exact emulation of 7400-series TTL family.

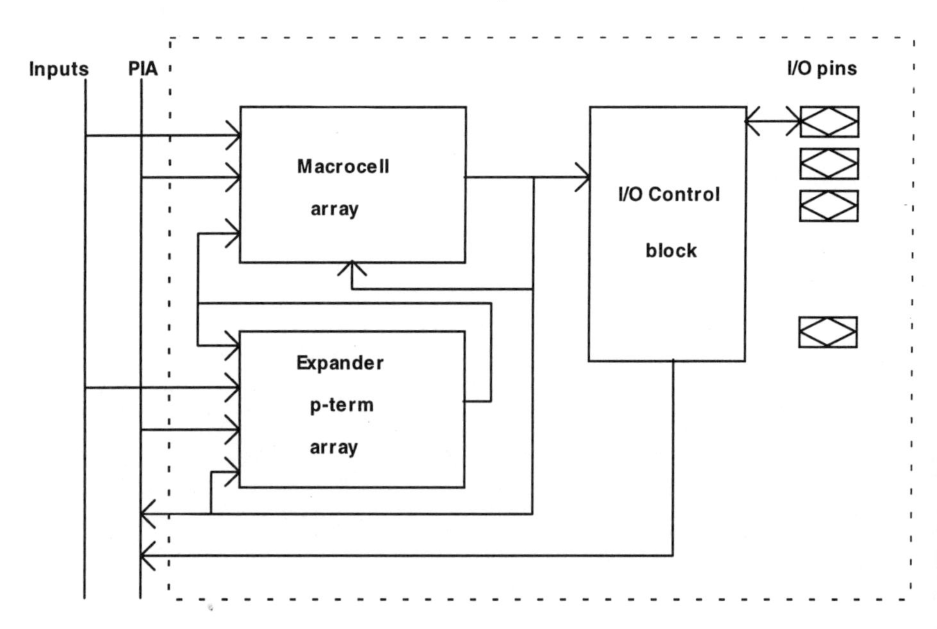


Figure 10.5 Logic Array Block architecture

Each LAB has two clocking modes: asynchronous and synchronous. During asynchronous clocking, each flip-flop is clocked by a p-term allowing that any input or internal logic to be used as a clock. Moreover, each flip-flop can be configured for positive- or negative-edge-triggered operation.

Synchronous clocking is provided by a dedicated system clock (CLK). Since each LAB has one synchronous clock, all flip-flop clocks within it are positive-edge-triggered from the CLK pin.

Altera EPLDs have an expandable, modular architecture, allowing several hundreds to tens of thousands of gates in one package. They are based on a logic matrix architecture that consists of matrix of LABs connected with a PIA, as it is shown in Figure 10.6. The PIA provides a connection path with a small fixed delay between all internal signal sources and logic destinations.

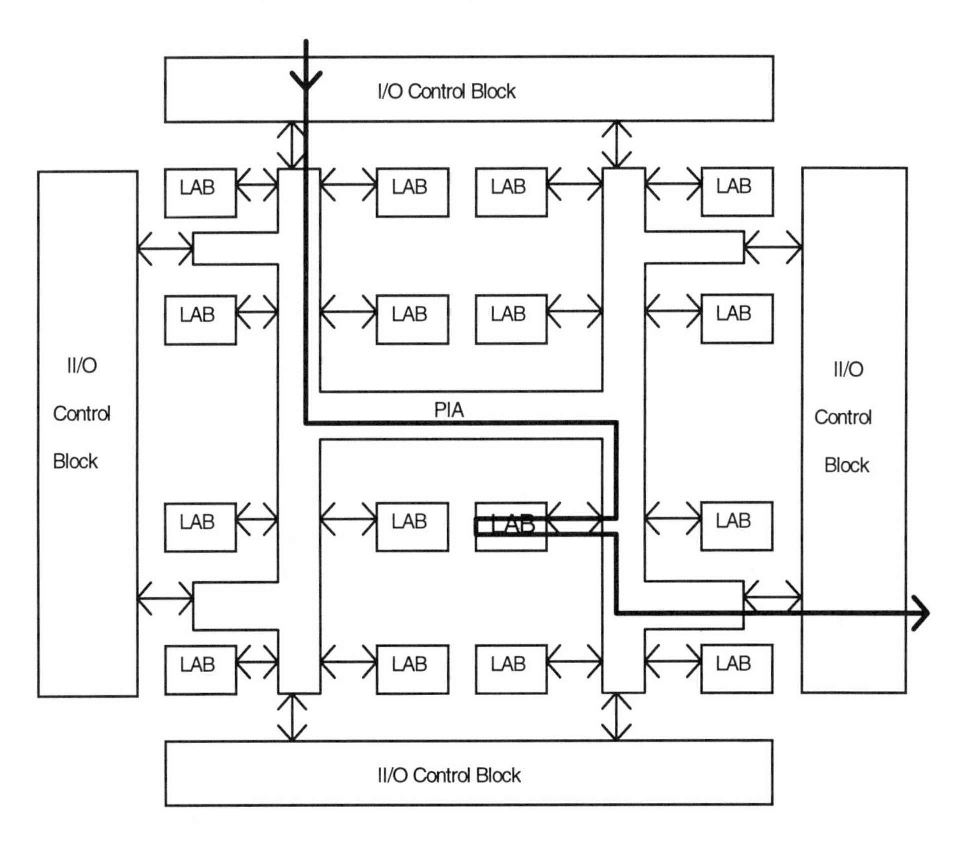


Figure 10.6 Altera EPLD entire block diagram

### *10.2.5 Programmable Interconnect Array*

Logic is routed between LABs on the Programmable Interconnect Array (PIA). This global bus is programmable and enables connection of any signal source to any destination on the device. All dedicated inputs, I/O pins, and macrocell outputs feed the PIA, which makes them available throughout the entire device. An EEPROM cell controls one input of a 2-input AND gate which selects a PIA signal to drive into the LAB, as shown in Figure 10.7. Only signals required by each LAB are actually routed into the LAB

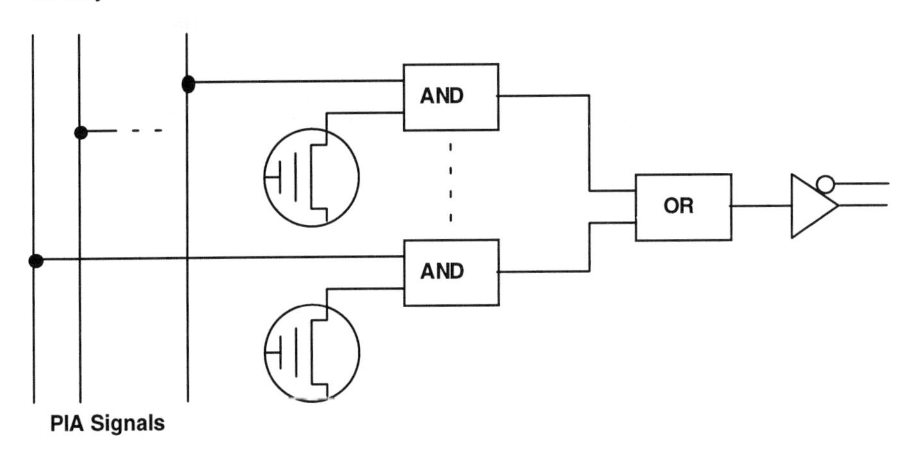


Figure 10.7 PIA routing

While the routing delays in channel-based routing schemes in MPGAs and FPLDs are cumulative, variable, and path-dependent, the MAX PIA has a fixed delay. Therefore, it eliminates skew between signals, and makes timing performance easy to predict. MAX 7000 devices have fixed internal delays that allow the user to determine the worst-case timing for any design.

### *10.2.6 Programming*

Programming of MAX 7000 devices consists of configuring EEPROM transistors as required by design. The normal programming procedure consists of the following steps:

1. The programming pin ($V_{pp}$) is raised to the super-high-input level (usually 12.5V).

2. Row and column address are placed on the designated address lines (pins).

3. Programming data is placed on the designated data lines (pins).

4. The programming algorithm is executed with a sequence of 100 microsecond programming pulses separated by program verify cycles.

5. Overprogram or margin pulses may be applied to double ensure EPLD programming.

The programming operation is typically performed eight bits at a time on specialised hardware. The security bit can be set to ensure EPLD design security.

Some of the devices from the MAX 7000 family have special features such as 3.3 V operation or power management. The 3.3 V operation offers power savings of 30% to 50% over 5.0 V operation. The power-saving features include a programmable power-saving mode and power-down mode. Power-down mode allows the device to consume near-zero power (typically 50 μA). This mode of operation is controlled externally by the dedicated power-down pin. When this signal is asserted, the power-down sequence latches all input pins, internal logic, and output pins preserving their present state.

The newest members of 7000S family provide another programming mechanism known as in-system programmability that enables programming of devices which are already placed on a PCB. Using this techniques does not require any special programming hardware, providing ease of modifications of design, as well as parallel development of design and printed circuit board.

## 10.3 Example2: Altera's CPLD Devices

Altera's Flexible Logic Element Matrix (FLEX) programmable logic combines the high register counts of FPGAs and the fast predictable interconnects of EPLDs. It is SRAM based providing low stand-by power and in-circuit reconfigurability. Logic is implemented with 4-input look-up tables (LUTs) and programmable registers. High performance is provided by a fast, continuous network of routing resources. FLEX 8000 devices, as a representative of this family, are configured at system power-up, with data stored in a serial configuration EPROM device or provided by a system controller. Configuration data can also be stored in an industry standard

EPROM or downloaded from system RAM. Because reconfiguration requires less than 100 ms, real-time "on the fly" changes can be made during system operation.

The FLEX architecture, presented in Figure 10.8, incorporates a large matrix of compact logic cells called logic elements (LEs). Each LE contains a 4-input LUT that provides combinatorial logic capability and a programmable register that offers sequential logic capability. LEs are grouped into sets of eight to create Logic Array

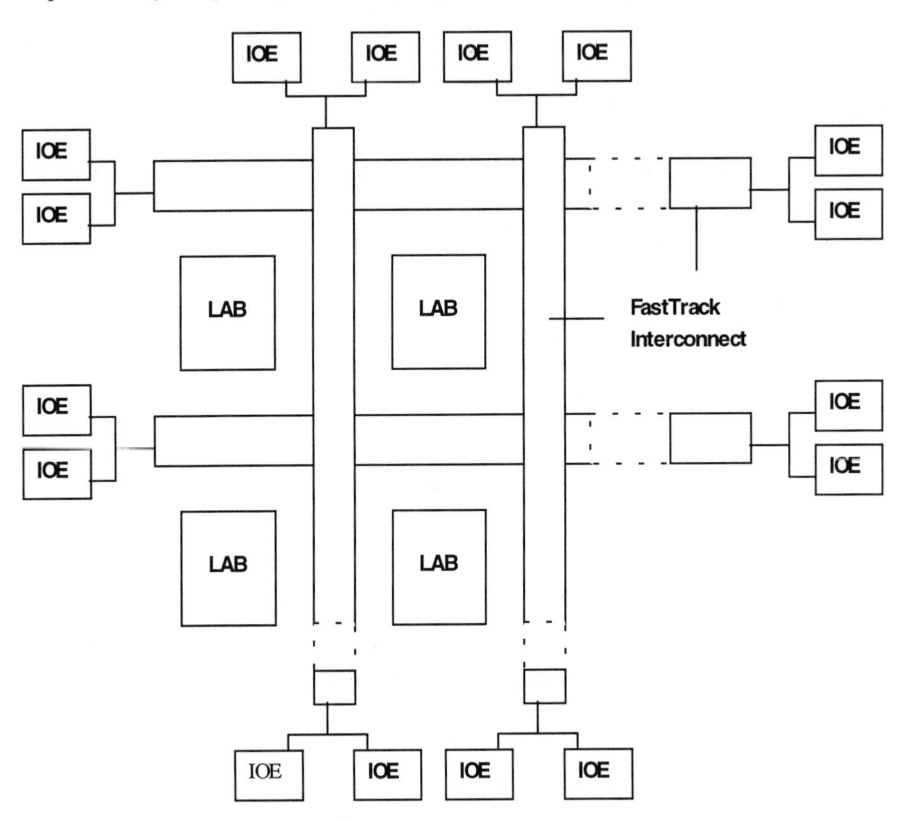


Figure 10.8 FLEX 8000 device architecture

Blocks (LABs). Each LAB is an independent structure with common inputs, interconnections, and control signals. LABs are arranged into rows and columns. The I/O pins are supported by I/O elements (IOEs) located at the ends of rows and columns. Each IOE contains a bi-directional I/O buffer and a flip-flop that can be used as either an input or output register. Signal interconnections within FLEX 8000 devices are provided by FastTrack Interconnect continuous channels that run the entire length and width of the device.

### *10.3.1 Logic Element*

The logic element (LE) is the basic logic unit in the FLEX 8000 architecture. Each LE contains a 4-input LUT, a programmable flip-flop, a carry chain, and a cascade chain as illustrated in Figure 10.9.

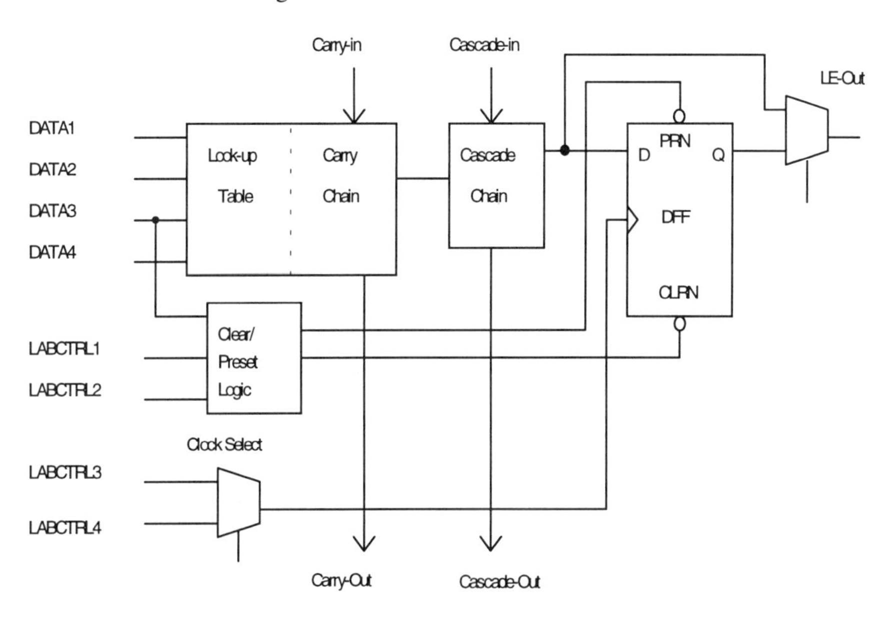


Figure 10.9 Logic element

The LUT quickly computes any Boolean function of four input variables. The programmable flip-flop can be configured for D, T, JK, or SR operation. The Clock, Clear, and Preset control signals can be driven by dedicated input pins, general-purpose I/O pins, or any internal logic. For purely combinational logic the flip-flop is bypassed and the output of the LUT goes directly to the output of the LE.

Two dedicated high-speed paths are provided in the FLEX 8000 architecture; the carry chain and cascade chain, that both connect adjacent LEs without using general-purpose interconnect paths. The carry chain supports high speed adders and counters. The cascade chain implements wide input functions with minimum delay. Carry and cascade chains connect all LEs in a LAB, and all LABs in the same row.

The carry chain provides a very fast (less than 1 ns) carry forward function between LEs. The carry-in signal from a lower order bit moves towards the higher-order bit via the carry chain, and also feeds both the LUT and a portion of the carry chain of the next LE. This feature allows implementation of high-speed counters and adders of practically arbitrary width. A 4-bit parallel full adder can be implemented in 4+1=5 LEs by using the carry chain as shown in Figure 10.10. The LE's look-up table is divided into two portions: one generates the sum of two bits using input signals and the carry-in signal, and the other generates the carry-out signal, which is routed directly to the carry-in input of the next higher-order bit. The final carry-out signal is routed to an additional LE, and can be used for any purpose.

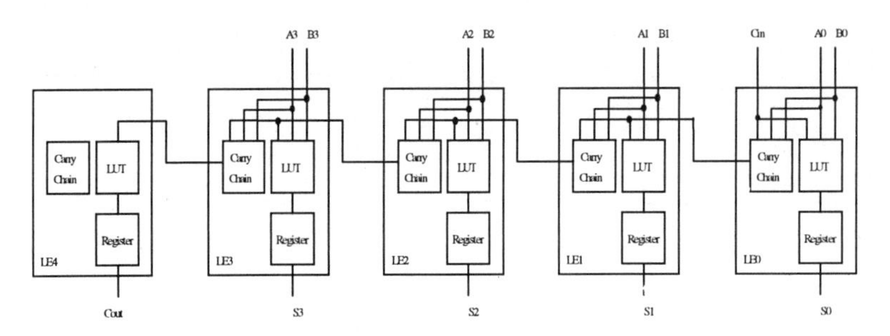


Figure 10.10 Carry chain illustration

With the cascade chain, the FLEX 8000 architecture can implement functions with a very wide fan-in. Adjacent LUTs can be used to compute portions of the function in parallel; while the cascade chain serially connects the intermediate values. The cascade chain can use a logical AND or logical OR to connect the outputs of adjacent LEs. Each additional LE provides four more inputs to the effective width of a function adding a delay of approximately 1 ns per LE. Figure 10.10 illustrates how the cascade function can connect adjacent LEs to form functions with wide fan-in. The LE can operate in the four different modes. In each mode, seven of the ten available inputs to the LE - the four data inputs from the LAB, local interconnect, the feedback from the programmable register, and the carry-in from the previous LE- are directed to different destinations to implement the desired logic function. The remaining inputs provide control for the register. The normal mode is suitable for general logic applications and wide decode functions that can take advantage of a cascade chain. The arithmetic mode offers two 3-input LUTs that are ideal for implementing adders, accumulators, and comparators. One LUT provides a 3-bit Boolean function, and the other generates a carry bit. The arithmetic mode also supports a cascade chain. The Up/Down counter mode offers counter enable, synchronous up/down control, and data loading options. Two 3-input LUTs are used: one generates the counter data, the other generates the fast carry bit. A 2-to-1 multiplexer provides synchronous loading. Data can also be loaded asynchronously with the Clear and Preset register control signals. The clearable counter mode is similar to the Up/Down counter mode, but supports a synchronous Clear instead of the up/down control. The Clear function is substituted for cascade-in signal in Up/Down Counter mode. Two 3-input LUTs are used: one generates the counter data, the other generates the fast carry bit.

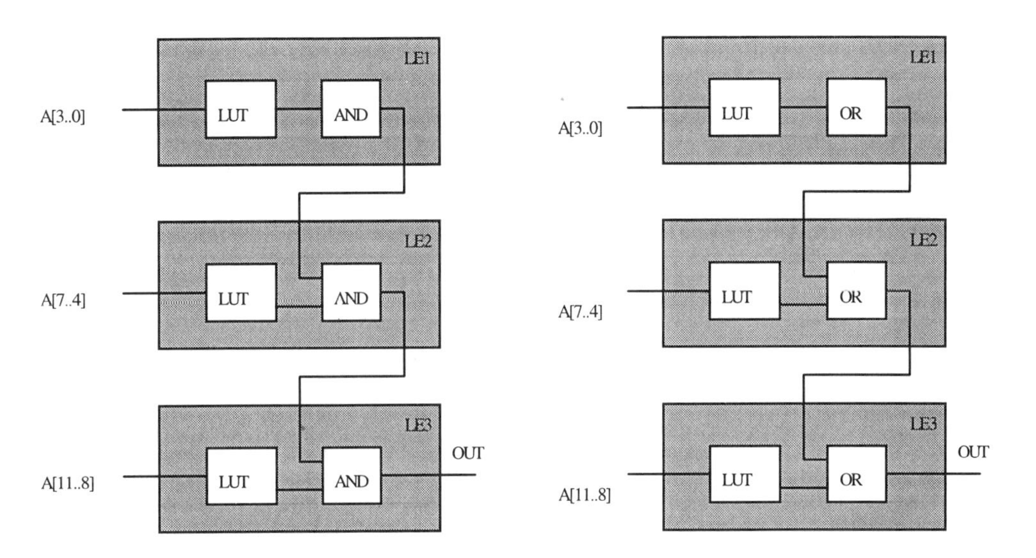


Figure 10.10 Cascade chain illustration

The Logic controlling a register's Clear and Preset functions is controlled by the DATA3, LABCTRL1, and LABCTRL2 inputs to LE. Default values for the Clear and Preset signals, if unused, are logic highs. If the flip-flop is cleared by only one of two LABCTRL signals, the DATA3 input is not required and can be used for one of the logic element operating modes.

### *10.3.2 Logic Array Block*

A Logic Array Block (LAB) consists of eight LEs, their associated carry and cascade chains, LAB control signals, and the LAB local interconnect. The LAB structure is illustrated in Figure 10.12. Each LAB provides four control signals that can be used in all eight LEs. Two of these signals can be used as Clocks and the other two for Clear/Preset control. The LAB control signals can be driven directly from a dedicated I/O pin, or any internal signal via the LAB local interconnect. The dedicated inputs are typically used for the global Clock, Clear, and Preset because they provide synchronous control with very low skew across the device. If logic is required on a control signal, it can be generated in one or more LEs in any LAB and driven into the local interconnect of the target LAB. Programmable inversion is available for all four LAB control signals.

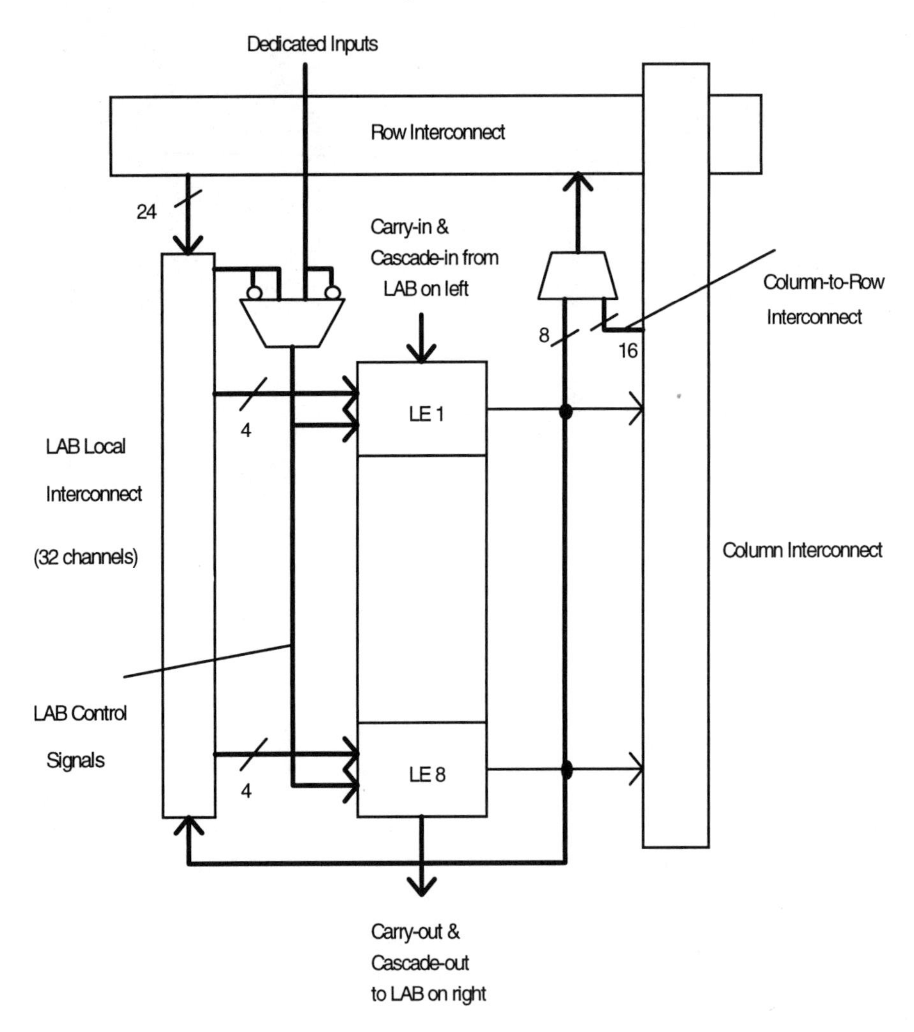


Figure 10.12 LAB Internal Architecture

### 10.3.3 FastTrack Interconnect

Connections between LEs and device I/O pins are provided by the FastTrack Interconnect mechanism represented by a series of continuous horizontal and vertical routing channels that traverse the entire device. The LABs within the device are arranged into a matrix of columns and rows. Each row of LABs has a dedicated row interconnect that routes signals both into and out of the LABs in the row. The row interconnect can then drive I/O pins or feed other LABs in the device. Figure 10.13 shows how an LE drives the row and column interconnect.

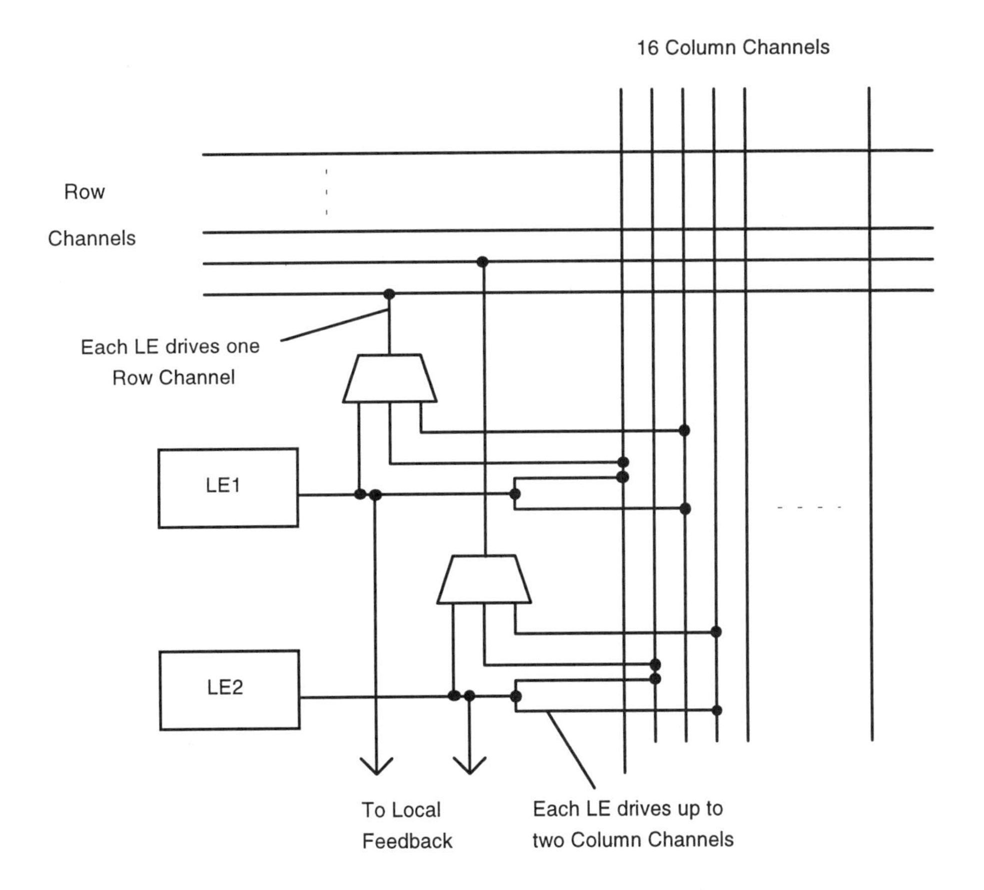


Figure 10.13 LAB Connections to Row and Column Interconnect

Each LE in a LAB can drive up to two separate column interconnect channels. Therefore, all 16 available column channels can be driven by the LAB. The column channels run vertically across the entire device, and LABs in different rows share access to them via partially populated multiplexers. A row interconnect channel can be fed by the output of the LE or by two column channels. These three signals feed a multiplexer that connects to a specific row channel. Each LE is connected to one 3-to-1 multiplexer. In a LAB, the multiplexers provide all 16 column channels with access to the row channels.

Each column of LABs has a dedicated column interconnect that routes signals out of the LABs in that column. The column interconnect can drive I/O pins or feed into the row interconnect to route the signals to other LABs in the device. A signal from the column interconnect, which can be either the output from an LE or an input from an I/O pin, must transfer to the row interconnect before it can enter a LAB. Figure 10.14 shows the interconnection of four adjacent LABs, with row, column, and local interconnects, as well as the associated cascade and carry chains.

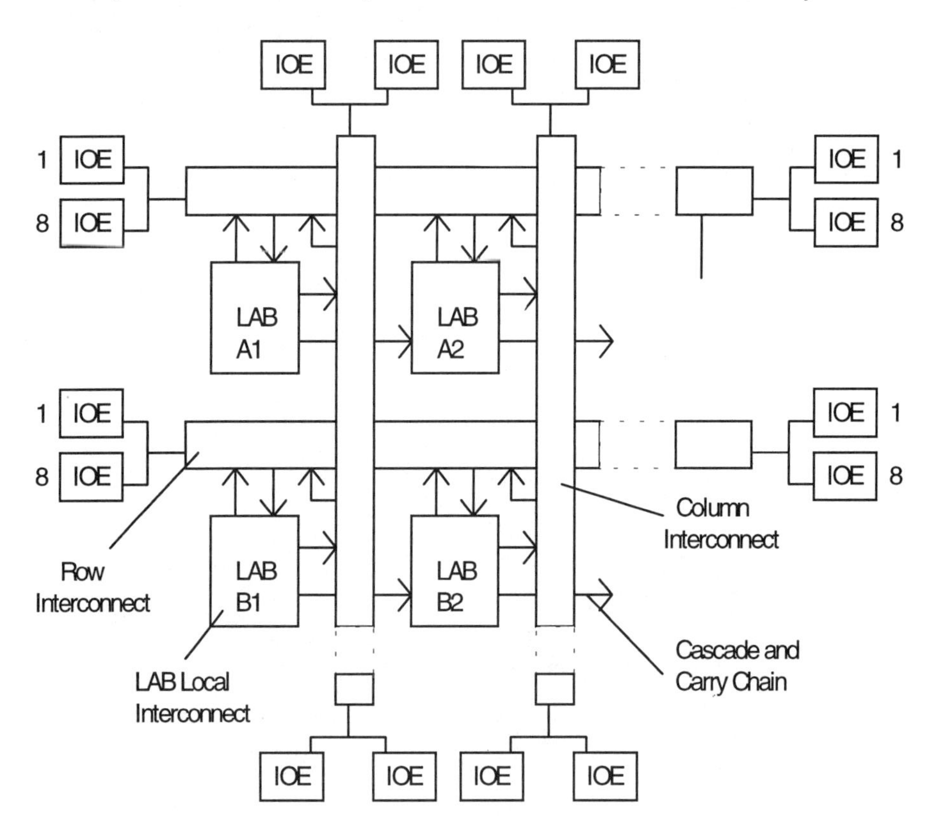


Figure 10.14 Device Interconnect Resources

In addition to general purpose I/O pins, four dedicated input pins are provided. They provide low-skew, device-wide signal distribution, and are typically used for global Clock, Clear, and Preset control signals. These signals are available for all LABs and IOEs in the device. The dedicated inputs can be used as general-purpose data inputs for nets with large fan-outs because they can feed the local interconnect of each LAB in the device.

### *10.3.4 Input/Output Element*

Input/Output Element (IOE) architecture is presented in Figure 10.15. IOEs are

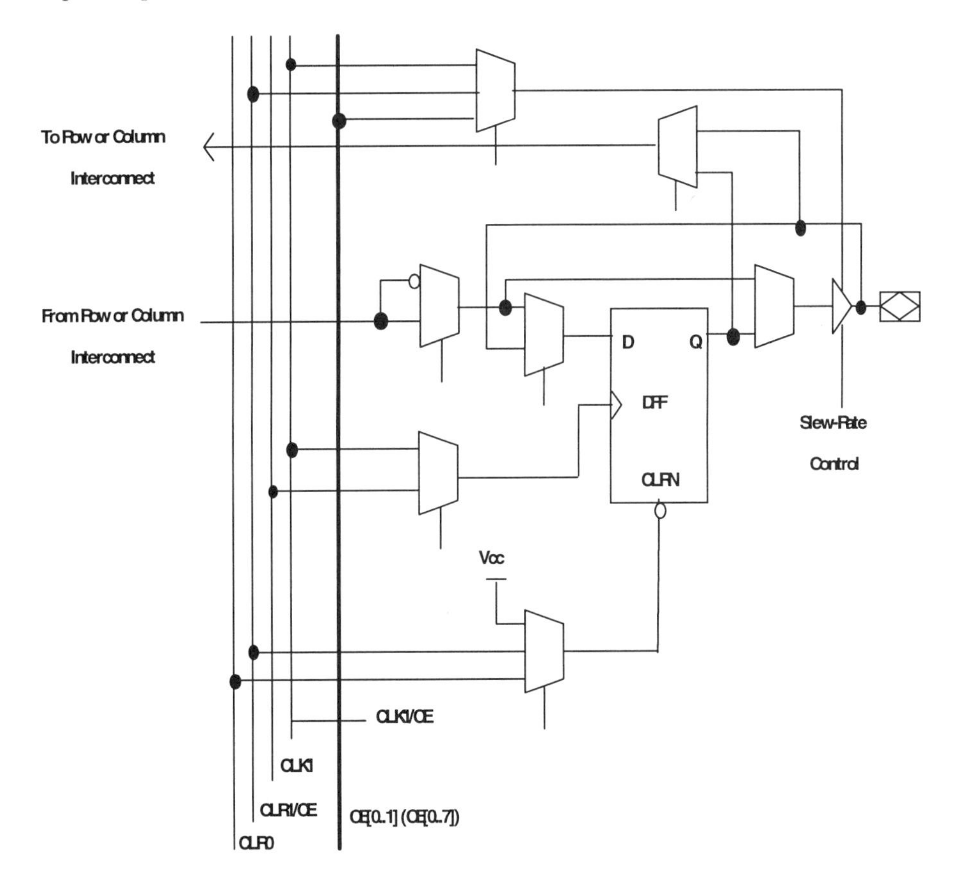


Figure 10.15 IOE Architecture

located at the ends of the row and column interconnect channels. I/O pins can be used as an input, output, or bi-directional pin. Each I/O pin has a register that can be used either as an input or output register in operations that require high performance (fast set-up time or fast Clock-to-output time). The output buffer in each IOE has an adjustable slew rate.

The fast slew rate should be used for speed-critical outputs in systems adequately protected against noise. The Clock, Clear, and Output Enable controls for the IOE are provided by a network of I/O control signals. These signals can be supplied by either the dedicated input pins or internal logic. All control signal sources are buffered onto high-speed drivers that drive the signals around the periphery of the device. This "peripheral bus" can be configured to provide up to four Output Enable signals, and up to two Clock or Clear signals.

### *10.3.5 Configuring FLEX Devices*

The FLEX family supports several different configuration schemes for loading the design into a chip on the circuit board. The FLEX architecture uses SRAM cells to store the configuration data for the device. These SRAM cells must be loaded each time the circuit powers up and begins operation. The process of physically loading the SRAM with programming data is called configuration. After configuration, the FLEX device resets its registers, enables I/O pins, and begins operation as a logic device. This reset operation is called initialization. Together, the configuration and initialization processes are called command mode. Normal in-circuit device operation is called user mode.

The entire process of configuration and initialization requires less than 100 ms, and can be used to dynamically reconfigure the device even during system operation. Device configuration can occur either automatically at system power-up or under control of external logic. The configuration data can be loaded into FLEX device with one of six configuration schemes, which is chosen on the basis of the target application. There are two basic types of configuration schemes: active, and passive. In an active configuration scheme, the device controls the entire configuration process and generates the synchronization and control signals necessary to configure and initialize itself from an external memory. In a passive configuration scheme, the device is incorporated into a system with an intelligent host that controls the configuration process. The host selects either a serial or parallel data source, and the data is transferred to the device on a common data bus. The best configuration scheme depends primarily on the particular application, and on such factors as the need to reconfigure in real-time, the need to periodically install new configuration data, etc.

Generally, active configuration schemes provide faster time-to-market because they require no external intelligence. The device is typically configured at system power-up, and reconfigured automatically if the device senses power failure. Passive configuration schemes are generally more suitable for fast prototyping and development (for example from development Max+Plus II software), or in applications that require real-time device reconfiguration. Reconfigurability allows reuse of the logic resources instead of designing redundant or duplicate circuitry into a system. A short description of some of the configuration schemes is presented in the following subsections.

### Active Serial Configuration

This scheme, with a typical circuit shown in Figure 10.16, uses Altera's serial configuration EPROM as a data source for FLEX devices. The nCONFIG pin is

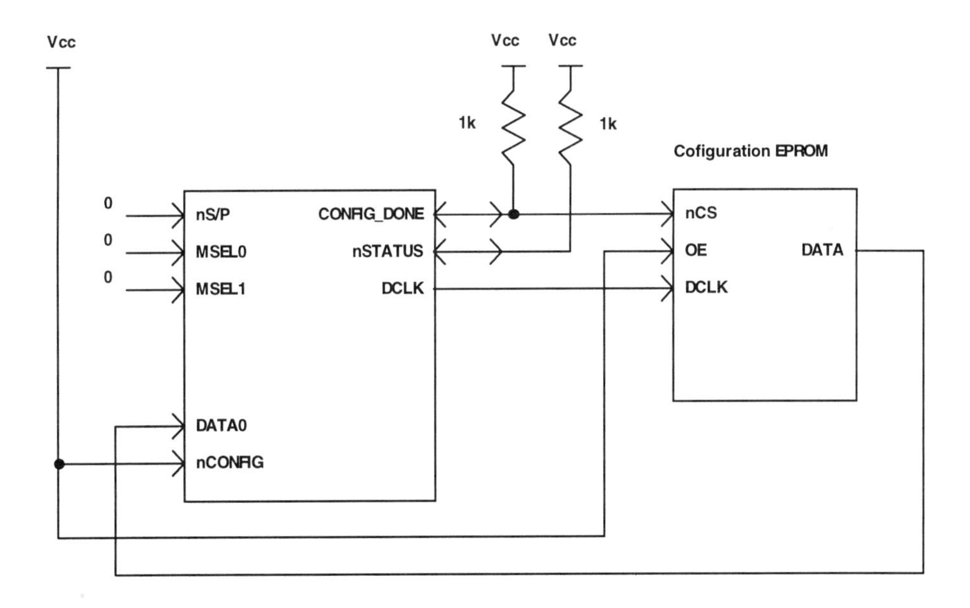


Figure 10.16 Active Serial Device Configuration

connected to Vcc, so the device automatically configures itself at system power-up. Immediately after power-up, the device pulls the nSTATUS pin low and releases it within 100 ms. The DCLK signal clocks serial data bits from the configuration

EPROM. When the configuration is completed, the CONF_DONE signal is released, causing at the same time nCS to activate and bring the configuration EPROM data output into a high-impedance state. After CONF_DONE goes high, the FLEX 8000 completes the initialization process and enters user mode. In the circuit shown in Figure 10.16, the nCONFIG signal is tied up to the Output Enable (OE) input of the configuration EPROM. External circuitry is necessary to monitor nSTATUS of the FLEX device in order to undertake appropriate action if configuration fails.

### Active Parallel Up (APU) and Active Parallel Down (APD) Configuration

In APU and APD configuration schemes, the FLEX 8000 device generates sequential addresses that drive the address inputs to an external EPROM. The EPROM then returns the appropriate byte of data on the data lines DATA[7..0]. Sequential addresses are generated until the device has been completely loaded. The CONF_DONE pin is then released and pulled high externally, indicating that configuration has been completed. The counting sequence is ascending (00000H to 3FFFFH) for APU or descending (3FFFFH to 00000H) for APD configuration. A typical circuit for parallel configuration is shown in Figure 10.17.

On each pulse of the RDCLK signal (generated by dividing DCLK by eight), the device latches an 8-bit value into a serial data stream. A new address is presented on the ADD[17..0] lines a short time after a rising edge on RDCLK. External parallel EPROM must present valid data before the subsequent rising edge of RDCLK, which is used to latch data based on address generated by the previous clock cycle.

Both active parallel configuration schemes can generate addresses in either an ascending or descending order. Counting up is appropriate if the configuration data is stored at the beginning of an EPROM, or at some known offset in an EPROM larger of 256 Kbytes. Counting down is appropriate if the low addresses are not available, for example if they are used by CPU for some other purpose.

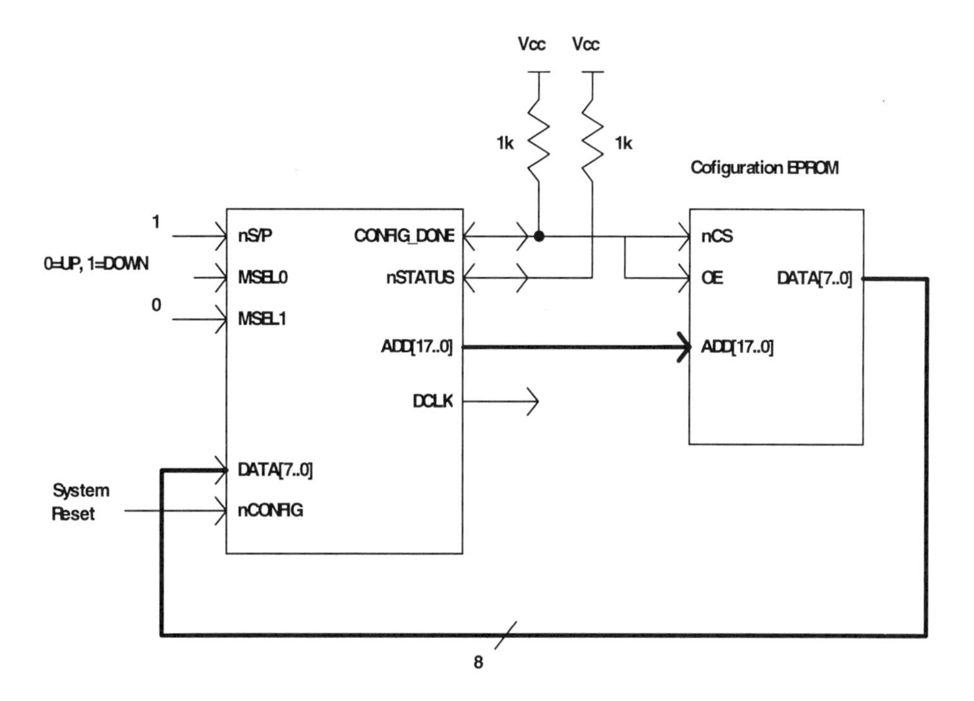


Figure 10.17 APU and APD Configuration with a 256 Kbyte EPROM

## Passive Parallel Synchronous Configuration

In this scheme the FLEX device is tied to an intelligent host. The DCLK, CONF_DONE, nCONFIG, and nSTATUS signals are connected to a port on the host, and the data can be driven directly onto a common data bus between the host and the FLEX 8000 device. New byte of data is latched on every eighth rising edge of DCLK signal, and serialized on every eight falling edge of this signal, until the device is completely configured.

A typical circuit for passive serial configuration is shown in Figure 10.17. The CPU generates a byte of configuration data. Data is usually supplied from a microcomputer 8-bit port. Dedicated data register can be implemented with an octal latch. The CPU generates clock cycles and data; eight DCLK cycles are required to latch and serialize each 8-bit data word. A new data word must be present at the DATA[7..0] inputs upon every eight DCLK cycle.

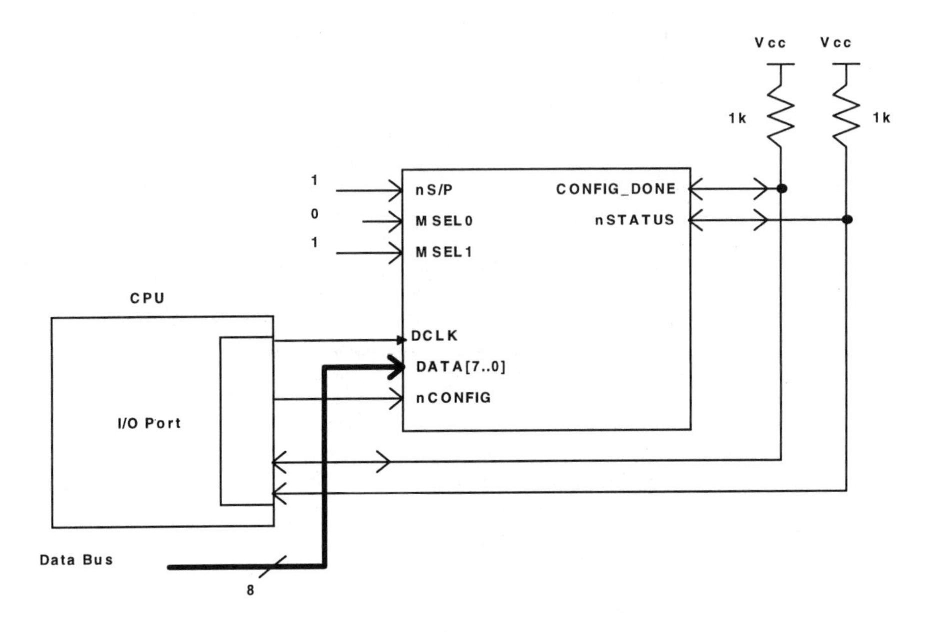


Figure 1.17 Parallel Passive Synchronous Configuration

## Passive Parallel Asynchronous Configuration

In this configuration, a FLEX device can be used in parallel with the rest of the board. The device accepts a parallel byte of input data, then serializes the data with its internal synchronization clock. The device is selected with nCS and CS chip select input pins. A typical circuit with a microcontroller as an intelligent host is shown in Figure 10.18. Dedicated I/O ports are used to drive all control signals and the data bus to the FLEX 8000 device. The CPU performs handshaking with a device by sensing the RDYnBUSY signal to establish when the device is ready to receive more data. The RDYnBUSY signal falls immediately after the rising edge of the nWS signal that latches data, indicating that the device is busy. On the eighth falling edge of DCLK, RDYnBUSY returns to Vcc, indicating that another byte of data can be latched.

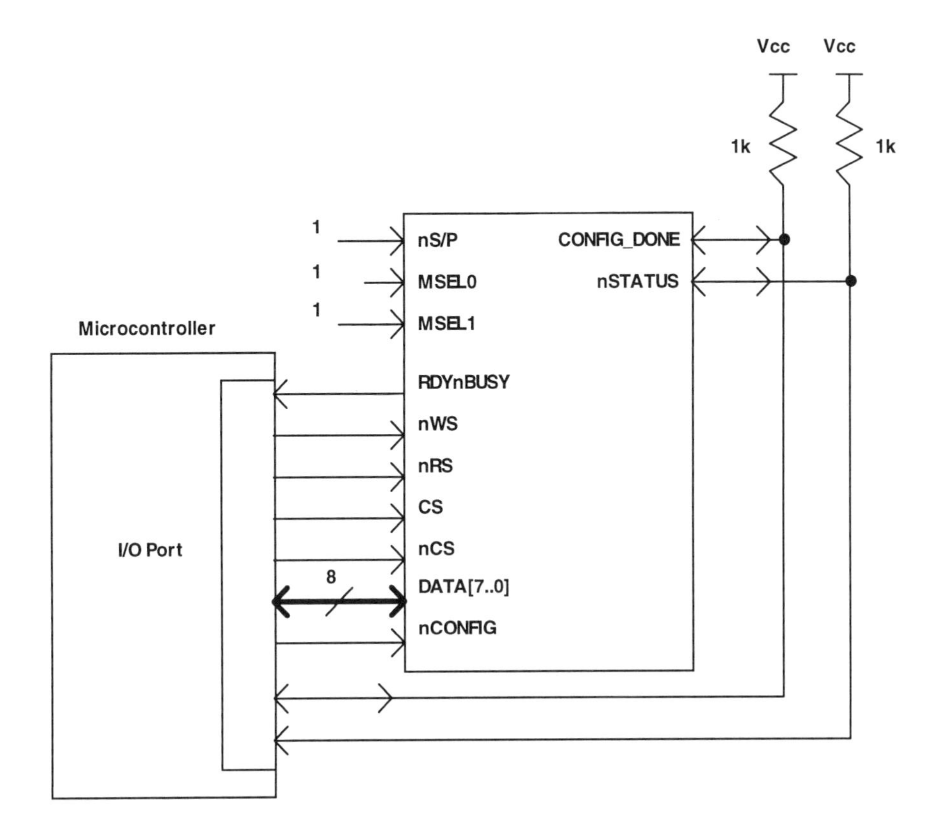


Figure 10.18 Passive Parallel Asynchronous Configuration

## Passive Serial Configuration

The passive serial configuration scheme uses an external controller to configure the FLEX 8000 device with a serial bit-stream. The FLEX device is treated as a slave, and no handshaking is provided. Figure 10.19 shows how a bit-wide passive configuration is implemented. Data bits are presented at the DATA0 input, with the least significant bit of each byte of data presented first. The DCLK is strobed with a high pulse to latch the data. The serial data loading continues until the CONF_DONE goes high, indicating that the device is fully configured. The data source can be any source that the host can address.

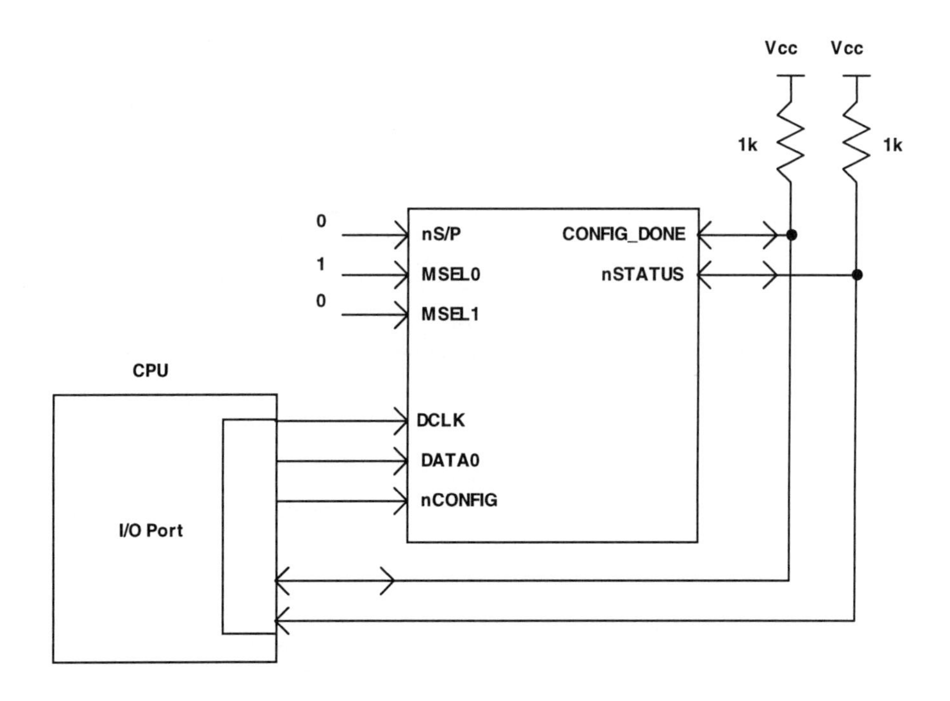


Figure 10.19 Bit-Wide Passive Serial Configuration

## 10.4 Example 3: Altera FLEX 10K Devices

The aim of this Section is to make a brief introduction to basic features of Altera's FLEX 10K devices which offer quite new design alternatives and solutions to existing problems than the other CPLDs and FPGAs. Altera's FLEX 10K devices are currently industry's most complex and most advanced CPLDs. Besides logic array blocks and their logic elements, which are with the same architecture as those in FLEX8000 devices, FLEX 10K devices incorporate dedicated die areas of embedded array blocks (EABs) for implementing large specialized functions providing at the same time programmability and easy design changes. The architecture of FLEX10K device family is illustrated in Figure 10.20. The EAB consists of memory array and surrounding programmable logic which can easily be configured to implement required function. Typical functions which can be implemented in EABs are memory functions or complex logic functions, such as microcontrollers, digital signal processing functions, data-transformations functions, and wide data path functions. The LABs are used to implement general logic.

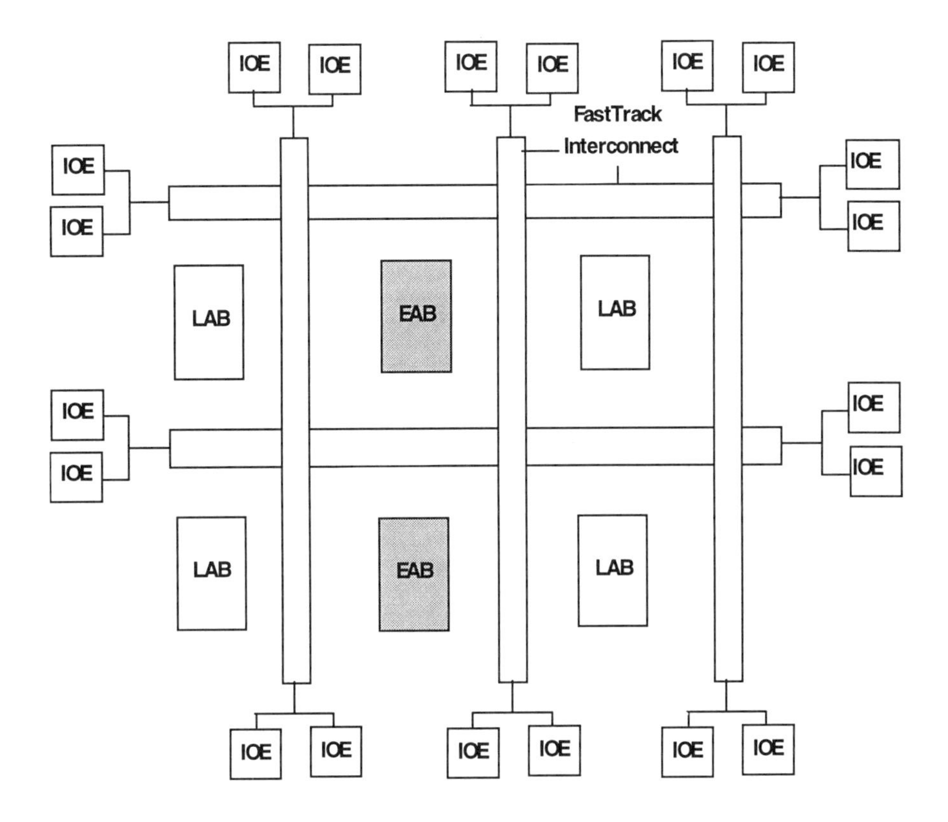


Figure 10.20 FLEX10K device architecture

### 10.4.1 Embedded Array Block

If the EAB is used to implement memory functions, it provides 2,048 bits, which are used to create single- or dual-port RAM, ROM or FIFO functions. When implementing logic, each EAB is equivalent to 100 to 600 gates for implementation of complex functions, such as multipliers, state machines, or DSP functions. One FLEX 10K device can contain up to 12 EABs. EABs can be used independently, or multiple EABs can be combined to implement more complex functions.

The EAB is a flexible block of RAM with registers on the input and output ports. Its flexibility provides implementation of memory of the following sizes: 2,048 x 1,

1,024 x 2, 512 x 4, or 2,048 x 1 as it is shown in Figure 10.21. This flexibility make it suitable for more than memory, for example by using words of various size as look-

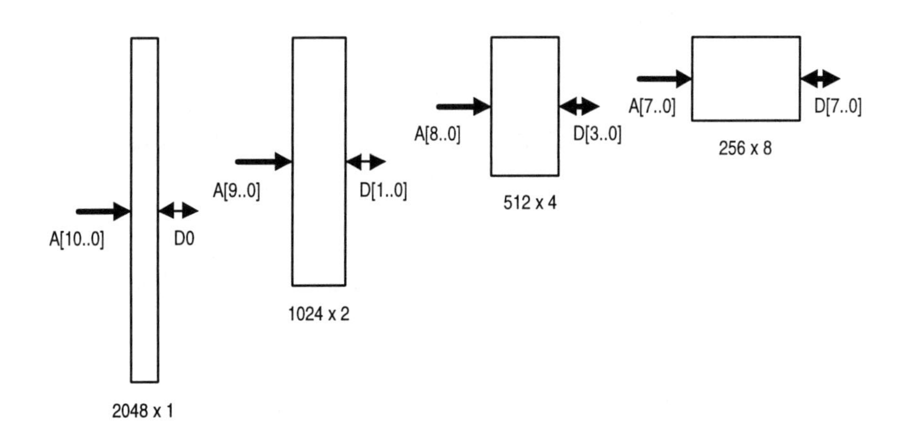


Figure 10.21 EAB Memory Configurations

up tables and implement functions such as multipliers, error correction circuits, or other complex arithmetic operations. For example, a single EAB can implement a 4 x 4 multiplier with eight inputs and eight outputs providing high performance by fast and predictable access time of the memory block. Dedicated EABs are easy to use, eliminate timing and routing concerns, and provide predictable delays. The EAB can be used to implement both synchronous and asynchronous RAM. In the case of synchronous RAM, the EAB generates its own "write enable" signal and is self-timed with respect to global clock. Larger blocks of RAM are created by combining multiple EABs in "serial" or "parallel" fashion. The global FLEX 10K signals, dedicated clock pins, and EAB local interconnect can drive the EAB clock signals. Because the LEs drive the EAB local interconnect, they can control the "write enable" signal or the EAB clock signal. The EAB architecture is illustrated in Figure 10.22.

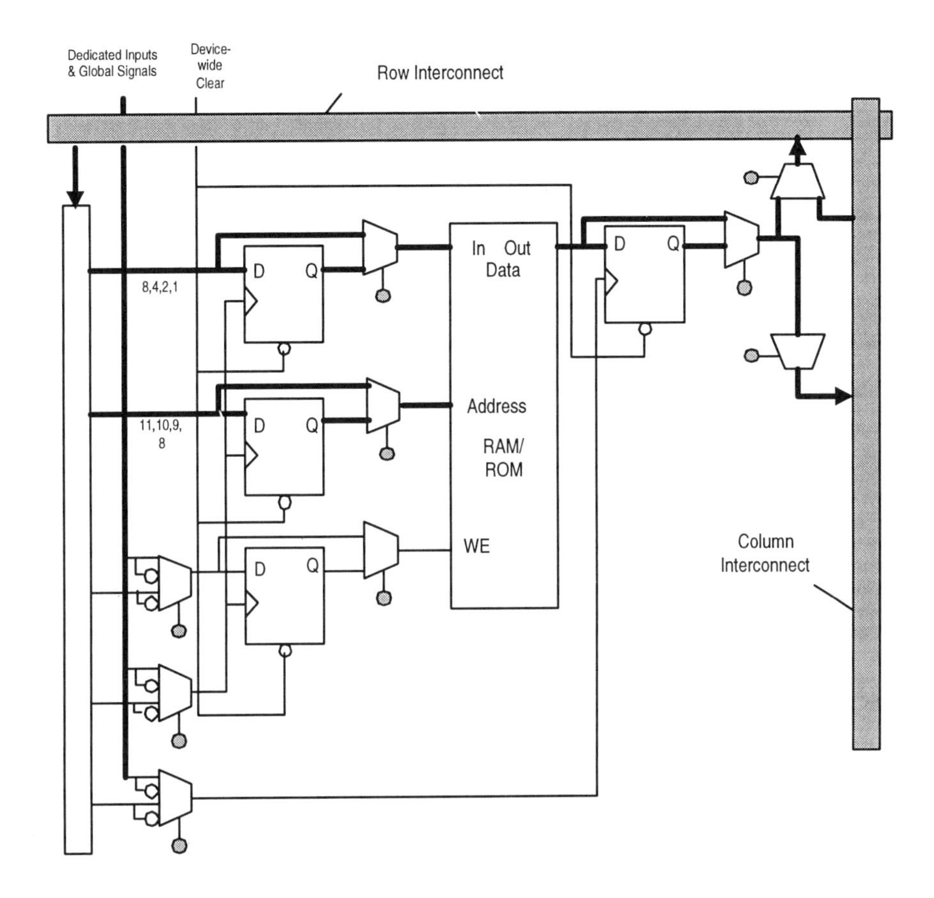


Figure 10.22 EAB architecture

In contrast to logic elements which implement very simple logic functions in a single element, and more complex functions in multi-level structures, the EAB implements complex functions, including wide fan-in functions, in a single logic level, resulting in more efficient device utilization and higher performance. The same function implemented in the EAB will often occupy less area on a device, have a shorter delay, and operate faster than functions implemented in logic elements. Depending on its configuration, an EAB can have 8 to 10 inputs and 1 to 8 outputs, all of which can be registered for pipelined designs. Maximum number of outputs depends on the number of inputs. For example, an EAB with 10 inputs can have only 1 output, and an EAB with 8 inputs can have 8 outputs.

### *10.4.2 Implementing Logic with EABs*

Logic functions are implemented by programming the EAB during configuration process with a read-only pattern, creating a large look-up table (LUT). The pattern can be changed and reconfigured during device operation to change the logic function. When a logic function is implemented in an EAB, the input data is driven on the address input of the EAB. The result is looked up in the LUT and driven out on the output port. Using the LUT to find the result of a function is faster than using algorithms implemented in general logic and LEs.

EABs make FLEX 10K devices suitable for a variety of specialized logic applications such as complex multipliers, digital filters, state machines, transcendental functions, waveform generators, wide input/wide output encoders, but also various complex combinatorial functions.

For example, in a 4-bit x 4-bit multiplier, which requires two 4-bit inputs and one 8-bit output, two data inputs drive address lines of the EAB, and the output of the EAB drives out the product. The contents of the EAB memory locations is product of input data (multiplicands) presented on address lines. Higher order multipliers can be implemented using multiple 4 x 4 multipliers and parallel adders.

Another interesting application is a constant multiplier which is found often in digital signal processing and control systems. The value of constant determines the pattern that is stored in the EAB. If the constant is to be changed at run-time, it can be easily done by changing the pattern stored in the EAB. The accuracy of the result of multiplication can be adjusted by varying width of the output data bus. This can be easily done by adjusting the EAB's configuration or connecting multiple EABs in parallel if the accuracy grater than 8 bits is required.

General multipliers, constant multipliers, and adders, in addition to delay liners implemented by D-type registers, are most frequently used in various data path applications such as digital filters. The EAB, configured as a LUT, can implement a FIR filter by coefficient multiplication for all taps. The required precision on the output determines the EAB configuration used to implement the FIR filter.

Another example is implementation of transcendental function such as sine, cosine and logarithms which are difficult to compute using algorithms. It is more efficient to implement transcendental functions using LUTs. The argument of the function drives the address lines to the EAB, and the output appears at the data output lines. After implementing functions such as sine and cosine, it is easy to use them in implementing waveform generators. The EAB is used to store and generate waveforms that repeat over time. Examples of such generators are presented in Chapter 14.

Similarly, large input full encoders can be implemented using LUTs stored in EABs. Number of input address line determines the number of combinations that can be stored in the EAB, and the number of data lines determines how many EABs are needed to implement encoder. For example, the EAB with 8 address lines can store 256 different output combinations. Using two EABs connected in parallel , enables encoding of input 8-bit numbers into up to 16-bit output numbers.

The contents of an EAB can be changed at any time without reconfiguring the entire FLEX 10K device. This enables the change of portion of design while the rest of device and design continues to operate. The external data source used to change the current configuration can be a RAM, ROM, or CPU. For example, while the EAB operates, a CPU can calculate a new pattern for the EAB and reconfigure the EAB at any time. The external data source then downloads the new pattern in the EAB. After this partial reconfiguration process, the EAB is ready to implement logic functions again. If we apply such design approach that some of the EABs are active and some dormant at the same time, on-the-fly reconfiguration can be performed on the dormant EABs and they can be switched into the working system. This can be accomplished using internal multiplexers to switch-out and switch-in EABs as it is illustrated in Figure 10.23.

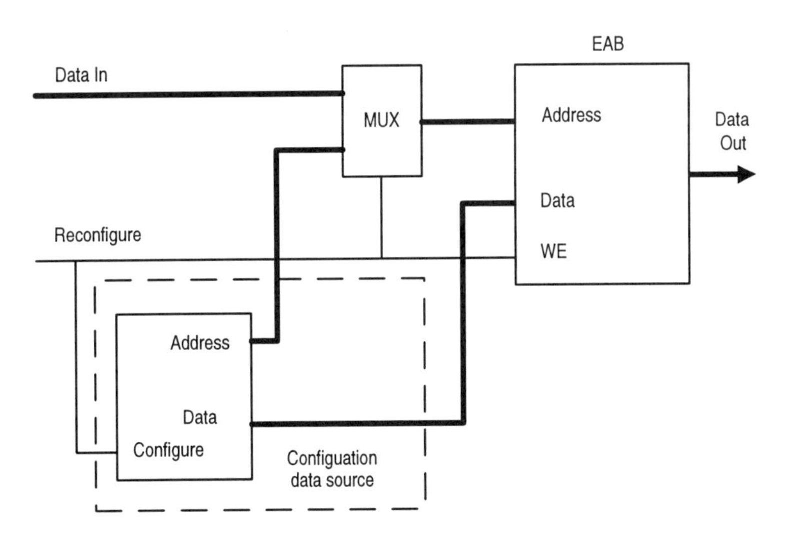


Figure 10.23 Implementation of reconfigurable logic in the EAB

If the new configuration is stored in external RAM, it has not to be defined in advance. It can be calculated and stored into RAM, and downloaded into the EAB when needed. For example, if the coefficients in an active filter are stored in an

EAB, the characteristics of the filter can be changed dynamically by modifying the coefficients. The coefficients are modified by writing to the RAM.

## 10.5 Questions and Problems

10.1. What are the major ways of classifying FPLDs?

10.2. What are the major programming technologies for FPLDs?

10.3. Why are FPLDs custom-configurable? Are all FPLDs dynamically reconfigurable?

10.4. Given the logic element of the Altera 8000 family. How would you design an 8-to-1 multiplexer using look-up tables present in the logic element? What is the content of all look-up tables that make the multiplexer.

10.5. Compare the synthesis of the multiplexer from the preceding example with the synthesized circuit based on elementary logic gates.

10.6. How can address decoders be implemented using different Altera logic families, namely MAX7000 and FLEX8000?

10.7. What is the In-Circuit Programmability?

10.8. Describe the external circuit that enables configuration of the FLEX8000 FPLD using an asynchronous passive parallel configuration scheme.

# 11 DESIGN AND PROTOTYPING ENVIRONMENTS FOR FPLDS

FPLD architectures provide identical logic cells (or some of their variations) and interconnection mechanisms as a base for the implementation of digital systems. These architectures can be used directly for the design of digital circuits, but the number of available resources and the complexity of designs that should be placed in a device require tools that are capable of translating the designer's functions into the exact cells and interconnections needed to form the final design. It is desirable to have such design software that will enable flexible design entry using varieties of methods, automatically translate designs for different FPLD architectures, support modular and hierarchical projects development, provide design verification using simulation, and finally provide data on physical aspects of the resulting design.

## 11.1 Design Frameworks

The complexity of FPLDs requires very sophisticated design tools that can efficiently handle complex designs. These tools usually integrate several different design steps into a uniform design environment enabling the designer to work with the different tools from within the same design framework. They enable design to be performed at a relatively high abstract level, but at the same time allowing the designer to see a physical relationship inside an FPLD device, and even change design details at the lowest, physical level.

### *11.1.1 Design Steps and Design Framework*

Design software must perform the following primary functions:

- Enable design entry in some of the commonly used and widely accepted formats - design entry software should provide an architecture-independent design environment that easily adapts to specific designer's needs. The most common design entries belong to the categories of graphic (schematic) design entry, hardware description languages, waveform editors, or some other appropriate tools to transfer designer's needs to a translator.

- Enable translation of design entered by any of the design entry tools or their combinations into the standard internal form that can be further translated for different FPLD architectures. Translation software performs functions such as logic synthesis, timing-driven compilation, partitioning, and fitting of design to a target FPLD architecture. Translation mechanisms also provide all of the needed information for the other design tools used in the subsequent design phases.
- Enable verification of a design using functional and timing simulation. In this way many design errors are discovered before actually programming devices, and can be easily corrected using design entry tools. Usually vendor-provided translators produce designs in the forms accepted by industry-standard CAE tools that provide very extensive verification models and procedures.
- Enable device programming consisting of downloading design control information into a target FPLD device.
- Enable reusability by providing the libraries of vendor- and user-designed units that have been proven to operate correctly.

All of the primary functions above are usually integrated into complex design environments or frameworks with the unified user interface. The common element of all these tools is some common circuit representation, most often described in the form of so-called netlists.

### *11.1.2 Netlists and Compiling*

The first step of every compiler is transformation of the design entered in user-provided form into the internal form which will be manipulated by the compiler and other tools. A compiler is faced with several problems, first being whether the design will fit into a target FPLD architecture at all. It obviously depends on the number of input and output pins, but also on the number of internal circuits needed to implement the desired functions. If the design is entered using a graphic editor and the usual schematic notation, a compiler must analyze the possible implementation of all logic elements in existing logic cells of the targeted FPLD. The design is dissected into known three- or four- input patterns that can be implemented in standard logic cells, and the pieces are subsequently added up. Initially, a compiler has to provide substitutions for the target design gates into equivalent FPLD cells, and also make the best use of substitution rules. Once substitution patterns are found, a sophisticated compiler eliminates all redundant circuitry increasing the probability that the design will fit to the target FPLD device. Compilers translate design from abstract form (schematic, equations, waveforms) to a concrete version- a bitmap forming functions and interconnections. As an intermediate form that unifies various design tools, a netlist appears. After the process of translating design into available cells provided by FPLD (sometimes also called the technology mapping phase), the cells are assigned to specific locations within the FPLD, and this is called cell placement. Once the cells are assigned to specific locations, the signals are assigned to specific interconnection lines. The portion of the compiler that performs placement and routing is usually called a fitter.

A netlist is a textfile representing logic functions and their input/output connections. Netlist can describe small functions like flip-flops, gates, inverters, switches, or even transistors, but also large units (building blocks) like multiplexers, decoders, counters, adders or even microprocessors. They are very flexible because the same format can be used at different levels of description. For example, a netlist with an embedded multiplexer can be rewritten to have the component gates comprising the multiplexer, as an equivalent representation. This is called netlist expansion. It simply specifies all gates with their input and output connections, including the inputs and outputs of the entire circuit.

A compiler uses traditional methods to simplify logic designs, but it also uses netlist optimization which represents design minimization after transformation to a netlist. Today's compilers include a large number of substitution rules and strategies in order to provide netlist optimization. The optimizer scans the multiplexer netlist first finding unused inputs, then eliminating gates driven by the unused inputs. In this way a new netlist, without unneeded inputs and gates, is created.

After netlist optimization, logic functions are translated into the available logic cells of the FPLD with the attempt to map as much as possible elementary gates into corresponding logic cells. The next step is to assign logic functions to specific locations within the FPLD device. The compiler usually attempts to place them into the simplest possible device if it is not specified in advance. The cell placement problem requires iteration. Sometimes, if compiler produces unsatisfactory results, manual cell placement may be necessary. The critical criteria for cell placement is that interconnections of the cells must be made in order to implement the required logic functions. Additional requirements may be minimum skew time paths or minimum time delays between input and output circuit pins. Usually, several attempts are necessary to meet all constraints and requirements.

Some compilers allow the designer to implement portions of a design manually. Any resource of the FPLD, such as an input/output pin (cell) or logic cell can perform a specific user-defined task. In this way some logic functions can be placed together in specific portions of the device, specific functions can be placed in specific devices (if the projects cannot fit into one device), or inputs or outputs of a logic function can be assigned to specific pins, logic cell or specific portion of the device. These assignments are taken as fixed by the compiler, and it then produces placement for the rest of the design.

Assuming that an appropriate placement of the cells and other resources occurred, the next step is to connect all resources. This step is called routing. Routing starts with the examination of the netlist that provides all interconnection information, and from inspection of the placement. The routing software assigns signals from resource outputs to destination resource inputs. As the connection proceeds, the interconnect lines become used, and congestion appears. In this case the routing software can fail to do further routing. Then, the software must replace resource placement into another arrangement and repeat routing again.

As the result of placement and routing design a file describing the original design in terms of the FPLD resources is obtained. The design file is then translated to a bitmap that can be passed to a device programmer to configure the FPLD. The type of device programmer depends on the type of the target FPLD in terms of the programming method used (RAM- or (E)EPROM-based devices).

Good interconnection architectures increase the probability that the placement and routing software will perform the desired task. However, bad routing software can waste a good connection architecture. Even in the case of total interconnectivity, when any cell could be placed at any site and connected to any other site, the software task is very complex. This complexity is increased by the addition of certain constraints that must be met. Such constraints are, for instance, a timing relationship or requirement that the flip-flops of some register or counter must be placed into adjacent logic cells within the FPLD. These requirements must be met first, and then the rest of the circuit is connected. In some cases, placement and routing become even impossible. This is the reason to keep the number of such requirements at the minimum.

## 11.2 Design Entry and High Level Modeling

Design entry can be performed in different levels of abstraction and in different forms. It represents different ways of design modeling, some of them being suitable for behavioral simulation of the system under the design, and some being suitable for circuit synthesis. Usually, two major design entry methods belong to:

- schematic entry systems that enable design to be described usual primitives in form of standard SSI and MSI blocks, or some more complex blocks provided by the FPLD vendor or designer or
- textual entry systems that use hardware description languages to describe system behaviour or structures and their interconnections.

Advanced design entry systems allow any combination of both design methods and the design of subsystems that will be interconnected with other subsystems at a higher level of design hierarchy. Usually, the highest current level of design hierarchy is called the project level. Current project can use and contain designs done in previous projects as its lower level design units.

In order to illustrate all design entry methods, we will use a small example of pulse distributor circuit that has Clock as an input and produces five non-overlapping periodic waveforms (clock phases) at the output as shown in Figure 11.1.

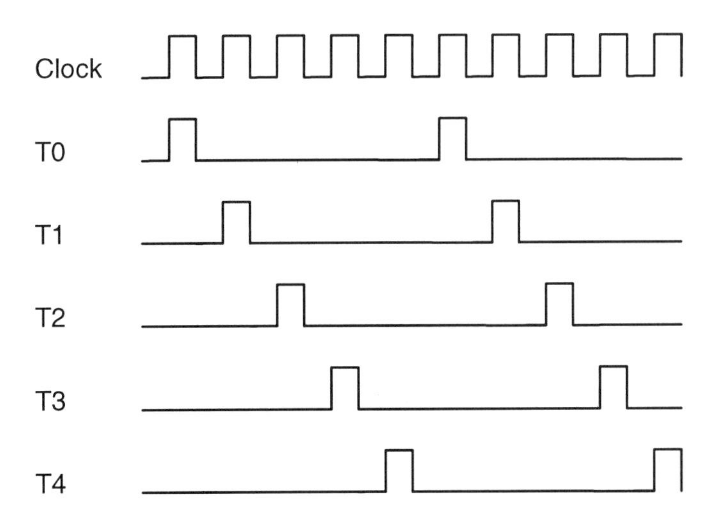


Figure 11.2 Waveforms produced by example circuit

The circuit has capabilities of asynchronous Clear, load initial parallel data, and Enable input which must be active when the circuit generates output waveforms.

### *11.2.1 Schematic Entry*

Schematic entry is one of the traditional ways to specify digital system design. A graphics editor is a schematic entry capture program that allows relatively complex designs to be entered quickly and easily. The built-in and extensible primitive and macrofunction libraries provide basic building blocks for constructing a design, while the symbol generation capability enables users to build libraries of custom functions. A graphic editor usually provides a WYSIWYG (What-You-See-Is-What-You-Get) environment. Typical vendor-provided primitives include input and output pins, elementary logic gates, buffers, and standard flip-flops. Vendor-provided libraries contain macrofunctions equivalent to standard 74- series digital circuits (SSI and MSI), with standard input and output facilities. In the translation process these circuits are stripped off the unused portions such as unused pins, gates and flip-flops.

The graphic editor enables easy connection of desired output and input pins, editing of new design, duplication of portions or complete design, etc. Symbols can be assigned to new designs and used in subsequent designs.

Usual features of graphic editor are:

- Symbols are connected with single lines or with bus lines. When the name is assigned to a line or bus, it can be connected to another line or bus either graphically or by name only.

- Multiple objects can be selected and edited at the same time.
- Complete areas containing symbols and lines can be moved around the worksheet while preserving signal connectivity. Any selected symbol or area can be rotated.
- Many resources can be viewed and edited in Graphic editor such as probes, ,pins, logic cells, blocks of logic cells, logic and timing assignments.

Our small example pulse distribution circuit represented by schematic diagram is shown in Figure 11.1.

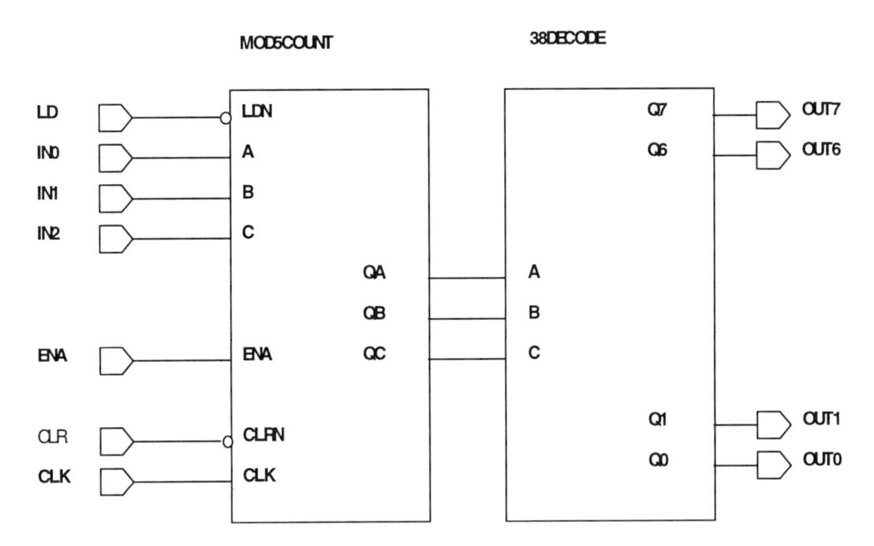


Figure 11.2 Pulse distribution circuit represented by schematic diagram

### *11.2.2 Hardware Description Languages*

Hardware description languages (HDLs) represent another tool for the description of digital system behaviour and structure at different abstraction levels. HDLs belong either to a category of vendor-designed languages, or general languages that are independent on the vendors of FPLDs.

VHDL has emerged as the standard tool for description of digital systems at various levels of abstraction optimized for transportability among many computer design environments. In order to compare different design tools on the same example, our small pulse distribution circuit is described in VHDL by the following description:

```
use work.mycomp.all;

entity pulsdist is
        port(d: in integer range 0 to 7;
                clk, ld, ena, clr: in bit;
                q: out integer range 0 to 255);
end pulsdist;

architecture puls_5 of pulsdist is
        signal a: integer range 0 to 7;
begin
```

```
        cnt_5: mod_5_counter port map (d,clk,ena,clr,a);
        dec_1: decoder3_to_8 port map (a, q);
end puls_5;
```

Some of the basic VHDL design units (entity and architecture) appear in this example. The architecture puls_5 of the pulse contains instances of two components from the library mycomp, cnt_5 of type mod_5_counter, and dec_1 of type decoder3_to_8. The complete example of the pulse distributor is presented in the next section, where the hierarchy of design units is introduced.

### *11.2.3 Hierarchy of Design Units - Design Example*

As mentioned earlier, all design entry tools usually allow not only using design units specified in the same tool but also the mixing of design units specified in other tools. This leads to the concept of project as the design at the highest level of hierarchy. The project itself consists of all files in a design hierarchy including some ancillary files produced during the design process. The top level design file can be schematic entry or textual design files, defining how previously designed units together with their design files are used.

Consider the design hierarchy of the example of our pulse distribution circuit. Suppose that the circuit consists of hierarchy of subcircuits as shown in the Figure 11.3. This figure also shows us how portions of the pulse distribution circuit are

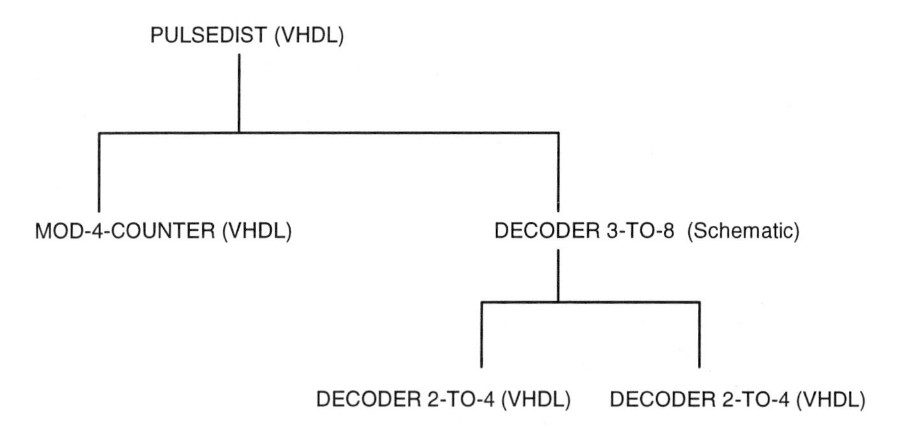


Figure 11.3 Design hierarchy of pulse distribution circuit

implemented. At the top of hierarchy we have the VHDL file (design) that represents the pulse distributor. It consists of modulo-5 counter circuit, denoted MOD-4-COUNTER, which is also designed in VHDL, and decoder 3-to-8, denoted DECODER 3-TO-8, designed using schematic editor. This decoder is designed, in turn, using two 2-to-4 decoders, denoted DECODER 2-TO-4, with enable inputs designed in VHDL. Any of the above circuits could be designed in AHDL, as well. This example just opens a window to the powerful integrated design environments which provide even greater flexibility than one shown in it. Many more complex designs which employ hierarchy of design units will be shown in Part 3.

Another important component in the hierarchical design of projects is the availability of libraries of primitives, standard 74-series, and application-specific macrofunctions, including macrofunctions that are optimized for the architecture of a particular FPLD device or family. Primitives are basic function blocks such as buffers, flip-flops, latches, input/output pins, and elementary logic functions. They are available in both graphical and textual form and can be used in schematic diagrams and textual files. Macrofunctions are high-level building blocks that can be used together with primitives and other macrofunctions to create new logic design. Unused inputs, logic gates and flip-flops are automatically removed by a compiler, ensuring optimum design implementation. Macrofunctions are usually given together with their detailed implementation, enabling the designer to copy them into their own library and edit them according to specific requirements.

The common denominator of all design entry tools is the netlist level at which all designs finally appear. If standard netlist descriptions are used, then further tools that produce actual programming data, or perform simulation, can even accept at their inputs designs specified by any other design specification tool provided that compiler produces standard netlist formats.

## 11.3 Design Verification and Simulation

Design verification is necessary due to the fact that there are bugs and errors in the translation, placement and routing processes, on one hand, and errors made by a designer, on another hand. Most of the verification tools are incorporated into design tools, examining netlists and analyzing properties of the final design. For instance, a design checker can easily identify the number of logic cells driven by any logic cell and determine how it contributes to a cumulative load resulting in a time delay attached to the driving cell's output. In the case that the delay is unacceptable, the designer must split the load among several identical logic cells. Similarly, a design checker can identify unconnected inputs of logic cells which float and can produce noise problems.

While some of the checks can be performed during design compilation, many checks can be done only during simulation which enables assessing functionality and the timing relationship and performance of an FPLD-based design.

Regardless of the type of logic simulation, a model of the system is created and it is driven by a model of inputs, called stimuli or input vectors, to generate a model of output signals called responses or output vectors. Simulation is useful not only for observing global behaviour of the system under design, but also because it permits observation of internal logic at various levels of abstraction, which is not possible in actual systems.

Two types of simulation are usually used in digital systems design:

- Functional simulation that enables observation of design units at the functional level, by combining models of logic cells with the models of inputs to generate response models taking into account only relative relationships among signals, and neglecting circuit delays. This type of simulation is useful because for quick analysis of system behaviour, but produces inaccurate results because propagation delays are not taken into account.
- Timing simulation, which takes into account additional element in association with each cell model output - a time delay variable. Time delay enables much more realistic modeling of logic cells and the system as the whole. Time delays consist of several components that can be, or not , taken into account, such as time delay of logic cells, without considering external connections, time delays associated with the routing capacitance of the metal connecting outputs with the inputs of logic cells, and time delay which is the function of the driven cell input impedance.

Timing simulation is the major part of verification process of the FPLD design. Timing simulator uses two basic information to produce output response:

- Input vectors which are given either in tabular or graphical form. Input timing diagrams represent a convenient form to specify stimuli of simulated design.
- Netlists, representing an intermediate form of the system modeled. Besides connectivity information, netlists contain information on delay models of individual circuits and logic cells, as well as logic models that describe imperfections in the behaviour of logic systems. These models increase complexity of used logic, but at the same time improve quality of model and system under design.

The simulator applies the input vectors to the model of the system under design, and, after processing according to the input netlists and models of individual circuits, produces resulting output vectors. Usually, outputs are also presented in the convenient form of timing diagrams. Later on, both input and output timing diagrams can be used by electronic testers to compare simulated with the real behaviour of the design.

In order to perform simulation, the simulator has to maintain several internal data structures that help easily and quickly to find the next event that requires simulation cycle to be started. A simulation event is the occurrence of a netlist node (gate, cell, output, etc) making a binary change from one value to another. The scheduler is a part of the simulator that keeps list of times and events, and does dispatching of events when their time comes. The process is initiated every simulated time unit regardless of existence of event (in that case we say that simulation is time-driven), or only at the time units at which there are some events (event-driven simulation). The second type of simulation is more popular and more efficient in today's simulators. The most important data structure is the list of events which must be ordered according to increased time of occurrence.

In the ideal case these changes are simply binary (from zero to one and vice versa), but more realistic models take into account imperfect or sometimes unspecified values of signals. The part of simulator called evaluation module is activated at each event, and it uses models of functions that describe behaviour of the subsystems of design. These models take into account more realistic electrical conditions of circuit behaviour, such as three-state outputs, unknown states, and time persistence, essentially introducing multi-valued instead of common binary logic. Some simulators use models with up to twelve values of the signals. This leads to much more complex truth tables even for very simple logic gates, very complex and time consuming simulation, but also produces more accurate simulation results.

Regardless of the type of logic simulation, a model of the system is created and it is driven by a model of inputs, called stimuli or input vectors, to generate a model of output signals called responses or output vectors. Simulation is useful not only for observing global behaviour of the system under design, but also because it permits observation of internal logic at various levels of abstraction, which is not possible in actual systems.

## 11.4 Integrated Design Environment Example: Altera Max+Plus II

An integrated design environment for the FPLD design represents a complete framework for all phases of the design process, starting with the design entry and ending with the device programming. An example of such an environment is Altera's Max+PLUS II development software which is represented in this section. It is a fully integrated software package for deigning with all Altera programmable devices. The same design can be retargeted to various devices without changes to the design itself. Max+PLUS II consists of a spectrum of logic design tools and capabilities such as a variety of design entry tools for hierarchical projects, logic synthesis algorithms, timing-driven compilation, partitioning, functional and timing simulation, linked multi-device simulation, timing analysis, automatic error location, and device programming and verification. It also has the capabilities of reading netlist files produced by other vendor systems or producing netlist files for other industry-standard CAE software.

The overall Max+PLUS II design environment is presented in Figure 11.4. In the heart of environment is a Max+PLUS II compiler capable to accept design specification in various design entry tools, and to produce files for two major purposes:

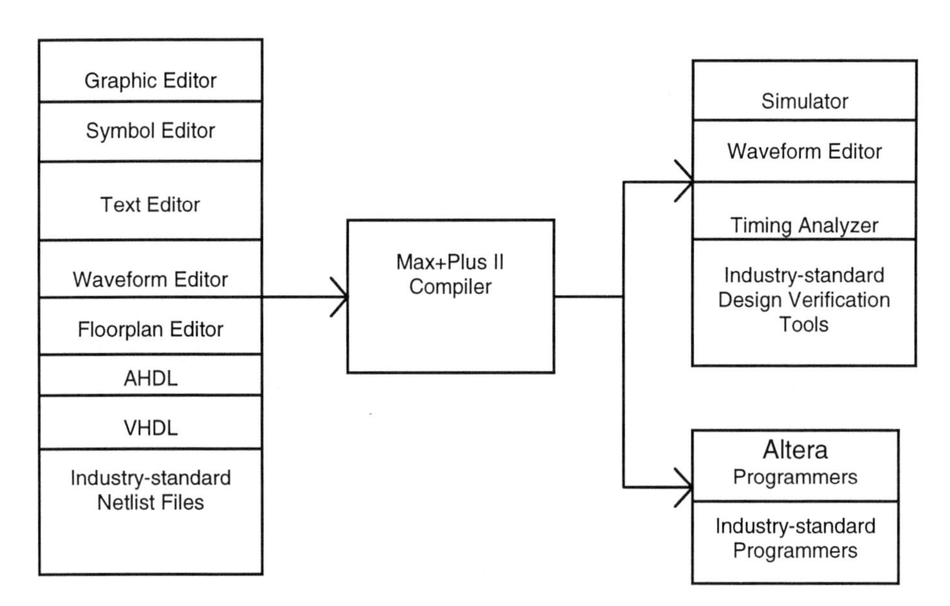


Figure 11.4 Max+Plus II Integrated Design Environment

- design verification, which is performed using functional or timing simulation, or timing analysis and
- device programming, which is performed on Altera's or other industry-standard programmers

Output files produced by the Max+Plus II compiler can be used by other CAE software tools.

Logic design is created as the entity called project. Project can include one or more subdesigns (previously designed projects). It allows combinations of different types of subdesigns (files) into a hierarchical project, choosing the design entry format that best suits each functional block. In addition, large libraries of Altera-provided macrofunctions simplify design entry. Macrofunctions are available in different forms and can be used in all design entry tools.

Project consists of all files in a design hierarchy. If needed, this hierarchy can be displayed at any moment, and the designer sees all files that make the project, including design and ancillary files. Design files represent a graphic, text, or waveform file created with a corresponding editor, or with another industry-standard schematic text editor or a netlist writer. The Max+PLUS II compiler can process the following files:

- Graphic design files (with .gdf extension)
- Text design files (with .tdf extension)
- Waveform files (with .wdf extension)
- VHDL files (with .vhd extension)
- OrCAD schematic files (with .sch extension)
- EDIF input files (with .edf extension)
- Xilinx netlist format files (with .xnf extension)
- Altera design files (with .adf extension)
- State machine files (with .smf extension)

Ancillary files are files associated with a project, but are not part of a project hierarchy tree. Most of them are generated by different Max+Plus II functions, and some of them can be entered or edited by a designer. Examples of ancillary files are assignment and configuration files (with .acf extension), and report files (with .rpt extension).

The Max+Plus II compiler provides powerful project processing and customization to achieve the best or desired silicon implementation of a project. Besides fully automated procedures, it allows a designer to perform some of the assignments or functions manually, and in that way to control resulting design. A designer can enter, edit, and delete the types of resource and device assignments that control project compilation, including logic synthesis, partitioning, and fitting.

The Max+Plus II design environment is Windows based. This means that all functions can be invoked using menus or simply by clicking on different buttons with the icons describing corresponding functions. Each function invoked is active in its own window which can be, if needed, minimized in order not to overload the screen surface.

### *11.4.1 Design Entry*

Max+Plus II provides three design entry editors: the graphic, text, and waveform editors. Two additional editors are included to facilitate design entry: the floorplan and symbol editor.

Design entry methods supported by Max+Plus II are:

- schematic - designs are entered in schematic form using Graphic editor.
- textual - AHDL or VHDL designs are entered using Altera's or any other standard text editor
- waveform - designs are specified with Altera's waveform editor
- netlist - designs in the form of netlist files or designs generated by other industry-standard CAE tools can be imported into Max+Plus II design environment
- pin, logic cell, and chip assignments for any type of design file in the current project can be entered in a graphical environment with the floorplan editor.
- graphic symbols that represent any type of design file can be generated automatically in any design editor. Symbol editor can be used to edit symbols or create own customized symbols.

The Assign menu, accessible from any Max+Plus II application, allows the user to enter, edit, and delete the types of resource and device assignments that control project compilation. This information is saved in the assignment and configuration

(.acf) file for the project. The assignment of device resources can be controlled by the following types of assignments:

- Clique assignment that specifies which logic functions must remain together in the same logic array block, row, or device
- Chip assignment that specifies which logic must remain together in a particular device when a project is partitioned into multiple devices
- Pin assignment that assigns the input or output of a single logic function to a specific pin, row, or column within a chip
- Logic cell assignment that assigns a single logic function to a specific location within a chip (to logic cell, I/O cell, LAB, row, or column)
- Probe assignment that assigns a specific name to an input or output of a logic function
- Connected pin assignment specifies how two or more pins are connected externally on the printed circuit board.
- Device assignment assigns project logic to a device (for example, maps chip assignments to specific devices in multi-device project)
- Logic option assignment that specifies logic synthesis style in logic synthesis (synthesis style can be one of three Altera-provided or specified by designer)
- Timing assignment that guides logic synthesis tools to achieve the desired performance for input to non-registered output delays ($t_{PD}$), clock to output delays ($t_{CO}$), clock setup time ($t_{SU}$), and clock frequency ($f_{MAX}$).

Max+Plus II allows the preservation of the resource assignment that the compiler made during the most recent compilation so that we can produce the same fit with subsequent compilation. This feature is called back-annotation. It becomes essential because after compiling, all time delays are known, and the design software can calculate a precise annotated netlist for the circuit, by altering the original netlist. The subsequent simulation using this altered netlist is very accurate and can show trouble spots in the design that are not observable from the outside world.

Some global device options can be specified in advance before compilation such as reservation of device capacity for future use, or some global settings such as automatic selection of global control signals like the Clock, Clear, Preset, and Output Enable signals. The compiler can be directed to automatically implement logic in I/O cell registers.

### *11.4.2 Design Processing*

Once the design is entered, it is processed by the Max+Plus II compiler to produce the different files that are used for circuit verification or programming. The Max+Plus II compiler consists of a series of modules that check a design for errors, synthesize the logic, fit the design into the needed number of Altera devices, and generate files for simulation, timing analysis, and device programming. It also provides a visual presentation of the compilation process showing which of the compilation modules is currently active and allowing this process to be stopped.

Besides design entry files, the inputs to the compiler are the assignment and configuration file of the project (.acf file), symbol files (.sym files) created with Symbol editor, include files (.inc files) imported into text design files containing function prototypes and constants declarations, and library mapping files (.lmf files) used to map EDIF and OrCAD files to corresponding Altera-provided primitives and macrofunctions.

The compiler netlist extractor first extracts information that defines hierarchical connections between a project's design files and checks the project for basic design entry errors. It converts each design file in the project into a binary Compiler Netlist File (.cnf file) and creates one or more Hierarchy Interconnect Files (.hif files), a Symbol File (.sym file) for each design file in a project, and a single Node Database File (.ndb file) that contains project node names for assignment node database.

If there are no errors all design files are combined into a flattened database for efficient further processing. Each CNF file is inserted into the database as many times as it is used within the original hierarchical project. The database preserves the electrical connectivity of the project.

The compiler applies a variety of techniques to implement the project efficiently in one or more devices. The logic synthesizer minimizes logic functions, removes redundant logic, and implements user-specified timing requirements.

If a project does not fit into a single device, the partitioner divides the database into the minimal number of devices from the same device family. A project is partitioned along logic cell boundaries, and the number of pins used for inter-device communication is minimized.

The fitter matches the requirements of the project with the known resources of one or more devices. It assigns each logic function to the specific logic cell location and tries to match specific resource assignments with the available resources. In the case that it does not fit, the fitter issues a message with the options of ignoring some or all of the required assignments.

Regardless of whether a fit is achieved, a report file (.rpt file) is created showing how a project will be implemented. It contains information on project partitioning, input and output names, project timing, and unused resources for each device in the project.

At the same time the compiler creates a functional or timing simulation netlist file (.snf file) and one or more programming files that are used to program target devices. The programming image can be in the form of one or more programmer object files (.pof files), or SRAM object files (.sof files). For some devices JEDEC files (.jed files) can be generated.

As an example, our pulse distributor circuit is compiled by the Max+Plus II compiler without any constraints and user-required assignments of resources. The tables 11.1 to 11.4 show some of the results of compilation as reported in report file. The compiler has placed the pulsdist circuit into the EPF8282LC84 device with the logic cell utilization of 5%. Other important information about the utilization of resources is available in the tables below.

Table 11.1 Device summary for pulsdist circuit

| Chip | Device | Input pins | Output pins | Bidir pins | Logic cells | % utilised |
|---|---|---|---|---|---|---|
| pulsdist | EPF8282 LC84 | 7 | 8 | 0 | 11 | 5% |
| User pins | | 7 | 8 | | | |

Table 11.2 Logic cell utilisation

| Column / Row | 01 | 02 | 03 | 04 | 05 | 06 | 07 | 08 | 09 | 10 | 11 | 11 | 13 | Total |
|---|---|---|---|---|---|---|---|---|---|---|---|---|---|---|
| A | 0 | 0 | 0 | 0 | 0 | 0 | 0 | 0 | 0 | 0 | 0 | 0 | 0 | 0 |
| B | 4 | 3 | 1 | 1 | 1 | 1 | 1 | 0 | 0 | 0 | 0 | 0 | 0 | 11 |
| Total | 4 | 3 | 1 | 1 | 1 | 1 | 1 | 0 | 0 | 0 | 0 | 0 | 0 | 11 |

Table 11.3 Interconnect mechanism usage

| LAB | Logic cells | Column Interconn. Driven | Row Interconn. Driven | Clocks | Clear/ Preset | External Interconn. |
|---|---|---|---|---|---|---|
| B1 | 4/8 (50%) | 0/8 (0%) | 3/8 (37%) | 1/2 | 1/2 | 6/24 (25%) |
| B2 | 3/8 (37%) | 0/8 (0%) | 3/8 (37%) | 0/2 | 0/2 | 3/24 (11%) |
| B3 | 1/8 (11%) | 0/8 (0%) | 1/8 (11%) | 0/2 | 0/2 | 3/24 (11%) |
| B4 | 1/8 (11%) | 0/8 (0%) | 1/8 (11%) | 0/2 | 0/2 | 3/24 (11%) |
| B5 | 1/8 (11%) | 0/8 (0%) | 1/8 (11%) | 0/2 | 0/2 | 3/24 (11%) |
| B6 | 1/8 (11%) | 0/8 (0%) | 1/8 (11%) | 0/2 | 0/2 | 3/24 (11%) |
| B7 | 1/8 (11%) | 0/8 (0%) | 1/8 (11%) | 0/2 | 0/2 | 3/24 (11%) |

Table 11.4 Other resources utilization

| | |
|---|---|
| Total dedicated input pins used | 1/4 (25%) |
| Total I/O pins used | 16/64 (25%) |
| Total logic cells used | 11/208 ( 5%) |
| Total input pins required | 7 |
| Total input registers required | 0 |
| Total output pins required | 8 |
| Total output registers required | 0 |
| Total burried I/O cell registers required | 0 |
| Total bidirectional pins required | 0 |
| Total reserved pins required | 2 |
| Total logic cells required | 11 |
| Total flip-flops required | 3 |
| Total logic cells in carry chains | 3 |
| Total number of carry chains | 1 |
| Total logic in cascade chains | 0 |
| Total number of cascade chains | 0 |
| Synthesised logic cells | 0/208 ( 0%) |

### *11.4.3 Design Verification*

The process of project verification is aided with two major tools: the simulator, and the timing analyzer. The simulator tests the logical operation and internal timing of a project. To simulate a project, a Simulator Netlist File (.snf file) must be produced by the compiler. An appropriate SNF file (for functional, timing, or linked multi-project simulation) is automatically loaded when the simulator is invoked.

The input vectors are given in the form of a graphical waveform Simulator Channel File (.scf file) or an ASCII Vector File (.vec file). The waveform editor can be used to create a default SCF file. The simulator allows the designer to check the outputs of the simulation against any outputs in SCF, such as user-defined expected outputs or outputs from a previous simulation. It can also be used to monitor glitches, oscillations, and setup and hold time violations.

An example of the simulator operation is given for our pulse distributor circuit in Figure 11.5. Input vectors are denote by capital letter I, and output vectors by capital letter O. The simulation ran for the simulated time of 800 ns. In Figure 11.5 only a portion of the simulation for the time interval of 280 ns is shown.

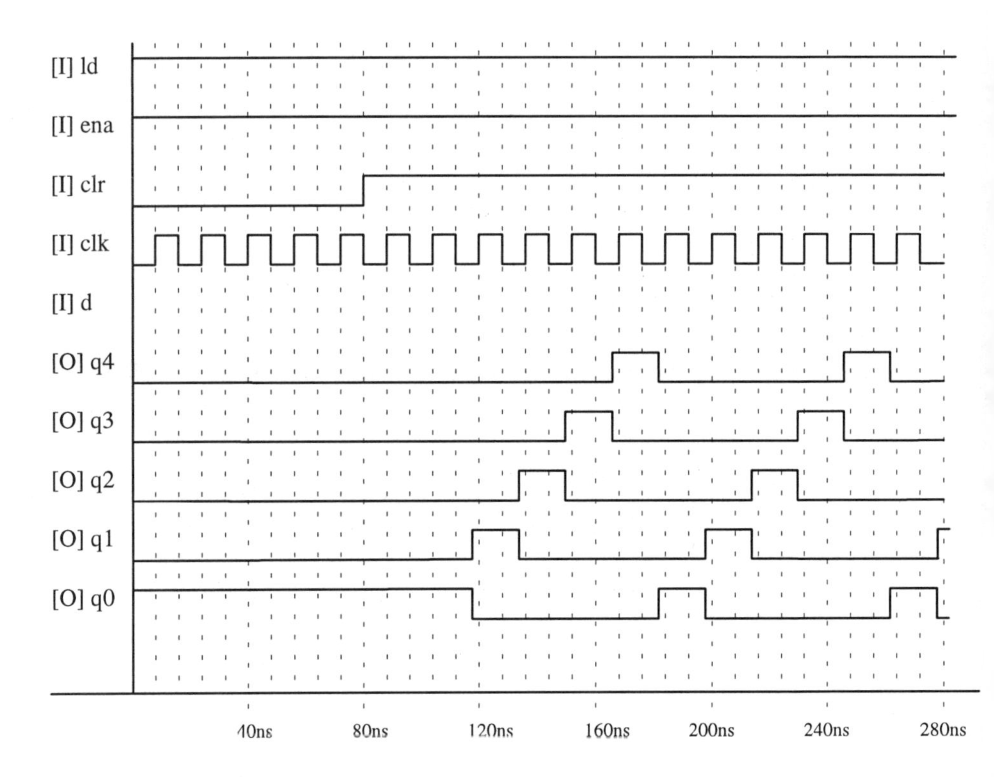


Figure 11.5 Simulation results for pulse distributor circuit

The input clock period is 16 ns. Further timing analysis has shown that the circuit can safely run to the minimum clock period of 15.7 ns, or frequency of about 63.69 MHz.

The Max+Plus II timing analyzer allows the designer to analyze the timing performance of a project after it has been optimized by the compiler. All signal paths in the project can be traced, determining critical speed paths and paths that limit the project's performance. The timing analyzer uses the network and timing information from a timing Simulator Netlist File (.snf file) generated by the compiler. It generates three types of analyses:

- The delay matrix shows the shortest and longest propagation delay paths between multiple source and destination nodes in a project.

- The setup/hold matrix shows the minimum required setup and hold times from input pins to the D, Clock and latch Enable inputs to flip-flops and latches.
- The registered performance display shows the results of a registered performance analysis, including the performance-limited delay, minimum Clock period, and maximum circuit frequency.

After the timing analyzer completes an analysis, it is possible to select a source or destination node and list all delay paths associated with it. Using the message processor it is easy to open and list the paths for the selected node, and locate a specific path in the original design file.

### *11.4.4 Device Programming*

The last portion of Altera's integrated design environment is the hardware and software necessary for programming and verifying Altera devices. The software part is called the Max+Plus II programmer. For EPROM-base devices Altera provides an add-on Logic Programmer card (for PC-AT compatible computers) that drives the Altera Master Programming Unit (MPU). The MPU performs continuity checking to ensure adequate electrical contact between the programming adapter and the device. With the appropriate programming adapter, the MPU also supports functional testing. It allows the application of simulation input vectors to a programmed device to verify its functionality.

For the FLEX 8000 family Altera provides the FLEX download cable and the BitBlaster. The FLEX download cable can connect any configuration EPROM programming adapter, which is installed on MPU, to a single target FLEX 8000 device in a prototype system. The BitBlaster serial download cable is a hardware interface to a standard RS-232 port that provides configuration data to FLEX 8000 devices on system boards. The BitBlaster allows the designer to configure the FLEX 8000 device independently from MPU or any other programming hardware.

## 11.5 Questions and Problems

11.1. What are the major steps in a digital system design process? Describe the role of each of the tools used in the design process.

11.2. What are the netlists? Generate a netlist for a 4-to-1 multiplexer described using a) elementary logic gates and b) logic elements using 4-input look-up tables.

11.3. What are the main features of functional and timing simulation?

11.4. How can we characterize Altera Max+Plus II design environment?

11.5. Analyze AHDL hardware description language present in the Max+Plus II design environment. What are the major features of AHDL and how can you compare it to VHDL? Knowing that all AHDL designs are synthesizable, is it possible to use VHDL in a AHDL-like manner to always provide synthesis of VHDL descriptions?

# 12 RAPID PROTOTYPING SYSTEMS FOR FPLDS

In this chapter we present hardware and software environments in which most of the examples shown in this book have been implemented. The first one is a microcontroller/FPLD prototyping system which combines flexibility of a standard microcontroller and an FPLD to a system suitable for hardware/software co-solutions and implementation of embedded applications. The second is a PC/FPLD prototyping system developed in the form of PC add-on card and additional software support that enables rapid prototyping in PC/Windows environment, or development of applications of FPLDs present on the add-on card as a kind of computation accelerator. Finally, a universal educationally oriented prototyping board from Altera Corporation is presented as an ideal tool for prototyping of medium complexity digital systems in the PC/Windows environment for educational and research purposes.

## 12.1 PROTOS - A Microcontroller/FPLD-based Prototyping System

In this section we describe PROTOS, a microcontroller/FPLD-based system used mainly for prototyping of embedded applications [11]. It consists of a standard Motorola 68HC11 microcontroller and one or more Altera FLEX 8000 FPLDs, as well as additional SRAM and EEPROM resources to accommodate a fairly wide class of embedded applications. The development tools which can be used for both software and hardware development for the system run on a PC with the Windows operating environment. They contain a standard suit of programming languages and compilers for 68HC11, Altera's Max+Plus II integrated hardware design environment for FPLDs, as well as additional software which facilitates the use of all resources in the system in both development and run-time stages.

### *12.1.1 The PROTOS Framework*

Some of the systems objectives the PROTOS system fulfills are listed below:

### Prototyping and Run-time Environment

The system should provide a platform for development of embedded applications that require not only application software but also hardware customized according to application requirements. The system has to provide easy and fast modification of both application programs and hardware structures associated to a standard microcontroller not only during prototyping process, but also in the working, operational system, without new hardware interventions (for example, without PCB re-design). Feasible interventions are mostly in the form of change of EEPROM contents in order to supply new versions of programs and/or hardware interfaces or other hardware structures. These hardware structures are represented by bitstreams, which are the result of hardware design process for FPLDs and are stored in EEPROM or SRAM before downloading into an FPLD.

### Standard Microcontroller

A Motorola 68HC11 microcontroller has been selected as a standard one providing a powerful 8-bit microprocessor with a rich instruction set and programming model, parallel and serial ports for interfacing, timers, and multichannel A/D converter.

### Standard FPLD Device

A standard Altera FLEX 8000 device is used as a major resource for implementation of application-specific hardware structures, which are a part of embedded system solution. In our case we implemented a PCB with a FLEX8282-84 devices, but it can be easily modified to accommodate any other FPLD from the FLEX 8000 family because they have the same architecture and reconfiguration mechanism.

### Memory

Existing 68HC11 on-the-chip memory resources are not sufficient for most of the intended applications. This is the reason we use the microcontroller in the expanded bus mode, and extend memory resources with external 8KB of SRAM and 32KB of EEPROM. Larger memory resources are needed to store programs and data, and also to store hardware configurations that are implemented in the FPLD chip.

### Serial Communication Link

A serial communication link is needed to provide communication with a personal computer, which is used as a software/hardware development platform. It enables us to download both programs, which run on microcontroller and hardware configurations from the PC to the prototyping board. It can also be used in the target application.

### System Clocks

The PROTOS system provides two system clocks. One is used to drive the microcontroller at 2MHz, and the other one to drive sequential circuits, which are implemented in FPLD at higher frequencies (up to 50MHz in our case).

### Access to the FPLD Using Memory-mapped I/O

Due to the fact that the 68HC11 supports memory-mapped I/O, our decision was to extend this I/O method to the FPLD. This enables access to the FPLD resources through a number of registers, implemented in an Altera MAX7000 EPLD, that appear in the address space of the 68HC11. However, this does not prevent a user from implementing more registers within the FPLD, as an application requires.

### Modes of Operation

The prototyping board supports two basic modes of operation:

- Configuration mode in which the hardware configuration of the FPLD is being downloaded. This mode is entered always at system power-up or system initialization whenever a new configuration has to be downloaded from memory or from the host development PC.
- System operation (or user) mode, in which the prototyping system carries out presumed function in embedded system.

### Hardware Reconfiguration

Hardware reconfiguration is performed in a passive parallel reconfiguration asynchronous scheme. It means that the reconfiguration process is controlled by the 68HC11 as host, which finds the configuration bitstream in external memory (SRAM or EEPROM) and downloads it into the FPLD using the appropriate control signals of FLEX8000 device as shown in Chapter 10. One configuration bitstream for FLEX8282 device requires 5KB of memory space, so several configurations can be stored at the same time in memory chips. This further allows dynamic reconfiguration of hardware in some embedded applications at a fairly high rate. The other type of reconfiguration is available during prototyping process. Under software control, a configuration bitstream can be directly downloaded from the host PC to the FPLD.

## *12.1.2 Global System View*

The overall structure of the prototyping environment is presented in Figure 12.1. As the Figure shows, the prototyping board is placed between the PC, which is used

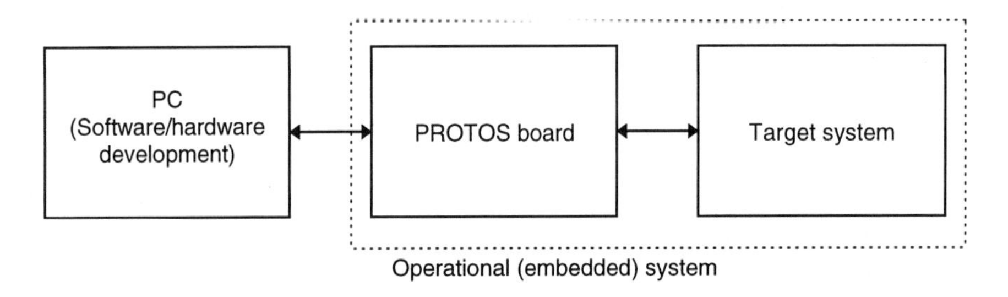


Figure 12.1 Prototyping system environment

as the software/hardware development environment, and the target system, which is controlled/supervised by the prototyping system. Once the application is finalized, an operational (embedded) system is used without a development environment. The system gives yet another possibility and degree of flexibility. It can be modified and changed remotely, using a modem link on both ends instead of a standard RS232 serial link of the PC and of the prototyping board. This means that both modifications of the system's software and hardware can be performed without any rewiring.

The structure of the hardware of the prototyping board is presented in Figure 12.2. Due to the fact that the 68HC11 must be used in expanded bus mode, proper address/data bus demultiplexing and mapping of the FPLD resources into the 68HC11 address space must be provided. It is all done in an Altera EPLD (MAX7096) which in fact extends the original 68HC11 capabilities and resources enabling simple addition of the other resources such as memories or FPLD chips. The current version of the design supports multi-FPLD structure, so the board can easily accommodate more than one FLEX8000 chip.

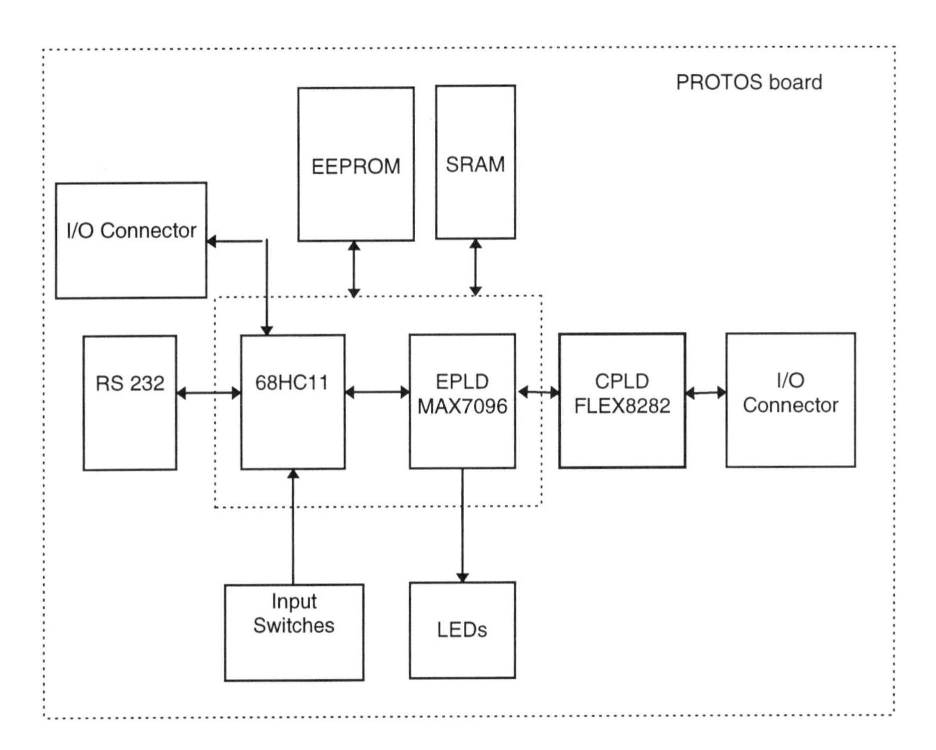


Figure 12.2. Prototyping board structure

Hardware design is supported by the Altera Max+Plus II design environment. As a result of the design process and compilation of the design entry file, a bitstream file is produced and stored on the hard disk. It can be downloaded and stored in any of the PROTOS memories. Starting address to which the configuration bitstream is downloaded is defined by a designer, and subsequently used during the configuration process. Hardware design entry may be performed by a number of entry tools supported by Max+Plus II, namely VHDL, AHDL (which stands for Altera Hardware Description Language), graphic design entry, or imported from another design tool. Prototyping program and design development, as well as the manipulations with the object program files and bitstream configuration files, are illustrated by the design process flow given in Figure 12.3.

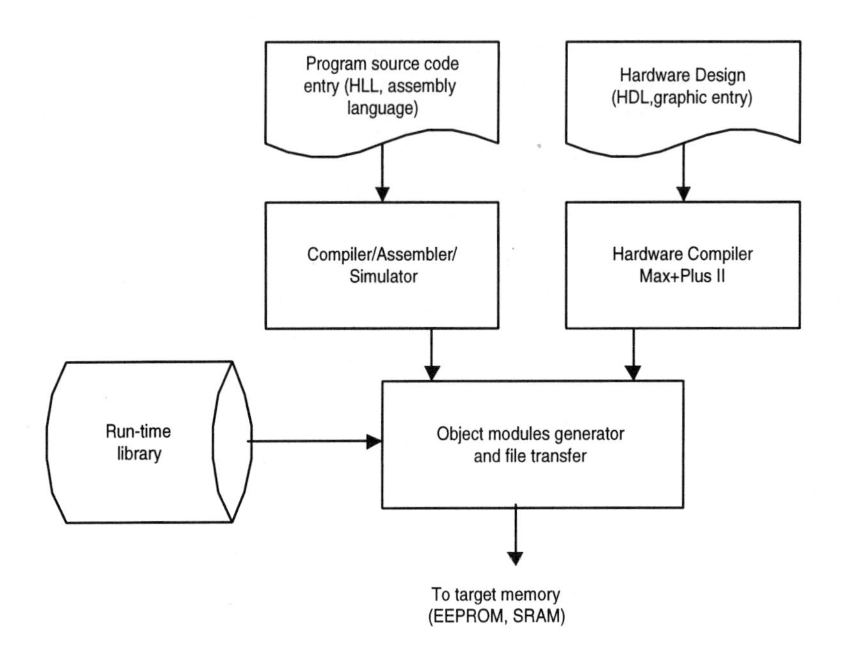


*Figure 12.3 Design process flow*

### *12.1.3 Hardware Design*

Only a brief description of hardware design, the details of which can be found in [11], is given below:

#### MC68HC11A1 Microcontroller

The 68HC11 microprocessor is the main core of this prototyping board. It controls the operations of all other components. The 68HC11 on the prototyping board is operated in the expanded bus mode in order to address external memories and the FPLD. In this mode, port B becomes the address bus upper byte. Port C becomes the multiplexed bus of address lower byte and data. Together with the signals R/W, AS and CLK, these lines are connected to the EPLD MAX7096. They are used to control the other components such as the memory chips. The 68HC11 is connected to a PC through the built-in serial communication interface (SCI) in port D and an RS232 interface. The SCI uses bits 1 and 0 of port D to transmit and receive data to and from PC. The remaining bits in port D are connected to the 4 single-bit switches. Two pins from port A are used to capture the control signals from the FLEX8282 during configuration. The rest of the pins in port A and port E are connected to an external connector. The users can make use of the timer system in port A or the A/D converter in port E in their applications. However, these ports can also be used as general purpose I/Os. The IRQ line is also connected to the external connector so the user can connect external devices that generate interrupts.

### EPLD MAX7096

The EPLD MAX7096 interconnects the 68HC11 and the FPLD and other components on the board. This EPLD receives the address, data and other control signals such as R/W and AS from the 68HC11. The main function of this chip is to decode the address from 68HC11 to enable the EPROM, RAM or one of six internal registers. It also demultiplexes the lower byte address and data from the 6811. The six internal registers in EPLD are used as buffers when the 68HC11 communicates with the FLEX8282. These registers appear in the address space of the 68HC11 providing easy use by user programs.

### FPLD FLEX 8282

The FLEX8282-84 is the lowest-capacity device in the Altera FLEX8000 family. There is a total of 208 logic elements in the device, and together with input/output blocks, there are 282 flip-flops available to the users. The whole device is equivalent to 2500 usable two-input logic gates to implement logic functions. The device has 84 pins. It uses a number of pins to connect to other devices on the prototyping board. There are 43 pins left unused. All these pins are connected to I/O connectors on the board. The users can use them for input and output purposes in their applications.

There are two clock sources provided on the prototyping board. One of the clocks is from 68HC11 and runs at 2MHz. The other source is the user clock. The user can plug in a separate clock to change the operating frequency of the FPLD. The users can select the clock source by using a fifth jumper on the prototyping board.

### LED Circuitry

The LED circuitry contains five LEDs. They can be used for any purpose to display the status of the board. For example, they can be used to indicate which set of configuration data is currently downloaded to the FPLD.

### Reset Circuitry

The Reset circuitry is designed such that the operation of the 68HC11 will be delayed by about half a second when the board is powered up. This ensures the power supply has reached a stable value of 5V. This delay only occurs when the board is powered up, but does not occur during normal reset.

## *12.1.4 Modes of Operation and User View*

The PROTOS board can be used in two operation modes: configuration and user mode, depending on the status of the FLEX8282 device. All other components on the board are used in the same way in both operating modes.

### Configuration Mode

The FLEX8282 is configured using passive parallel asynchronous (PPA) scheme. The 68HC11 reads configuration bitstream from the EEPROM and transfers it to the FPLD. The control signals are generated by a configuration program and transferred to the FPLD through registers implemented in MAX7096. Status signals are transferred from FLEX8282 to the 68HC11 through another status registers or directly through the 68HC11 port A. Transfer of individual configuration bytes is carried out asynchronously using simple program controlled handshaking.

The data for configuring the FLEX8282 in the PPA scheme is stored in Raw Binary File which is generated by Max+Plus II compiler. It has a fixed size of 5120 bytes for any design using FLEX8282. The program for configuring the FLEX8282 was written as a subroutine for 68HC11. It implements all timing and other requirement for configuration protocol. The user can start the configuration process at any time by calling this subroutine. In this way, in-circuit reconfiguration is performed. The subroutine is located in EEPROM at fixed location. Actual configuration data set is selected by passing its starting address to this subroutine.

## User Mode

After the FLEX8282 had been properly configured, the PROTOS board enters the user mode. Except dedicated pins, all other pins used by the FLEX8282 for configuration are now used for communication between the 68HC11 and the FPLD. Data transfers between 68HC11 and FLEX8282 are achieved using 2 8-bit registers implemented in MAX7096. Other register implemented in MAX7096 controls the FLEX8282 lines. The user can make use of bit 2 of input port A, PA2, as of input capture input to generate interrupt in the 68HC11 if necessary.

## PROTOS Memory Map

The memory map is important for both hardware and software design. The global memory map is shown in Figure 12.4, without presenting the details of EEPROM

| Address | |
|---|---|
| $0000<br>$00FF | Internal RAM (256 bytes) |
| $0100 | Reg1 (Data to FPGA) |
| $0101 | Reg2 (Data From FPGA) |
| $0102 | Reg3 (Control to FPGA) |
| $0103 | Reg4 (CONF_DONE) |
| $0104 | Reg5 (Data to LEDs) |
| $0105 | Reg6(nCONFIG) |
| $1000<br>$103F | Internal registers of 68HC11 |
| $2000<br>$3FFF | External SRAM (8KB) |
| $8000<br>$FFFF | External EEPROM |

Figure 12.4. PROTOS memory map

map. All internal 68HC11 RAM and control registers are left at their default locations without remapping. Registers Reg1-6 which are used for communication with the FPLD are implemented in MAX7096 and appear in the 68HC11 address space. The external RAM is placed at addresses $2000 to $3FFF, while external EEPROM is placed at addresses $8000 to $FFFF. The internal 68HC11 EEPROM is disabled by software. In the user mode some of the port A and B input and output pins are available to the user.

### Clocks

The globe clock pin of the FLEX8282 is connected to the E-clock pin of the 68HC11. However, the FLEX8282 can operate at a much higher clock frequency than the 68HC11. Therefore, a clock socket is reserved on the prototyping board to allow the user to use a higher clock frequency for the FLEX8282. A jumper is used to switch between E-clock and user-supplied clock.

### Other Software Support

The PROTOS board can be connected to the host PC by using SCI port of the 68HC11 and RS-232 interface on the board. Two basic system subroutines are written that transmit and receive data from the PC and are stored in the EEPROM. The subroutine to receive data stores them in the external SRAM. If this data, for instance, is a configuration bitstream for the FPLD, the user can configure the FPLD using configuration subroutine subsequently. Also, the user can reprogram EEPROM if a new configuration will be needed during the later development. The transmitting subroutine is used to transmit a string to the PC. It is useful when the user wants to display messages on the PC screen to show the status of the prototyping board, transfer results of its operation, etc.

## 12.2 RAPROS - Prototyping System for PC/FPLD Applications

RAPROS stands for Rapid PROtotyping System for hardware/software solutions on PC-compatible platforms. The system contains a number of field-programmable devices, memory chips, and pre-wired interconnect structures, that can be custom-configured by downloading a design which is in the form of a configuration file. It offers a very flexible interface to the designer in the prototyping phase, but also flexible run-time support for integration of the target system into an application, which is implemented partly in software and partly in hardware on a PC platform.

The main goals of developing a RAPROS [13] have been to:

- Provide a prototyping system which can be easily used in PC environments for low
- and medium complex hardware solutions that also require moderate external memory resources.
- Support dynamic hardware reconfigurability.
- Support easy integration of hardware solutions with software solutions for PC-based systems.
- Integrate existing hardware designs and program development tools and environments.
- Use standard devices and tools that can be easily replaced by new versions as they appear.
- Provide an inexpensive solution that can be easily used in educational and research environment.
- Provide some of the standard features found in the PC add-on cards (such as 8-bit and 16-bit data transfers, use of interrrupt and DMA channels) to be easily integrated into the user designs.

### *12.2.1 RAPROS System Features*

The inherent feature of SRAM based CPLDs and the design process for them is that they support a flexible system for rapid prototyping of digital systems, which is especially suitable for PC-compatible hardware/software solutions. The RAPROS system consists of two major parts:

- An ISA-bus compatible RAPROS prototyping card used for implementation of digital hardware in CPLDs, and
- The RAPROS software environment used to enable easy integration of the designed hardware into the target application.

RAPROS can be considered as an addition to already existing hardware and software design environments, but also as an application implementation system once the design is completed and included into an application. It supports downloading of design files into CPLDs, software-based control of the card operation once the configurations are downloaded, and run-time reconfiguration of CPLDs. Altera's CPLD FLEX 8282 devices are used in the first implementation of the card. More complex devices from the FLEX8000 family can easily replace them because they support the same (re)configuration schemes. Altera's Max+Plus II design environment is used to provide design of digital hardware using schematic entry, hardware description languages, or waveform design entry. A powerful design compiler fits the design into one or more CPLDs, and simulates the design before actually downloading it into the CPLDs. Once the design is downloaded, it implements the desired system which can be accessed and used by application programs. Programs can be designed using any program development tool running under MS Windows, either high-level or assembler language. The RAPROS system provides a programming model, which fits into standard PC programming model, but also allows the system designer to change it as the application requires.

The RAPROS card is an ISA-bus compatible card containing four FLEX8282 FPLD devices each having around 2,500 equivalent usable logic gates, four static RAM memory chips, an interface to the ISA-bus, and an interface to external devices/processes. Some of the device interconnections are pre-wired to enable connection/communication between CPLD devices and between these devices and memory chips, as well as the connections to the external world. The pre-wired connections can be used to implement simple bus structures between CPLDs and memories. The memory chips, each with a capacity of 32Kbytes can be used in two groups (banks) of 16-bit memories resulting in a capacity of 64K 16-bit words. The ISA-bus interface, besides providing standard bus signals, also provides dual-port access to on-card memory chips, as well as mapping these memory addresses to the address space of the PC. In this way, both CPLDs and the PC's CPU can access the same memory chips, thereby providing a number of applications with temporary data storage on the RAPROS card. The RAPROS card can be used in two basic operating modes:

- Configuration mode in which the hardware configuration of the CPLD is being downloaded; this mode is entered whenever a new configuration has to be downloaded from the configuration file which is present either in RAPROS memory or from the hard disk of the host PC, and
- System operation (or user) mode, in which the RAPROS card carries out user specified function(s) in an integrated hardware/software environment.

The RAPROS software provides downloading of new designs into the selected CPLD or the reconfiguration of the design. Access to the individual CPLDs for reconfiguration enables more advanced use of the card such as *dynamic* reconfiguration. A part of the CPLDs can be reconfigured "on the fly" while the rest of the card is performing the function specified by the designer. The RAPROS software also enables access to the memory chips at the programming level. The whole software support is provided in two forms:

- A user friendly interface that enables access to the resources on the RAPROS card for downloading of configurations and/or testing purposes, and
- A Library of low-level routines integrated into the Windows environment in the form of a dynamic link library with the corresponding application programming interface, which can be used from any software development environment, and as the run-time support.

### *12.2.2 RAPROS Card*

The RAPROS card consists of devices and interconnect structures used for prototyping purposes and interface and control logic that enables access to prototyping resources from the PC-bus.

#### Global view

The global organisation of the card is presented in Figure 12.5. Prototyping resources comprise:

- Four CPLD prototyping devices (four FLEX8282's are used in the first implementation.).
- Four static (S)RAM chips (which can be replaced with pin-compatible EPROM chips).
- A number of interconnects that can be used to form user bus structures, or any other type of interconnections between devices.
- A card control unit.
- A PC ISA-bus interface.

The prototyping devices, together with pre-wired interconnect, make user platform to implement specific designs. A data path for communication with external modules through the attached connector is also included. Both the prototyping FPLDs and the PC can access on-card SRAMs.

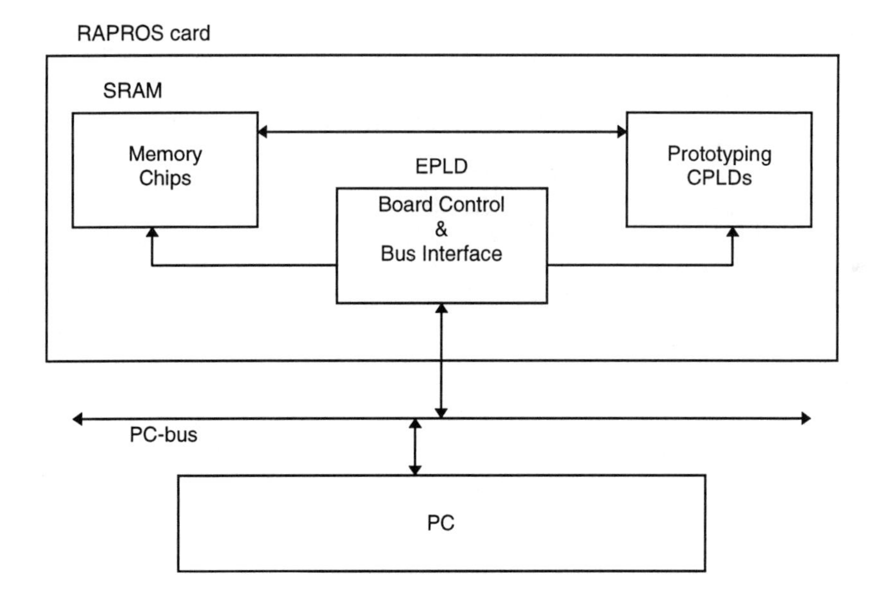


Figure 12.5 RAPROS card global structure

The role of card control unit and PC-bus interface is:

- To interpret, decode, and redirect signals between the PC and prototyping devices. These signals include interrupt related signals, DMA related signals, address and data bus signals, and the I/O and memory access commands
- To generate control signals for the data bus transceivers to provide different modes of memory and I/O access. It also resolves contention on address and data bus.
- To generate control signals for accessing SRAMs according to instructions from the PC or prototyping devices.

## Programming model

The RAPROS card memory is viewed as two 32k-word memory banks, high and low. Each bank consists of two 32k-byte sections, left and right. When accessed in

16-bit mode, both sections are used to store data. If the system is operated in 8-bit mode, only one section in each bank will be chosen to store data.

## More Detailed Card Structure

The card control unit resolves conflicts if both PC and FPLDs try to access on-card RAM at the same time. It also controls the access and configuration of FPLDs when the configuration files are downloaded from PC. The card control unit and PC-bus ISA interface are both implemented in Altera's MAX7128 EPLD 84-pin device. Bus buffers and transceivers, which are not shown in Figure 12.5, and are implemented in external chips.

More detailed structure of the prototyping subsystem on the RAPROS card is illustrated in Figure 12.6. Details describing connections of buffers, transceivers and DIP switches to their corresponding buses are not shown. It should be noted that the four FPLDs are grouped into two partitions (devices 0 and 1 form one, and devices 2 and 3 form another partition). Each partition has the same set of control signals connected to the card control unit and can be operated individually. The following buses provide point-to-point interconnections between two FPLDs:

- Gray bus: There are two, 17-bit gray buses which provide connections within each partition of prototyping devices. They can be treated as internal buses within a partition.
- Purple bus: This is a 10-bit bus between prototyping device 1 and 3, and can be considered as the bus between partitions.
- Green bus: This is a 14-bit bus between prototyping device 0 and 2, and can be considered as the bus between partitions.
- Red bus: This is a 15-bit address bus, which interconnects prototyping devices (1 and 3), DB37 connector and four SRAMs.
- Yellow bus: This is 16-bit bus, which interconnects all four FPLDs, connector, SRAMs and PC.

The last two buses can also be used as an additional connection between four FPLDs if they are not used to access SRAMs.

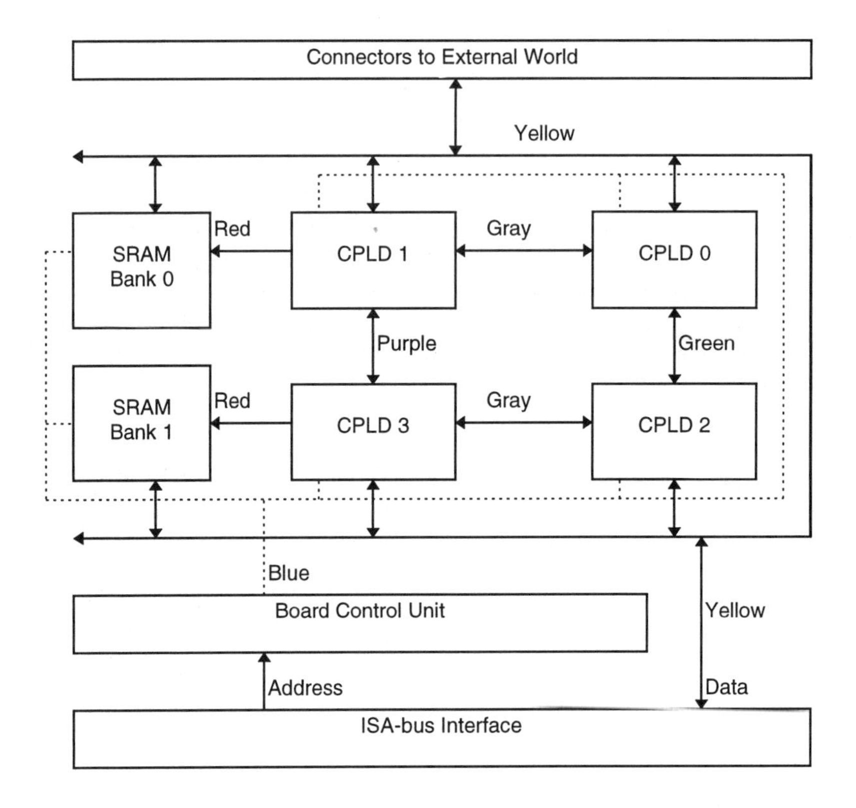


Figure 12.6 Prototyping resources.

There are two 4-bit identical sets of interrupt and direct memory access (DMA) control signals connected from prototyping devices 0 and 2 to PC to the ISA-bus. The application of RAPROS card may require service or help from the PC through the interrupt control signals. One of four alternative interrupt request lines- IRQ3, 5, 7, and 11 and one of the 3 DMA channels – DRQ 1,2, and 6 can be set by a DIP switch. It gives users adequate choice, so that the application will not have conflict with entire PC system. The remaining buses are related to the card control unit. The blue control bus has two types of signals. The first type is for FPLDs to access SRAMs through card control unit. There will be no access conflict because only either PC or FPLD may access the SRAMs at the same time. The second type is related to DMA. The PRDACK from the ISA-bus is used to acknowledge the FPLDs that DMA is going to take place.

The 22-bit address from the ISA-bus is used by card control unit to decode where SRAMs will be placed in PC address space and to decode to which part of I/O space the card will placed.

The 10MHz PC-bus clock signal, which can used as a prototype clock, is connected from the PC host to each FPLD. Users can implement ripple counter within FPLD to reduce the clock speed. This represents a limitation on the speed of the user design. In the case that the design requires higher speed, separate clock source must be added on the card.

### *12.2.3 Configuration of FPLDs*

The prototyping device is configured using the multiple-device passive parallel asynchronous (MD-PPA) scheme. This scheme allows the PC host to control and monitor the whole configuration process through the ISA-bus using simple I/O operations. There are four software controllable chip select signals for each CPLD for configuration. Signals to monitor and control the configuration process are provided and shown in Figure 2 as the 9-bit black CPLD configuration bus. The data for configuring each CPLD is stored in an ASCII text file on the hard disk. After a successful configuration, the device is in user mode and some of the control signals can be used to initiate and terminate operations on the card.

### *12.2.4 RAPROS Software*

The RAPROS software is used for communication between the RAPROS card hardware and software development environments used on the PC platform. It is also used to support run-time operation of the target application system. Its place in the overall prototyping process and run-time environment is illustrated in Figure 12.7.

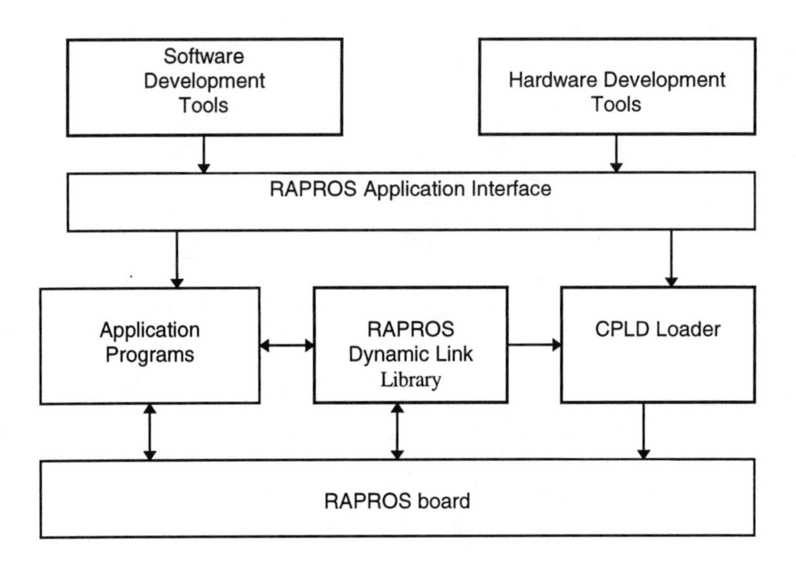


Figure 12.7 RAPROS software support

There are two main tasks RAPROS software performs. First, it allows users to easily download configuration files of their own design to the selected devices on the RAPROS card. To fulfill the first requirement, a windows based FPLD loader is implemented. The user is only required to select the design files which will appear in the listbox. This software provides monitoring of the configuration process, read-out of the contents of SRAM devices, and supports the testing of operation of FPLDs.

Different CPLD applications require different ways of accessing the on-card resources. To achieve the second task, a set of dynamic link library functions is provided, as summarized in Table 12.1. Parameters of the functions are not shown in the table. The users can integrate these functions into their own software for specific applications. Obviously, these functions can be used to perform reconfiguration of CPLDs dynamically, as well as to access main on-card resources. If the access to the user design is needed, then either low-level access features have to be used, or the designer has to write new functions that interact with the design.

Table 12.1. Summary of the dynamic link library functions

| Function | Meaning |
|---|---|
| Reset() | Reset the interface devicefor configuration |
| DataSize() | Set 8- or 16-bit accessing mode |
| ChipSelect() | Select FPLD for configuration |
| WriteIOB() | Write data to FPLD during configuration mode |
| ReadIOB() | Read data from FPLD during configuration. |
| SETRAM() | To select which memory location of the card memory appears in the PC memory map |
| Download() | Download configuration configuration file to FPLD device(FLEX8282) |
| ReadRamB() | Read 8-bit data from a specific memory location |
| WriteRamB() | Write 8 bit data to a specific memory location |
| ReadRam() | Read16 bit data from specific memory location |
| WriteRam() | Write 16-bit data to a specific memory location |

## 12.3 Altera UP1 Prototyping Board

Altera UP prototyping package was designed specifically to meet the needs of educational purposes of digital design at the University level. The package includes prototyping board, ByteBlaster download device that enables downloading FPLDs from the PC computer, and the Max+Plus II design environment. These three components provide all of the necessary tools for creating and implementing digital logic designs. The whole UP prototyping environment is illustrated in Figure 12.8.

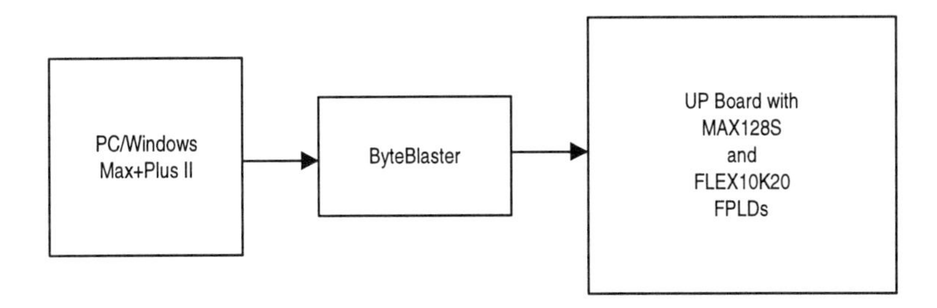


Figure 12.8 Overall UP Prototyping Environment

The Max+Plus II design environment has already been introduced in preceding Chapter. The ByteBlaster represents an interface with download cable that plugs in into PC parallel port and enables downloading of configuration bitstreams into two types of FPLDs present on the UP prototyping board: one is a product-based MAX7128S device with built-in In-System Programmability feature, and the other one is look-up-table based FLEX10K20 device that uses SRAM for programming purposes. An external EPROM can be alo used for an on-board configuration. In this section we provide description of the UP board, which has been used in verification of many of examples presented in this book.

The UP board functional organisation is presented in Figure 12.9. It contains two parts dedicated to two types of FPLDs present on the board:

- MAX7000S part. This part contains the EPM128S device with 128 macrocells and an equivalent of 2500 gates for designs of medium complexity,. The device is suitable for introductory designs which include larger combinatorial and sequential functions. It is mounted on 84-pin socket, and all pins are accessible via on-board connectors. Associated with this device are 2 octal DIP switches, 16 LEDs, dual digit 7-segment display, 2 momentary push buttons, on-board oscillator with 25.175MHz crystal, and an expansion port with 42 I/O pins and dedicated Global CLR, OE1 and OE2 pins. The switches and LEDs are not pre-wired to the device pins, but they are broken out to female connectors providing flexibility of connections using hook-up wires.
- FLEX10K part. This part contains the EPF10K20-240 device with 1152 logic elements, 6 embedded array blocks of 2048 bits of SRAM each, and total of 240 pins. With a typical gate count of 20,000 gates, this device is suitable for advanced designs, including more complex computational, communication, and DSP systems. This part contains a socket for an EPC1 serial configuration EPROM, an octal DIP switch, 2 momentary push buttons, dual digit 7-segment display, on-board oscillator 25.175MHz crystal, a VGA port, a mouse port, and 3 expansion ports each with 42 I/O pins and 7 global pins. The VGA interface allows the FLEX10K device to control an external monitor according to the VGA standard. The FLEX device can send signals to an external monitor through the diode-resistor network and a D-sub connector that are designed to generate voltages for the VGA standard. Information about the color of the screen, and the row and column indexing of the screen are sent from the FLEX device to the monitor via 5 signals (3 signals for red, green and blue, and 2 signals for horizontal and vertical synchronization). With the proper manipulation of these signals, images can be written to the monitor's screen. The mouse interface allows the FLEX10K device to receive data from a PS/2

mouse or PS/2 keyboard. The FLEX10K device outputs the data clock signal to the external device and receives data signal from the device.

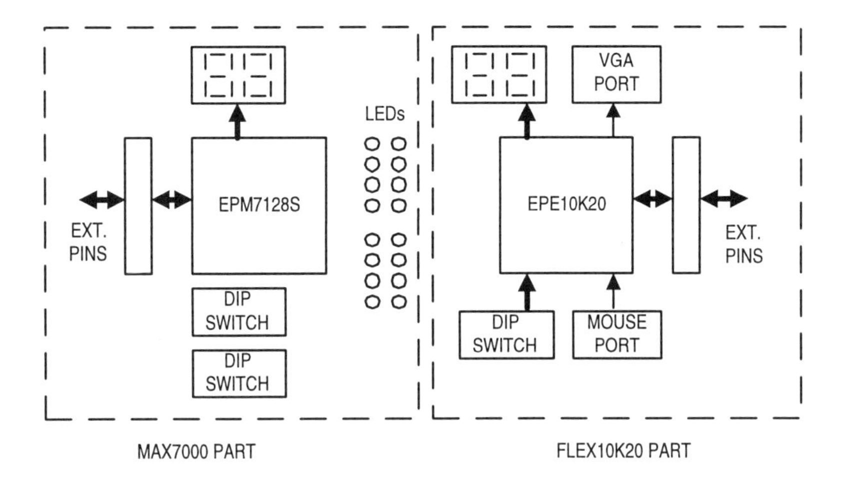


Figure 12.9 UP Prototyping board structure

Configuration of the devices on the UP prototyping board is performed simply by selecting menus and options in the Max+Plus II design environment.

## 12.4 Questions and Problems

12.1. Analyze the operation of 68HC11 in extended mode of operation. Design an interface to the 68HC11 that demultiplexes and separates address and data bus and "opens" six read/write registers at adjacent locations in the 68HC11 address space starting at any feasible address. These registers enable access to external world. For interface and registers' description use VHDL.

12.2. Using the interface from the preceding example implemented in a separate chip design a connection to an external FLEX8000 type FPLD that enables passive asynchronous configuration of the FPLD controlled by a program from the 68HC11. For description of the circuit use VHDL.

12.3. Implement a simple logic analyzer circuit in the external FPLD of preceding question. It captures external data on four groups of 16 lines each and stores them in internal registers before being transferred to an external RAM memory. The logic analyzer controls transfers of data to memory using DMA. Start and end of data capture is controlled by a program through the registers implemented in the interface circuit from problem 1.

12.4. Analyze the operation of ISA-bus. Using VHDL describe the interface that enables a) 8-bit transfers between a PC and logic implemented on a card similar to RAPROS board and b) 16-bit transfers between a PC and logic implemented on a card similar to RAPROS board. Open a window to the PC I/O address space with 16 ports at adjacent addresses.

12.5. Analyze the operation of Altera UP prototyping board. Design a circuit that enables demonstration of input and output functions using switches, push buttons and 7-segment displays.

12.6. Analyze the operation of VGA interface on Altera UP prototyping board. Design a circuit that enables the access to a VGA monitor.

12.7. Analyze the operation of mouse interface on Altera UP prototyping board. Design a circuit that provides data input from a mouse.

12.8. Using circuits from 12.6 and 12.7 design a circuit that will enable interaction between the user and VGA monitor using mouse.

# 13 USING VHDL TO DESIGN FOR ALTERA'S FPLDS

This Chapter represents a quick tour through VHDL in the Max+Plus II design environment, and to a certain extent a repetition of the VHDL basic features presented in Part 1. Further examples of VHDL descriptions which result in valid synthesized circuits for Altera FPLDs are also presented. More complex examples and features of Altera's VHDL implementation and its use in design for FPLDs can be found in the later chapters of Part 3.

## 13.1 Specifics of Altera's VHDL

VHDL is fully integrated into the Max+Plus II design environment. VHDL designs can be entered with any text editor, and subsequently compiled with the compiler to create output files for simulation, timing analysis, and device programming. Max+Plus II supports a subset of IEEE 1076-1987/1993 VHDL as described in the corresponding Altera documentation. VHDL design files, with the extension .vhd, can be combined with the other design files into a hierarchical design, called project. The other types of files include Altera specific AHDL (Altera Hardware Description Language) design files (Text Design Files or TDF files), schematic entry files (Graphic Design Files or GDF files), waveform design files (WDF files), altera design files (ADF files), and state machine files (SMF files), as well as industry standard EDIF files, or Xilinx Netlist format files (XNF files). Each file in a project hierarchy, i.e., each macrofunction, is connected through its input and output ports to one or more design files at the next higher hierarchy level.

Max+Plus II environment allows a designer to create a symbol that represents a VHDL design file and incorporate it into graphic design file. The symbol contains graphical representation of input and output ports, as well as some parameters which can be used to customize design to the application requirement. Custom functions, as well as Altera-provided macrofunctions, can be incorporated into any VHDL design file. Max+Plus II Compiler automatically processes VHDL design files and optimizes them for Altera FPLD devices. The Compiler can be directed to create a VHDL Output File (.vho file) that can be imported into an industry standard environment for simulation. On the other side, after a VHDL project has compiled successfully, optional simulation and timing analysis with Max+Plus II can be done, and then Altera device programmed.

Altera provides ALTERA library that includes the maxplus2 package, which contains all Max+Plus II primitives and macrofunctions supported by VHDL. Besides that, Altera provides several other packages located in subdirectories of the \maxplus2\max2vhdl directory, as it is shown in Table 1.

In addition, Altera provides the STD library with the STANDARD and TEXTIO packages that are defined in the IEEE standard VHDL Language Reference Manual. This library is located in the \maxplus2\max2vhdl\std directory.

Table 13.1 Max+Plus II Packages

| Package | Library | Contents |
|---|---|---|
| maxplus2 | altera | Max+Plus II primitives and macrofunctions supported by VHDL |
| std_logic_1164 | ieee | Standard for describing interconnection data types for VHDL modeling, and the std_logic and std_logic_vector types |
| std_logic_arith | ieee | Signed and unsigned types, arithmetic and comparison functions for use with signed and unsigned types, and the conversion functions conv_integer, conv_signed, and conv_unsigned |
| std_logic_signed | ieee | Functions that allow Max+Plus II to use std_logic_vector types as if they are signed types |
| std_logic_unsigned | ieee | Functions that allow Max+Plus II to use std_logic_vector types as if they are unsigned types |

## 13.2 Combinational Logic Implementation

Combinational logic is implemented in VHDL with concurrent signal assignment statements or with process statements that describe purely combinatorial behaviour.

Both of these statements should be placed in the architecture body of a VHDL design file, as shown in the following template:

```
architecture arch_name of and_gate is
begin
        [concurrent_signal_assignments]
        [process_statements]
        [other concurrent statements]
end arch_name;
```

### *13.2.1 Basic Combinational Functions*

Concurrent signal assignment statements assign values to signals, directing the compiler to create simple gates and logical connections. Three basic types of concurrent signal assignment statements are available:

- Simple signal assignment
- Conditional signal assignment
- Selected signal assignment

The simple signal assignment statements shown in the example below create an and gate and connect two nodes, respectively. These statements are executed concurrently:

```
library ieee;
use ieee.std_logic_1164.all;
entity simpsig is
port (a, b: in std_logic;
        c, d: out std_logic);
end simpsig;

architecture arch1 of simpsig is
begin
        c <= a and b;
        d <= b;
end arch1;
```

Figure 13.1 shows a graphic design file that is equivalent to the simple signal assignment statements of the preceding example.

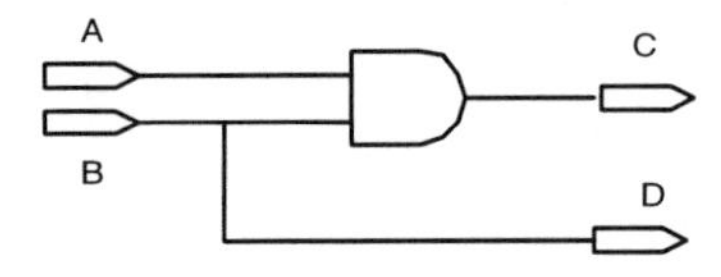


Figure 13.1 GDF equivalent of simple signal assignment statements

Conditional signal assignment statements list a series of expressions that are assigned to a target signal after the positive evaluation of one or more Boolean expressions. The following example shows a basic 2-to-1 multiplexer in which the input value a is assigned to output y when s equals '0'; otherwise, the value b is assigned to output y:

```
library ieee;
use ieee.std_logic_1164.all;
entity mux21 is
          port (a, b, s: in std_logic;
                    y: out std_logic);
end mux21;

architecture arch1 of mux21 is
begin
          y <= a when s = '0' else
                    b;
end arch1;
```

A GDF equivalent of this example is shown in Figure 13.2. When a conditional signal assignment statement is executed, each Boolean expression is tested in order in which it is written. The value of the expression preceding **when** keyword is assigned to the target signal for the first Boolean expression that is evaluated as true. If none of the Boolean expressions are true, the expression following the last **else** keyword is assigned to the target.

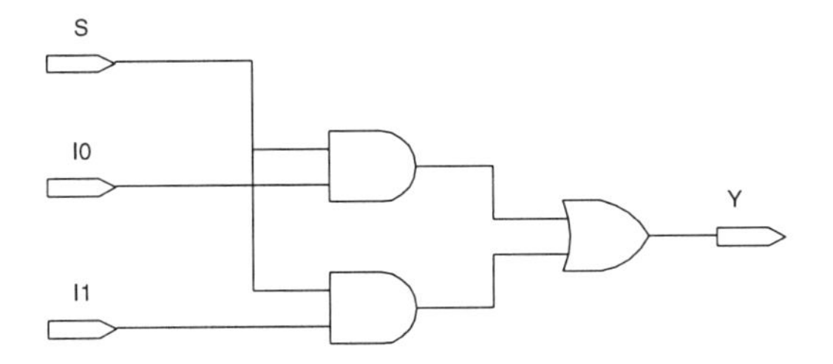


Figure 13.2 GDF equivalent of conditional assignment statement with one alternative

An example of priority encoder is shown below:

```
library ieee;
use ieee.std_logic_1164.all;

entity prior3 is
        port (high, mid, low: in bit;
        q: out integer);
end prior3;

architecture arch1 of prior3 is
begin
        q <=    3 when high = '1' else
                2 when mid = '1' else
                1 when low = '1' else
                0;
end arch1;
```

Figure 13.3 shows GDF equivalent of priority encoder as synthesized by the Compiler.

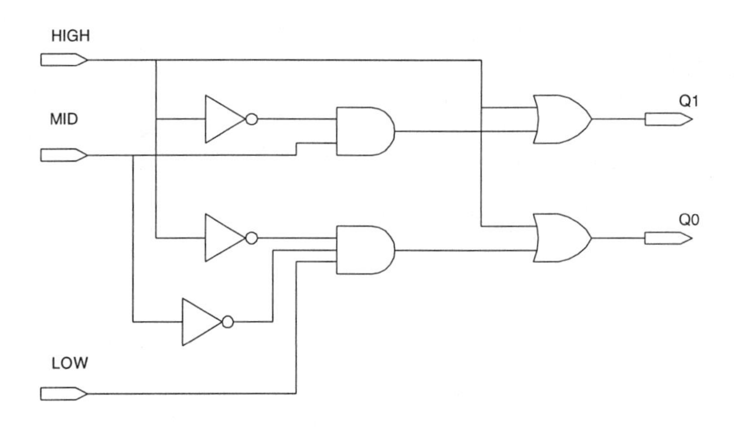


Figure 13.3 GDF equivalent of priority encoder

Selected signal assignment statement lists alternatives that are available for each value of an expression, then selects a course of action based on the value of expression. In the following example selected signal assignment statement is used to create multiplexer:

```
library ieee;
use ieee.std_logic_1164.all;

entity mux41 is
        port ( a, b, c, d: in bit;
                s: in integer range 0 to 3;
                y: out bit);
end mux41:

architecture arch1 of mux41 is
begin
with s select
        y <=    a when 0,
                b when 1,
                c when 2,
                d when 3;
end arch1;
```

The select expression, found between with and select keywords, selects the signal that is to be assigned to the target signal. Figure 13.4 shows GDF equivalent of the selected signal assignment statement of example above.

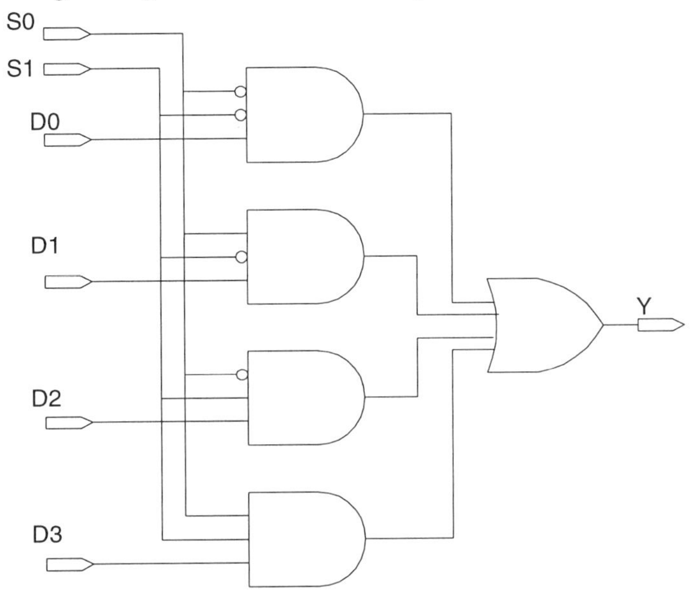


Figure 13.4 GDF equivalent of 4-to-1 multiplexer

Process statements include a set of sequential statements that assign values to signals. Process statements that describe purely combinatorial behaviour can be used to create combinatorial logic. To ensure that a process is combinatorial, its sensitivity list must contain all signals that are read in the process. The following is an example of the process that counts the number of bits with the value 1 in a 4-bit word represented by signal d:

```
library ieee;
use ieee.std_logic_1164.all;

entity counter is
        port ( d: in std_logic_vector (3 downto 0);
         q: out integer range 0 to 4);
end counter;

architecture arch1 of counter is
```

```
begin
process (d)
        variable n: integer;
begin
        n := 0;

        for i in d'range loop
                if d(i) = '1' then
                  n = n + 1;
                end if;
        end loop;

        q <= n;
end process;
end arch1;
```

Signal d is the only signal contained in the sensitivity list that follows the process keyword. This signal is declared as an array in the entity declaration. If d(i) equals 1, the if statement increments internal counting variable n. The n variable is then assigned to the signal q, which is also declared in the entity declaration.

### *13.2.2 Examples of Standard Combinational Blocks*

In this subsection we will present a number of examples of standard combinational blocks and various ways to describe them in VHDL. These blocks usually represent units that are used to form data paths in more complex digital designs. All of these designs are easily modifiable to suit the needs of specific application. Different approaches to modeling are used to demonstrate both versatility and power of VHDL.

#### Example: 8-bit 4 to 1 multiplexer - behavioral model

Multiplexer is modeled with two concurrent statements: one is producing an intermediate internal signal sel (of integer type) which selects the input that is forwarded to the output of multiplexer in another concurrent statement. This example demonstrates the use of various data types, and conversion between those types.

```
library ieee;
use ieee.std_logic_1164.all;
```

```
entity mux41beh is
port(a, b, c ,d: in std_logic_vector(7 downto 0);
        s0, s1: in std_logic; -- select input lines
        y: out std_logic_vector(7 downto 0));
end mux41beh;
architecture beh of mux41beh is
        signal sel: integer;
begin
        sel <=  0 when (s1='0' and s0 ='0') else
                1 when (s1='0' and s0 ='1') else
                2 when (s1='1' and s0 ='0') else
                3 when (s1='1' and s0 ='1') else
                4;

        with sel select
        y <= a when 0,
                b when 1,
                c when 2,
                d when 3,
                "XXXXXXXX" when others;

end beh;
```

## Example: 8-bit 4 to 1 multiplexer - structural model

In order to model the same multiplexer structurally, we first design elementary logic gates (minv inverter, *mand3* 3-input and gate and mor4 4-input or gate) and include them in the package my_gates. Components from this package are used in the structural model. Because of the multi-bit inputs and outputs to the multiplexer, components are instantiated using for-generate statement. The whole VHDL model including the order in which design units are compiled (first individual components, then package, and at the end multiplexer unit) is shown below.

```
library ieee;       -- must be declared before each design unit
use ieee.std_logic_1164.all;

entity minv is      -- inverter
        port (a: in std_logic; y: out std_logic);
end minv;
```

```
architecture dflow of minv is
begin
        y <= not a;
end dflow;

library ieee;
use ieee.std_logic_1164.all;
entity mand3 is   -- 3-input and
        port (a, b, c: in std_logic; y: out std_logic);
end mand3;
architecture dflow of mand3 is
begin
        y <= a and b and c;
end dflow;

library ieee;
use ieee.std_logic_1164.all;
entity mor4 is    -- 4-input or
        port (a, b, c, d: in std_logic; y: out std_logic);
end mor4;
architecture dflow of mor4 is
begin
        y <= a or b or c or d;
end dflow;

library ieee;       -- separately compiled package
use ieee.std_logic_1164.all;
use work.all;  -- all previously declared components are in work library

package my_gates is
component minv
        port(a: in std_logic; y: out std_logic);
end component;

component mand3
        port(a, b, c: in std_logic; y: out std_logic);
end component;

component mor4
        port(a, b, c,d: in std_logic; y: out std_logic);
end component;
```

```
end my_gates;

library ieee;
use ieee.std_logic_1164.all;
use work.my_gates.all;  -- package used in structural model

entity mux41str is
        port(a, b, c, d: in std_logic_vector(7 downto 0);
        s1, s0:   in std_logic;
        y: out std_logic_vector(7 downto 0));
end mux41str;

architecture struct of mux41str is
        signal s1n, s0n: std_logic;  -- internal signals to interconnect
                                     -- components
        signal ma, mb, mc, md: std_logic_vector(7 downto 0);
begin
        u_inv0: minv port map(s0, s0n);
        u_inv1: minv port map(s1, s1n);

f1:  for i in 0 to 7 generate
        u_ax: mand3 port map (s1n, s0n, a(i), ma(i));
        u_bx: mand3 port map (s1n, s0, b(i), mb(i));
        u_cx: mand3 port map (s1, s0n, c(i), mc(i));
        u_dx: mand3 port map (s1, s0, d(i), md(i));

        u_ex: mor4 port map (ma(i), mb(i), mc(i), md(i), y(i));
end generate f1;
end struct;
```

### Example: 8-to-3 Encoder

This example shows two different behavioral architectures of 8-to-3 encoder. The first architecture uses if statement while the second architecture uses a case statement within a process. The use of the if statements introduces delays because the circuit inferred will evaluate expressions in the order in which they appear in the model (the expression at the end of the process is evaluated last). Therefore, the use of the case statement is recommended. It also provides a better readability.

```
library ieee;
use ieee.std_logic_1164.all;
```

```
entity encoder83 is
        port(a: in std_logic vector (7 downto 0);
                y: out std_logic_vector(2 downto 0));
end encoder83;

architecture arch1 of encoder83 is
begin
        process(a)
        begin
                if (a=”00000001”) then
                        y <= “000”;
                elsif (a=”000000010”) then
                        y <= “001”;
                elsif (a=”00000100”) then
                        y <= “010”;
                elsif (a=”00001000”) then
                        y <= “011”;
                elsif (a=”00010000”) then
                        y <= “100”;
                elsif (a=”00100000”) then
                        y <= “101”;
                elsif (a=”01000000”) then
                        y <= “110”;
                elsif (a=”10000000”) then
                        y <= “111”;
                else
                        y <= “XXX”;
                end if;
        end process;
end arch1;

architecture arch2 of encoder83 is
begin
        process(a)
        begin
                case a is
                        when “00000001” => y <= “000”;
                        when “00000010” => y <= “001”;
                        when “00000100” => y <= “010”;
                        when “00001000” => y <= “011”;
                        when “00010000” => y <= “100”;
                        when “00100000” => y <= “101”;
```

```
                        when "01000000" => y <= "110";
                        when "10000000" => y <= "111";
                end case;
        end process;
end arch2;
```

## Example: 3-to-5 Binary Decoder with Enable Input

This example is straightforward. However, it is important to notice that the behavior for unused combinations is specified to avoid generation of unwanted logic or latches.

```
library ieee;
use ieee.std_logic_1164.all;

entity decoder35 is
        port(a: in integer; en: in std_logic;
                y: out std_logic_vector(4 downto 0));
end encoder83;

architecture arch1 of encoder83 is
begin
        y <=    1 when (en='1' and a=0) else
                2 when (en='1' and a=1) else
                4 when (en='1' and a=2) else
                8 when (en='1' and a=3) else
                16 when (en='1' and a=1) else
                0;
end arch1;
```

## Example: A Simple Arithmetic and Logic Unit

This example is introduced just to illustrate an approach to the description of a simple arithmetic and logic unit (ALU) as a more complex, but still common, combinational circuit. However, most of the issues in the design of the real ALUs are related to efficient implementation of basic operations (arithmetic operations such as addition, subtraction, multiplication, and division, shift operations, etc.). The ALU in this example performs operations on one or two operands that are received on two 8-bit busses (a and b) and produces output on 8-bit bus (f). Operation performed by the ALU is specified by operation select (opsel) input lines. Input and output carry are not taken into account.

```
library ieee;
use ieee.std_package_1164.all;

type ops is (add, nop, inca, deca, loada, loadb, op_nega, op_negb, op_and,
        op_or, shl, shr);

entity alu is
        port (a, b: in std_logic_vector(7 downto 0); opsel: in ops;
        clk: in std_logic;
        f: out std_logic_vector(7 downto 0));
end alu;

architecture beh of alu is
begin
process
procedure "+" (a, b: std_logic_vector) return std_logic_vector is
variable sum: std_logic_vector (0 to a'high);
variable c: std_logic:= '0';
begin
for i in 0 to a'high loop
        sum(i) := a(i) xor b(i) xor c;
        c := (a(i) and c) or (b (i) and c) or(a(i) and b(i));
end loop;
return sum;
end;
function shiftl(a: std_logic_vector) return std_logic_vector is
variable shifted: std_logic_vector (0 to a'high);
begin
        for i in 0 to a'high -1 loop
                shifted(i + 1) := a(i);
        end loop;
return shifted;
end;
function shiftr(a: std_logic_vector) return std_logic_vector is
constant highbit: integer := a'high;
variable shifted: bit_vector (0 to highbit);
begin
        shifted(0 to highbit - 1) := a(1 to highbit);
```

```
return shifted;
end;

begin
wait until clk;
case opsel is
        when add => f <= a + b;
        when inca => f <= a + 1;
        when deca => f <= a -1;
        when nop => null; -- A null statement,
        when op_nega => f <= not a;
        when op_negb => f <= not b;
        when op_and => f <= a and b;
        when op_or => f <= a or b;
        when shl => f <= shiftl(a);
        when shr => f <= shiftr(a);
        when loada => f <= a;
        when loadb => f <= b;
end case;
end process;
end beh;
```

## 13.3 Sequential Logic Implementation

Sequential logic is implemented in VHDL with process statements. Process statements direct the Compiler to create logic circuitry that is controlled by the process statement's clock signal.

### *13.3.1 Registers and Counters Synthesis*

A register is implemented implicitly with a register inference. Register inferences in Max+Plus II VHDL support any combination of clear, preset, clock, enable, and asynchronous load signals. The Compiler can infer memory elements from the following VHDL statements which are used within a process statement:

- If statements can be used to imply registers for signals and variables in the clauses of the if statement
- Wait statements can be used to imply registers in a synthesized circuit. The Compiler creates flip-flops for all signals and some variables that are assigned

values in any process with a wait statement. The wait statement must be listed at the beginning of the process statement.

Registers can be also implemented with component instantiation statement. However, register inferences are technology-independent.

The following example shows several ways to infer registers that are controlled by a clock and asynchronous clear, preset, and load signals:

```
library ieee;
use ieee.std_logic_1164.all;

entity register_inference is
        port ( d, clk, clr, pre, load, data: in std_logic;
        q1, q2, q3, q4, q5: out std_logic);
end register_inference;

architecture arch1 of register_inference is

begin

-- register with active-low clock
process
begin
        wait until clk = '0';
        q1 <= d;
end process;

-- register with active-high clock and  asynchronous clear
process
begin
        if clr = '1' then
                q2 <= '0';
        elsif clk'event and clk = '1' then
                q2 <= d;
        end if;
end process;

-- register with active-high clock and asynchronous preset
process
begin
        if pre = '1' then
```

```
                q3 <= '1';
        elsif clk'event and clk = '1' then
                q3 <= d;
        end if;
end process;

-- register with active-high clock and asynchronous load
process
begin
        if load = '1' then
                q4 <= data;
        elsif clk'event and clk = '1' then
                q4 <= d;
        end if;
end process;

-- register with active-low clock and asynchronous clear and preset
process
begin
        if clr = '1' then
                q5 <= '0';
        elsif pre = '1' then
                q5 <= '1';
        elsif clk'event and clk = '0' then
                q5 <= d;
end process;
end arch1;
```

All above processes are sensitive only to changes on the control signals (clk, clr, pre, and load) and to changes on the input data signal called data.

A counter can be implemented with a register inference. A counter is inferred from an if statement that specifies a clock edge together with logic that adds or subtracts a value from the signal or variable. The If statement and additional logic should be inside a process statement. The following example shows several 8-bit counters controlled by the clk, clear, ld, d, enable, and up_down signals that are implemented with if statements:

```
library ieee;
use ieee.std_logic_1164.all;
entity counters is
        port (d : in integer range 0 to 255;
```

```
                clk, clear, ld, enable, up_down: in std_logic;
            qa, qb, qc, qd, qe, qf: out integer range 0 to 255);
end counters;

architecture arch of counters is
begin

-- an enable counter

process (clk)
        variable cnt: integer range 0 to 255;
begin
        if (clk'event and clk = '1') then
                if enable = '1' then
                        cnt := cnt + 1;
                end if;
        end if;

        qa <= cnt;
end process;
-- a synchronous load counter
process (clk)
        variable cnt: integer range 0 to 255;
        if (clk'event and clk = '1') then
                if ld = '0' then
                        cnt := d;
                else
                        cnt := cnt +1;
                end if;
        end if;

        qb <= cnt;
end process;

-- an up_down counter

process (clk)
        variable cnt: integer range 0 to 255;
        variable direction: integer;
begin
        if (up_down = '1') then
                direction := 1;
```

```
        else
                direction := -1;
        end if;

        if (clk'event and clk ='1') then
                cnt := cnt + direction;
        end if;

        qc <= cnt;
end process;

-- a synchronous clear counter

process (clk)
        variable cnt: integer range 0 to 255;
begin
        if (clk'event and clk = '1') then
                if clear = '0' then
                        cnt := 0;
                else
                        cnt := cnt + 1;
                end if;

        qd <= cnt;
end process;
-- a synchronous load clear counter

process (clk)
begin
        if (clk'event and clk = '1') then
                if clear = '0' then
                        cnt := 0;
                else
                        if ld = '0' then
                                cnt := d;
                        else
                                cnt := cnt +1;
                end if;
        end if;

        qe <= cnt;
end process;
```

```
                    -- a synchronous load enable up_down counter

process (clk)
        variable cnt: integer range 0 to 255;
        variable direction: integer;
begin
        if up_down = '1' then
                direction := 1;
        else
                direction := -1;
        end if;

        if (clk'event and clk = '1') then
                if ld = '0' then
                        cnt := d;
                else
                        if enable = '1' then
                                cnt := cnt + direction;
                        end if;
                end if;
        end if;
        gf <= cnt;
end process;
end arch;
```

All processes in this example are sensitive only to changes on the clk input signal. All other control signals are synchronous. At each clock edge, the cnt variable is cleared, loaded with the value of d, or incremented or decremented based on the value of the control signals.

Register inference also allows implementation of a latch. The latch is inferred from the incompletely specified if statement inside a process statement. The following VHDL design file shows an implementation of a latch:

```
library ieee;
use ieee.std_logic_1164.all;

entity latch is
        port (enable, data: in std_logic;
         q: out std_logic);
end latch;

architecture arch of latch is
```

```
begin

        latch: process (enable, data)
        begin
                if (enable = '1') then
                        q <= data;
                end if;
        end process latch;
end arch;
```

If enable equals '1', the value data is assigned to q. However, the if statement does not specify what happens if enable = '0'. In this case, the circuit maintains the previous state, creating a latch.

### *13.3.2 State Machines Synthesis*

To describe a state machine, an enumeration type for states, and a process Statement for the state register and the next-state logic can be used. The VHDL design file that implements a 2-state state machine from the Figure 13.5 is shown below:

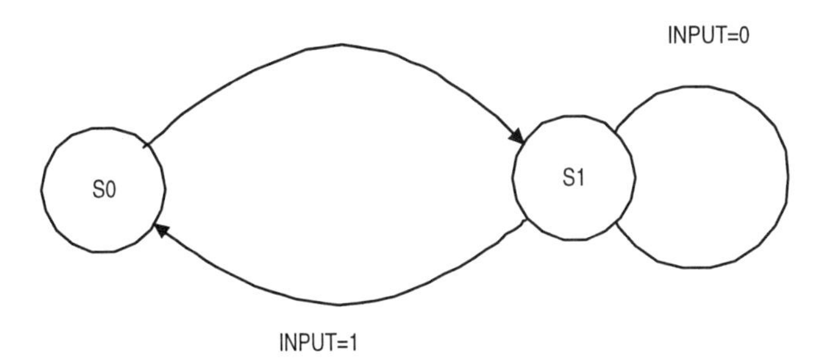


Figure 13.5 An example 2-state state machine

```
library ieee;
use ieee.std_logic_1164.all;

entity state_machine is
        port (clk, reset, input: in std_logic;
```

```
        output: out std_logic);
end state_machine;

architecture arch of state_machine is

        type state_typ is (s0, s1);
        signal state: state_typ;

begin
process (clk, reset)
begin
        if reset = '1' then
                state <= s0;
        elsif (clk'event and clk = '1') then
                case state is
                        when s0 =>
                                state <= s1;
                        when s1 =>
                                if input = '1' then
                                        state <= s0;
                                else
                                        state <= s1;
                                end if;
                end case;
        end if;
end process;

        output <= '1' when state = s1 else '0';

end arch;
```

The process statement in this example is sensitive to the clk and reset control signals. An if statement inside the process statement is used to prioritize the clk and reset signals, giving reset the higher priority. GDF equivalent of the state machine from the preceding example is shown in Figure 13.6.

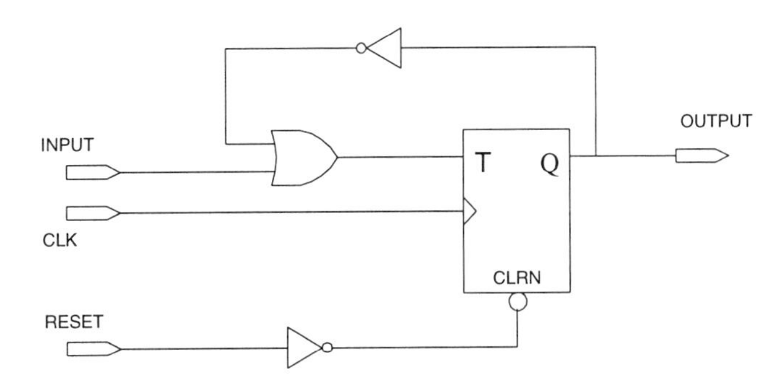


Figure 13.6 GDF equivalent of 2-state state machine

In general, the VHDL Compiler assigns the value 0 to the first state, the value 1 to the second state, the value 2 to the third state, and so on. This state assignment can be overridden by manual state assignment using enum_encoding attribute which follows the associated Type declaration. The following example shows the manual state assignment:

```
library ieee;
use ieee.std_logic_1164.all;

entity state_machine is
          port (up_down, clock: in std_logic;
           lsb, msb: out std_logic);
end state_machine;

architecture enum_state_machine is
          type state_typ is (zero, one, two, three);
          attribute enum_encoding: string;
          attribute enum_encoding of state_typ: type is "11 01 10 00";
          signal present_state, next_state: state_typ;
begin
process (present_state, up_down)
begin
          case present_state is
                    when zero =>
                              if up_down = '0' then
                                        next_state <= one;
```

```
                        lsb <= '0';
                        msb <= '0';
                else
                        next_state <= three;
                        lsb <= '1';
                        msb <= '1';
                end if;
        when one =>
                if up_down = '0' then
                        next_state <= two;
                        lsb <= '1';
                        msb <= '0';
                else
                        next_state <= zero;
                        lsb <= '0';
                        msb <= '0';
                end if;
        when two =>
                if (up_down = '0') then
                        next_state <= three;
                        lsb <= '0';
                        msb <= '1';
                else
                        next_state <= one;
                        lsb <= '1';
                        msb <= '0';
                end if;
        when three =>
                if (up_down = '0') then
                        next_state <= zero;
                        lsb <= '1';
                        msb <= '1';
                else
                        next_state <= two;
                        lsb <= '0';
                        msb <= '1';
                end if;
        end case;
end process;

process
begin
        wait until clock'event and clock = '1';
```

```
        present_state <= next_state;
    end process;

    end enum_state_machine;
```

The enum_encoding attribute must be a string literal that contains a series of state assignments. These state assignments are constant values that correspond to the state names in the enumeration type declaration. The states in the example above are encoded with following values:

zero = '11'
one = '01'
two = '10'
three = '00'

The enum_encoding attribute is Max+Plus II specific, and may not be available with other vendors' VHDL tools.

### *13.3.3 Examples of Standard Sequential Blocks*

#### Example: 16-bit Counter wit Enable Input and Additional Controls

This example demonstrates design of 16-bit counter which allows initialization to zero value (reset), and control of the counting by selection of counter step: incrementing for 1 or 2 and decrementing for 1. It also demonstrates the use of various data types.

```
library ieee;
use ieee.std_logic_1164.all;
use ieee.std_logic_arith.all;          -- use of numeric standard
use ieee.std_logic_unsigned.all;

entity flexcount16 is
port (up1, up2, down1, load, enable, clk, clr: in std_logic;
        q: out unsigned(15 downto 0));
end flexcount16;

architecture beh of updncnt8 is
begin
        process(clk, clr)
```

```
            variable dir: integer range -1 to 2;
            variable cnt: unsigned(15 downto 0);
    begin
            if up1 = '1' and up2='0' and down1 ='0' then
                    dir :=1;
            elsif up1 = '0' and up2 ='1' and down1 = '0' then
                    dir := 2;
            elsif up1 = '0' and up2 ='0' and down1 = '1' then
                    dir := -1;
            else
                    dir :=0;
            end if;

            if clr = '1' then
                    cnt := "0000000000000000";
            elsif clk'event and clk ='1' then
                    cnt := cnt + dir;
            end if;
            q <= cnt;
    end process;
end beh;
```

## Example: Frequency Divider

This example demonstrates how a frequency divider (in this case divide by 11) can be designed using two different architectures. The output pulse must occur at the $11^{th}$ pulse received to the circuit. The first architecture is purely structural and uses decoding of a 4-bit counter implemented with toggle flip-flops at the value 9 (which is reached after 10 pulses received). When this value is detected it is used to toggle an additional toggle-flip flop which will produce the output pulse at the next clock transition, but also will be used to reset the 4-bit counter. The relationship between 4-bit counter and toggle flip-flop is presented in Figure 13.7. It uses separately designed and compiled toggle flip-flops that are also presented within this example.

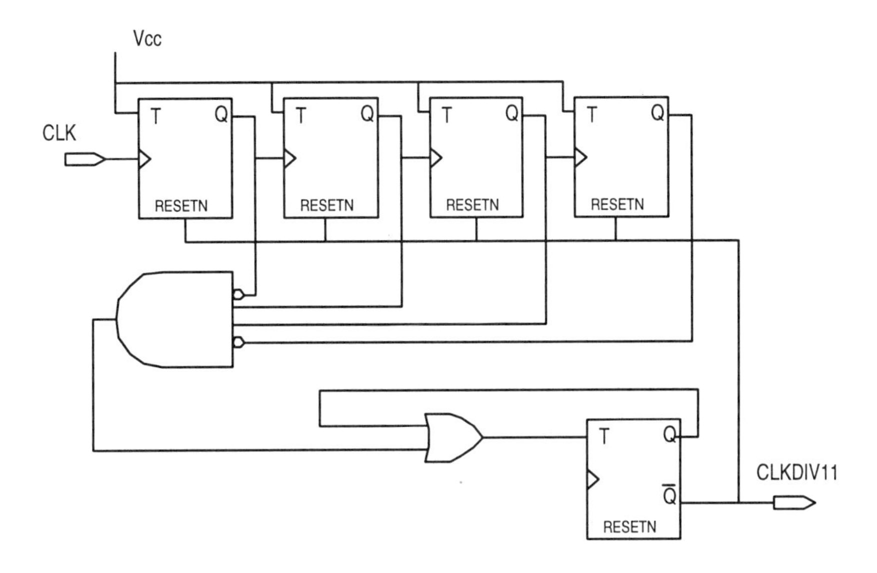


Figure 13.7 Structural Architecture of Divider by 11

```
library ieee;
use ieee.std_logic_1164.all;

entity t_ff is
        port(t, clk, resetn: in std_logic;
        q: out std_logic);
end t_ff;

architecture beh of t_ff is
        signal mq: std_logic;
begin
        process(clk, resetn)
        begin
                if resetn = '0' then
                        mq <= '1';
                elsif clk'event and clk= '1' then
                        mq <= mq xor t;
```

```
                end if;
                          q <= mq;
        end process;
end beh;

use work.all;

entity divider11 is
        port (clk: in std_logic;
        clkdiv11: out std_logic);
end divider11;

architecture struct of divider11 is
component t_ff
port (t, clk, resetn: in std_logic;
        q: out std_logic);
end component;

signal vcc: std_logic;
signal z: std_logic_vector(0 to 4);
signal m0, m1, m2, m3: std_logic;

begin
        vcc <='1';
        z(0) <= clk;
        out_ff: t_ff port map(m1, clk, vcc, m2); -- 5th toggle flip-flop

        m0 <= not(z(1)) and (z(2)) and (z(3)) and not (z(4)); -- detect 9
        m1 <= m0 or m2;
        m3 <= not(m2);

        f1: for i in 0 to 3 generate
                ux: t_ff port map (vcc, z(i), m3, z(i+1));
        end generate f1;
        clkdiv11<= m2;
end struct;
```

The second architecture is a behavioral one using two cooperating processes to describe counting and the process which detects the boundary conditions, produces a resulting pulse, and resets the counter to start a new counting cycle. It is illustrated in Figure 13.8.

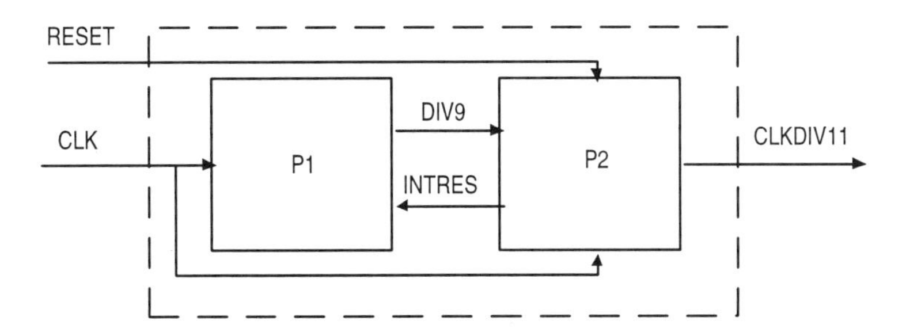


Figure 13.8 Behavioral Architecture of Divider by 11

```
library ieee;
use ieee.std_logic_1164.all;
use ieee.std_logic_arith.all; -- additional packages used in behavioral
                              -- architecture
use ieee.std_logic_unsigned.all;

entity divider11 is
        port (clk, reset: in std_logic;
        clkdiv11: out std_logic);
end divider11;

architecture beh of divider11 is
        signal div9: std_logic;      -- indication that the contents of
                                     -- counter is 9
        signal intres: std_logic;    -- internal reset signal used to reset
                                     -- counter

begin
        p1: process(clk, intres)
        variable m: integer range 0 to 9; -- state of the counter
        begin
                if intres = '1' then
                        m:= 0;
                elsif clk'event and clk = '1' then
                        m:=m + 1;
                else
```

```
                    m:=m;
            end if;
            if (m = 9) then
                    div9 <= '1';
            else
                    div9 <= '0';
            end if;
    end process p1;

    p2: process(clk, reset)
            variable n: std_logic;
    begin
            if reset = '1' then
                    n:='0';
                    intres <= '1';
            elsif clk'event and clk = '1' then
                    if div9='1' then
                            n:='1';
                    else
                            n:='0';
                    end if;
                    intres <= n;
            end if;
            clkdiv11 <= n;
    end process p2;
end beh;
```

## Example: Timer

Timer is a circuit that is capable of providing very precise time intervals based on the frequency (and period) of external clock (oscillator). Time interval is obtained as a multiple of clock period. The initial value of the time interval is stored into internal register and then by counting down process decremented at each either positive or negative clock transition. When the internal count reaches value zero, the desired time interval is expired. The counting process is active as long as external signal enable controlled by external process is active. Block diagram of the timer is presented in Figure 13.9.

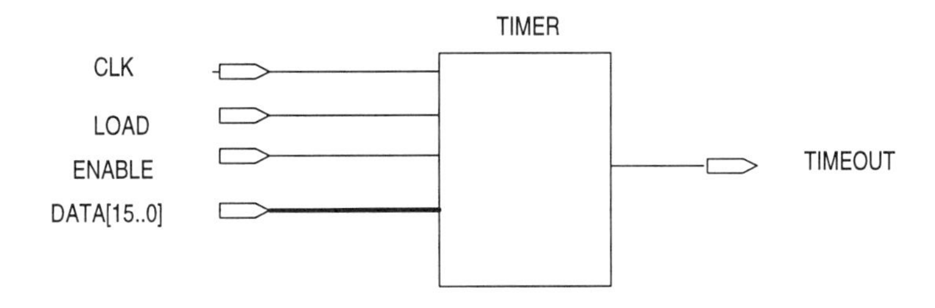


Figure 13.9 Illustration of the timer ports

VHDL description of the timer is given below:

```
library ieee;
use ieee.std_logic_1164.all;
use ieee.std_logic_arith.all;
use ieee.std_logic_unsigned.all;

entity timer is
        port(clk, load, enable: in std_logic;
                data: in std_logic_vector (15 downto 0);
                timeout: out std_logic);
end timer;

architecture beh of timer is
begin
        process
                variable cnt: unsigned (15 downto 0);
        begin
                if clk'event and clk = '1' then
                        if load = '1'  then
                                cnt:= data;
                        elsif enable = '1' then
                                cnt := cnt + "0000000000000001"
                        else
                                cnt:=cnt;
                        end if;
                end if;
                if cnt = "0000000000000000" then
```

```
                    timeout <= '1' ;
            else
                    timeout <= '0';
            end if;
        end process;
end beh;
```

## 13.4 Hierarchical Projects

VHDL design file can be combined with the other VHDL design files, and other design files from various tools (AHDL Design Files, GDF Design files, OrCAD Schematic Files, and some other vendor specific design files into a hierarchical project at any level of project hierarchy.

Max+Plus II design environment provides a number of primitives and bus, architecture-optimized, and application-specific macrofunctions. The designer can use component instantiation statements to insert instances of macrofunctions and primitives. register inference shown in preceding sections can be used to implement registers.

### *13.4.1 Max+Plus II Primitives*

Max+Plus II primitives are basic functional blocks used in circuit designs. Component Declarations for these primitives are provided in the maxplus2 package in altera library in the maxplus2\max2vhdl\altera directory. Table 13.2 shows primitives that can be used in VHDL Design Files.

Table 13.2 Primitives supported by Max+Plus II VHDL

| Primitive Type | Primitive Name |
|---|---|
| Buffer | CARRY, CASCADE, EXP, GLOBAL, LCELL, SOFT, TRI |
| Flip-Flop | DFF, DFFE, JKFF, JKFFE, SRFF, SRFFE, TFF, TFFE |
| Latch | LATCH |

### *13.4.2 Max+Plus II Macrofunctions*

Max+Plus II macrofunctions are collections of high-level building blocks that can be used in logic designs. Macrofunctions are automatically installed in the \maxplus2\max2lib directory. Component declarations for these macrofunctions are provided in the maxplus2 package in the Altera library in the \maxplus2\max2vhdl\altera directory. The Compiler analyses logic circuit and automatically removes all unused gates and flip-flops. All input ports have default signal values, so the designer can simply leave unused inputs unconnected. From the functional point of view all macrofunctions are the same regardless of target architecture. However, implementations take advantage of the architecture of each device family, providing higher performance and more efficient implementation. Examples of Max+Plus II macrofunctions supported by VHDL are shown in Table 13.3, and the rest can be found in corresponding Altera literature. Macrofunction usual names have the prefix a_ due to the fact that VHDL does not support names that begin with digits.

Table 13.3 Max+Plus II Macrofunctions supported by VHDL

| Macrofunction Type | Macrofunction Name | Description of Operation |
|---|---|---|
| Adder | a_8fadd<br>a_7480<br>a_74283 | 8-bit full adder<br>Gated full adder<br>4-bit full adder with fast carry |
| Arithmetic Logic Unit | a_74181<br>a_74182 | Arithmetic logic unit<br>Look-ahead carry generator |
| Application specific | ntsc | NTSC video control signal generator |
| Buffer | a_74240<br>a_74241 | Octal inverting 3-state buffer<br>Octal 3-state buffer |
| Comparator | a_8mcomp<br>a_7485<br>a_74688 | 8-bit magnitude comparator<br>4-bit magnitude comparator<br>8-bit identity comparator |
| Converter | a_74184 | BCD-to-binary converter |
| Counter | gray4<br>a_7468<br>a_7493<br>a_74191<br>a_74669 | Gray code counter<br>Dual decade counter<br>4-bit binary counter<br>4-bit up/down counter with asynch. load<br>Synchr. 4-bit up/down counter |

| | | |
|---|---|---|
| Decoder | a_16dmux<br>a_7446<br>a_74138 | 4-to-16 decoder<br>BCD-to-7-segment decoder<br>3-to-8 decoder |
| EDAC | a_74630 | 16-bit parallel error detection &correction circuit |
| Encoder | a_74148<br>a_74348 | 8-to-3 encoder<br>8-to-3 priority encoder with 3-state outputs |
| Frequency divider | a_7456 | Frequency divider |
| Latch | inpltch<br>a_7475<br>a_74259<br>a_74845 | Input latch<br>4-bit bistable latch<br>8-bit addresable latch with Clear<br>8-bit bus interface D latch with 3-state outputs |
| Multiplier | mult4<br>a_74261 | 4-bit parallel multiplier<br>2-bit parallel binary multiplier |
| Multiplexer | a_21mux<br>a_74151<br>a_74157<br>a_74356 | 2-to-1 multiplexer<br>8-to-1 multiplexer<br>Quad 2-to-1 multiplexer<br>8-to-1 data selector/multiplexer/register with 3-state outputs |
| Parity generator/checker | a_74180 | 9-bit odd/even parity generator/checker |
| Register | a_7470<br><br>a_7473<br>a_74171<br>a_74173<br>a_74396 | AND-gated JK flip-flop with Preset and Clear<br>Dual JK flip-flop with Clear<br>Quad D flip-flops with Clear<br>4-bit D register<br>Octal storage register |
| Shift register | barrelst<br>a_7491<br>a_7495<br>a_74198<br>a_74674 | 8-bit barrel shifter<br>Serial-in serial-out shift register<br>4-bit parallel-access shift register<br>8-bit bidirectional shift register<br>16-bit shift register |
| Storage register | a_7498 | 4-bit data selector/storage register |
| SSI Functions | inhb<br>a_7400<br>a_7421<br>a_7432<br>a_74386 | Inhibit gate<br>NAND2 gate<br>AND4 gate<br>OR2 gate<br>Quadruple XOR gate |

| True/Complement I/O Element | a_7487<br>a_74265 | 4-bit true/complement I/O element<br>Quadruple complementary output elements |
|---|---|---|

The component instantiation statement can be used to insert an instance of a Max+Plus II primitive or macrofunction in circuit design. This statement also connects macrofunction ports to signals or interface ports of the associated entity/architecture pair. The ports of primitives and macrofunctions are defined with component declarations elsewhere in the file or in referenced packages. Consider the following example:

```
library ieee;
use ieee.std_logic_1164.all;
library altera;
use altera.maxplus2.all;

entity example is
        port (data, clock, clearn, presetn: in std_logic;
                q_out:  out std_logic;
                a, b, c, gn: in std_logic;
                d: in std_logic_vector(7 downto 0);
                y, wn:  out std_logic);
end example;

architecture arch of example is
begin

dff1: dff port map (d=>data, q=>q_out, clk=>clock,
                            clrn=>clearn, prn=>presetn);
        mux: a_74151b port map (c, b, a, d, gn, y, wn);
end arch;
```

Component instantiation statements are used to create a DFF primitive and a 74151b macrofunction. The library altera is declared as the resource library. The use clause specifies the maxplus2 package contained in the **altera** library. Figure 13.10 shows a GDF equivalent to the component instantiation statements of the preceding example.

Besides using Max+Plus II primitives and macrofunctions, a designer can implement the user-defined macrofunctions with one of the following methods:

- Declare a package for each project-containing component declaration for all lower-level entities in the top-level design file.
- Declare a component in the architecture in which it is instantiated.

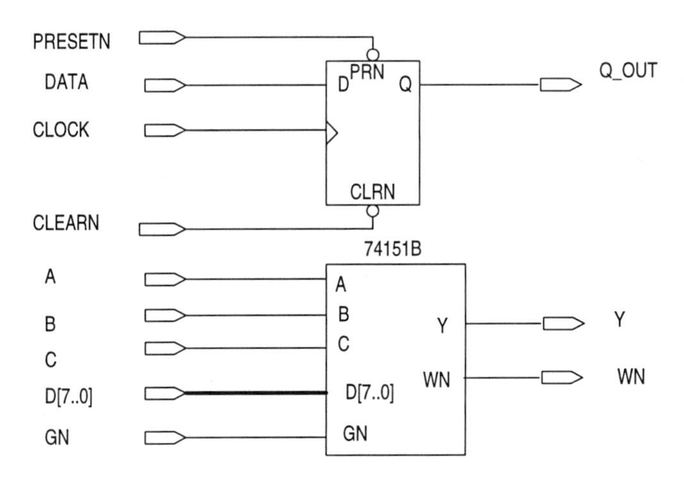


Figure 13.10 A GDF Equivalent of Component Instantiation Statement

The first method is described below. The following example shows reg12.vhd, a 12-bit register that will be instantiated in a VHDL Design File at the higher level of design hierarchy:

```
library ieee;
use iee.std_logic_1164.all;

entity reg12 is
          port (d: in std_logic_vector (11 downto 0);
                    clk: in std_logic;
          q:        out std_logic_vector (11 downto 0));
end reg12;

architecture arch of reg12 is
begin
process
```

```
begin
        wait until clk'event and clk = '1';
        q <= d;
end process;
end arch;
```

Figure 13.11 shows a GDF File equivalent to the preceding VHDL example.

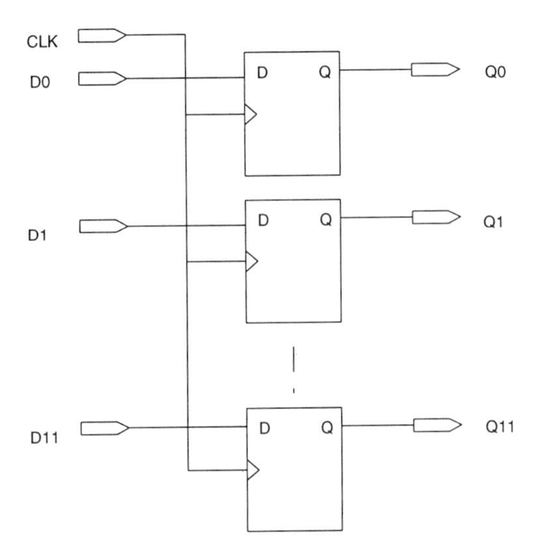


Figure 13.11 GDF Equivalent of reg12 Register

The following example declares reg24_package, identifies it with a use clause, and uses reg12 register as a component without requiring an additional component declaration:

```
library ieee;
use iee.std_logic_1164.all;

package reg24_package is
        component reg12
        port (d: in std_logic_vector(11 downto 0);
                clk: in bit;
                q: out std_logic_vector(11 downto 0));
```

```
        end component;
end reg24_package;

library work;
use work.reg24_package.all;

entity reg24 is
        port( d: in std_logic_vector(23 downto 0);
                clk: in std_logic;
                q: out std_logic_vector(23 downto 0));
end reg24;

architecture arch of reg24 is
begin
        reg12a: reg12 port map (d => d(11 downto 0),
                clk => clk, q => q(11 downto 0));
        reg12b: reg12 port map (d => d(23 downto 13),
                clk => clk, q => q(23 downto 13));
end arch;
```

From the preceding example we see that the user-defined macrofunction is instantiated with the ports specified in a component declaration. In contrast, Max+Plus II macrofunctions are provided in the maxplus2 package in the altera library. The architecture body for reg24 contains two instances of reg12. A GDF example of the preceding VHDL file is shown in Figure 13.12.

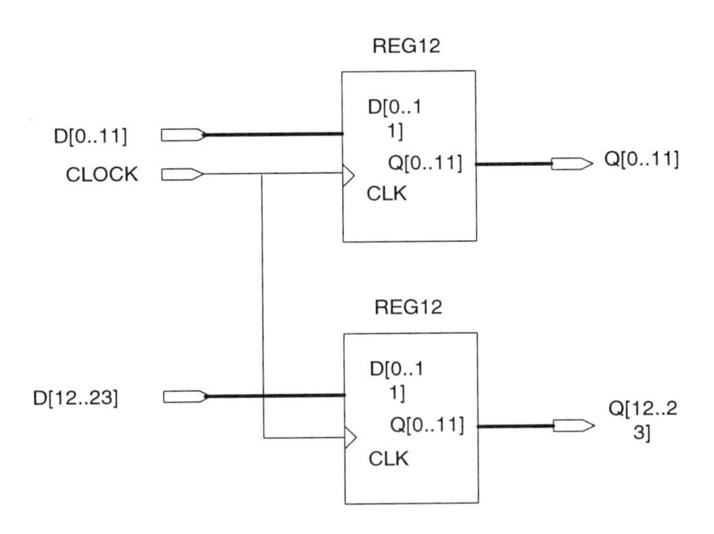


Figure 13.12 A GDF Equivalent of reg24

All VHDL libraries must be compiled. In Max+Plus II compilation is performed either with the Project Save and Compile command in any Max+Plus II application, or with the START button in the Max+Plus II Compiler window.

## 13.5 Using Parametrized Modules and Megafunctions

Altera provides another abstraction in the form of library design units which use parameters to achieve scalability, adaptability, and efficient silicon implementation. By changing parameters a user can customize design unit for a specific application. They belong to two main categories:

- Library of Parametrized Modules or LPMs, and
- Megafunctions

Moreover, the designer can create in VHDL and use parameterized functions, including LPM functions supported by MAX+PLUS II design environment. To create a parameterized logic function in VHDL, the generic clause in the entity declaration must list all parameters used in the architectural description and optional default values. An instance of a parameterized function is created with a component instantiation statement in the same way as unparameterized functions, with a few additional steps. The logic function instance must include a generic map aspect that lists all parameters for the instance. The generic map aspect is based on the generic clause in the component declaration, which is identical to the generic map in the component entity declaration.

The designer assigns values to parameters in the component instance. If no value is specified for a parameter, the Compiler searches for a default value in the parameter value search order. If a parameterized VHDL design file is the top-level file in a project, the Compiler takes parameter values from the global project parameters dialog box as the "instance" values, or, if values are not entered there, from the default values listed in the generic clause.

Parameter information cannot pass between functions that are defined in the same file. If an entity contains a generic clause, it must be the only entity in the file. Since parameterized functions do not necessarily have default values for unconnected inputs, the designer must ensure that all required ports are connected.

The following example shows reg24lpm.vhd, a 24-bit register which has an entity declaration and an architecture body that use two parameterized lpm_ff megafunctions. The generic map aspect for each instance of lpm_ff defines the register width by setting the lpm_width parameter value to 12.

```
library ieee;
use ieee.std_logic_1164.all;
library lpm;
use lpm.lpm_components.all;

entity reg24lpm is
port( d: in  std_logic_vector(23 downto 0);clk: in  std_logic;
          q: out std_logic_vector(23 downto 0));
end reg24lpm;

architecture arch of reg24lpm is
begin
          reg12a  : lpm_ff
          generic map (lpm_width => 12)
          port map (data => d(11 downto 0), clock => clk, q =>
                         q(11 downto 0));
          reg12b  : lpm_ff
          generic map (lpm_width => 12)
          port map (data => d(23 downto 12), clock => clk, q =>
                         q(23 downto 12));
```

```
end arch;
```

The following file, reggen.vhd, contains the entity declaration and architecture body for reggen, a parameterized register function. The generic clause defines the reg_width parameter.

```
entity reggen is
        generic(reg_width : integer);
        port(d: in std_logic_vector(reg_width - 1 downto 0);
        clk : in std_logic;
        q: out std_logic_vector(reg_width - 1 downto 0));
end reggen;

architecture arch of reggen is
begin
        process
        begin
                wait until clk = '1';
                q <= d;
        end process;
end arch;
```

The following example, reg24gen.vhd, instantiates two copies of reggen, reg12a and reg12b. The package declaration specifies the value of the top_width constant as the integer 24; the half_width constant is half of top_width. In the generic map aspect of each instance of reggen, the constant half_width is explicitly assigned as the value of the reg_width parameter, thereby creating two 12-bit registers.

```
package reg24gen_package is
        constant top_width : integer := 24;
        constant half_width : integer := top_width / 2;
end reg24gen_package;

use work.reg24gen_package.all;

entity reg24gen is
        port(d : in std_logic_vector(23 downto 0);
                clk : in std_logic;
                q: out std_logic_vector(23 downto 0));
end reg24gen;
```

```
architecture arch of reg24gen is

component reggen
        generic(reg_width : integer);
        port(d : in std_logic_vector(reg_width - 1 downto 0);
                clk : in std_logic;
                q : out std_logic_vector(reg_width - 1 downto 0));
end component;

begin
reg12a : reggen
        generic map (reg_width => half_width)
        port map (d => d(half_width - 1 downto 0), clk => clk,
        q => q(half_width - 1 downto  0));
reg12b : reggen
        generic map (reg_width => half_width)
        port map (d => d(half_width*2 - 1 downto half_width), clk =>
                clk, q => q(half_width * 2 - 1 downto half_width));

end arch;
```

In functions with multiple parameters, parameter values can also be assigned with positional association in a generic map aspect. The order of parameter values must be given in the same order as the parameters in the generic clause of the function's component declaration.

A list of LPMs supported by Altera for use in VHDL and other tools within MAX+Plus II design environment is shown in Table 13.4.

Table 13.4 Library of Parametrized Modules

| Name | Description |
|---|---|
| Gates | |
| lpm_and | Multi-bit and gate for bit-wise and operation |
| lpm_bustri | Multi-bit three-state buffer for unidirectional and bidirectional buffer implementation |
| lpm_clshift | Combinatorial Logic Shifter or Barrel Shifter |
| lpm_constant | Constant Generator |
| lpm_decode | Decoder |
| lpm_or | Multi-bit or gate for bit-wise or operation |

| | |
|---|---|
| lpm_xor | Multi-bit xor gate for bit-wise xor operation |
| lpm_inv | Multi-bit inverter |
| busmux | Two-input multi-bit multiplexer (can be derived from lpm_mux) |
| lpm_mux | Multi-input multi-bit multilexer |
| mux | Single input multi-bit multiplexer (can be derived from lpm_mux) |
| Arithmetic Components | |
| lpm_abs | Absolute value |
| lpm_add_sub | Multi-bit Adder/Subtractor |
| lpm_compare | Two-input multi-bit comparator |
| lpm_counter | Multi-bit counter with various control options |
| lpm_mult | Multi-bit multiplier |

Altera's VHDL supports several LPM functions and other megafunctions that allow the designer to implement RAM and ROM devices. The generic, scaleable nature of each of these functions ensures that you can use them to implement any supported type of RAM or ROM.

Table 13.5 lists the megafunctions that can be used to implement RAM and ROM in in Altera's VHDL:

Table 13.5 Megafunctions to implement RAM and ROM

| Name | Description |
|---|---|
| lpm_ram_dq | Synchronous or asynchronous RAM with separate input and output ports |
| lpm_ram_io | Synchronous or asynchronous RAM with a single I/O port |
| lpm_rom | Synchronous or asynchronous ROM |
| csdpram | Cycle-shared dual-port RAM |
| csfifo | Cycle-shared first-in first-out (FIFO) buffer |

In these functions, parameters are used to determine the input and output data widths; the number of data words stored in memory; whether data inputs, address/control inputs, and outputs are registered or unregistered; whether an initial memory content file is to be included for a RAM block; and so on. The designer must declare parameter names and values for RAM or ROM function by using generic map aspects. The following example shows a 256 x 8 bit lpm_ram_dq function with separate input and output ports.

```
library ieee;
use ieee.std_logic_1164.all;
library lpm;
use lpm.lpm_components.all;
library work;
use work.ram_constants.all;

entity ram256x8 is
port( data: in std_logic_vector (data_width-1 downto 0);
        address: in std_logic_vector (addr_width-1 downto 0);
        we, inclock, outclock: in std_logic;
        q: out std_logic_vector (data_width - 1 downto 0));
end ram256x8;

architecture arch of ram256x8 is

begin
inst_1: lpm_ram_dq
generic map (lpm_widthad => addr_width, lpm_width => data_width)
port map (data => data, address => address, we => we,
          inclock => inclock, outclock => outclock, q => q);
end arch;
```

The lpm_ram_dq instance includes a generic map aspect that lists parameter values for the instance. The generic map aspect is based on the generic clause in the function's component declaration. The designer assigns values to all parameters in the logic function instance. If no value is specified for a parameter, the Compiler searches for a default value in the parameter value search order.

## 13.6 Questions and Problems

13.1. Write a VHDL description which implements 8-input priority encoder.

13.2. Given a VHDL description:

```
library ieee;
use ieee.std_logic_1164.all;
use ieee.numeric_std.all;

entity example1 is
port (a, b in std_logic;
        clk: in std_logic;
        y1, y2: out std_logic);
end example1;

architecture arch1 of example1 is
begin
        p1: process(clk)
                if (clk = '1' and clk'event) then
                        y1 <= a;
                end if;
        end process;
        p2: process
                wait until (clk = '0' and clk'event);
                y2 <= b;
        end process;

end arch1;
```

Draw a schematic diagram which represents the result of synthesis of this description.

13.3. Given a VHDL description:

```
library ieee;
use ieee.std_logic_1164.all;
use ieee.numeric_std.all;

entity example2 is
port ( clk, a, b, c, d: in std_logic;
        y: out std_logic);
end example2

architecture arch1 of example2 is
        signal p: std_logic;
begin
        process(clk)
```

```
                    variable q: std_logic;
          begin
                    if (clk =  '0' and clk'event) then
                              p <= (a and b);
                              q :=  (c xor d);
                              y  <= (p or q);
                    end if;
          end process;
end arch1;
```

Draw a schematic diagram which represents the result of synthesis of this description.

13.4. Describe a generic synchronous n-bit up/down counter that counts up-by-p when in up-counting mode, and counts down-by-q when in down-counting mode. Using this model instantiate 8-bit up-by-one, down-by-two counter.

13.5. Describe an asynchronous ripple counter that divides an input clock by 32. For the ripple stages the counter uses a D-type flip-flop whose output is connected back to its D input such that each stage divides its input clock by two. For description use behavioral-style modeling. How would you modify the counter to divide the input clock by a number which is between 17 and 31 and cannot be expressed as $2^k$ (k is an integer).

# PART3

# Examples of Applications and Case Studies

# 14 EXAMPLES OF VHDL DESIGNS

The purpose of this Chapter is to present examples of relatively simple designs using VHDL as a high level design tool and Altera's FPLDs as the target technology. The first example is a sequence classification and recognition circuit which receives a sequence of binary coded characters, classifies them according to some decision criteria until a specific sequence of input characters (code or password) is detected, when the classification process stops. The purpose of this example is to demonstrate a hierarchical approach to the design of circuit that contains several subcircuits. The second example is a simple, but full traffic light controller. It is assumed for an intersection of a main road and a relatively infrequent side road traffic. Although simple, this example demonstrates the use of VHDL to describe a realistic state machine that controls a number of other input, output and timing circuits. The last example, a frequency generator/modulator, represents a different type of digital system with an emphasis on implementation of sinusoidal functions of required frequency using look-up tables, and the use of these functions in frequency modulation of digital signal. Because of its complexity, the whole frequency modulator is only conceptually shown, and the rest of design is left to the readers as an exercise.

## 14.1 Sequence Recognizer and Classifier

The aim of the sequence classification and recognition circuit is to receive the sequence of binary coded decimal numbers, compare the number of zeros and ones in each code, and, depending on that, increment one of two counters: the counter that stores the number of codes in which the number of ones has been greater than or equal to the number of zeros, and the counter that stores the number of codes in which the number of ones has been less than the number of zeros. The counting, and classification of codes in the input sequence continues until a specific five digit sequence is received, in which case the counting process stops. However, the recognition process continues in order to recognize another sequence of input numbers which will restart classification and counting process.

The overall sequence classifier and recognizer is illustrated in Figure 14.1. The input sequence appears on both inputs of the classifier and recognizer. As a result of classification, one of two outputs that increment counters is activated. The recognizer is permanently analyzing the last five digits received in sequence. When a specific sequence, given in advance, is recognized, the output of the recognizer is activated. This output stops counting on both counters. The counters are of the BCD type providing three BCD-coded values on their outputs. These outputs are used to drive 7-segment displays, so that the current value of each counter is continuously displayed. In order to reduce the display control circuitry, three digits are multiplexed to the output of the display control circuitry, but also a 7-segment LED enable signal is provided that determines to which seven segment display output value is directed. Before displaying, values are converted into a 7-segment code.

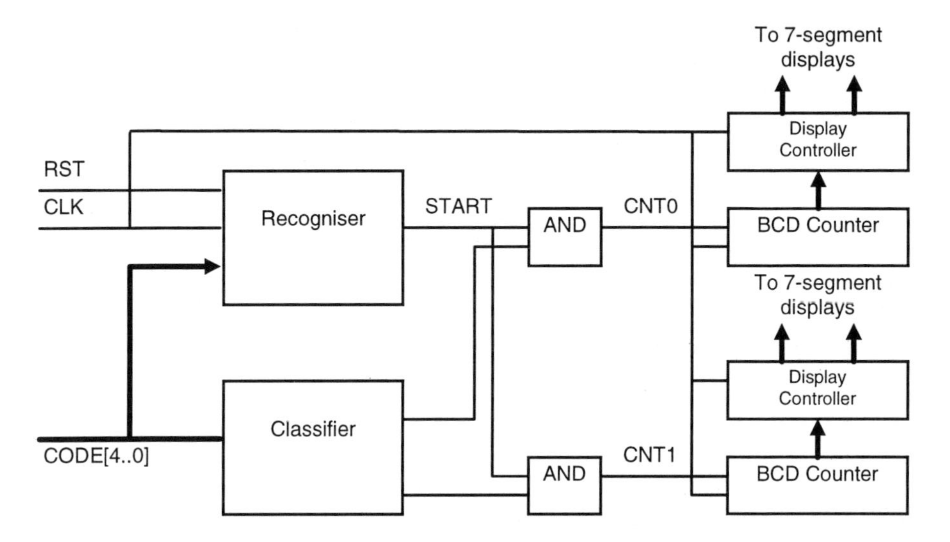


Figure 14.1 Block diagram of sequence classifier and recognizer circuit

From the above diagram we can establish the first hierarchical view of components which will be integrated into overall design. It is presented in Figure 14.2. Further decomposition is not necessary in this case. It is obvious that two instances of sequence recognizer are needed (one for recognition of start and another for recognition of stop sequence), two BCD counters, and two display controllers. Depending on the approach to BCD counter and display controller design, further decomposition is possible, but it will be discussed in the following subsections.

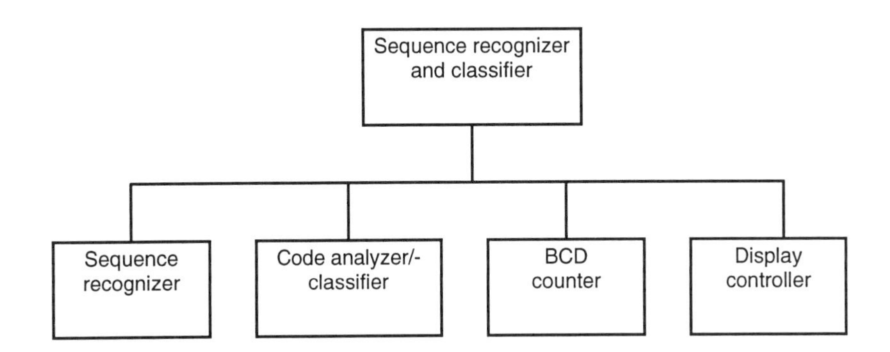


Figure 14.2 Hierarchy of design units in overall design

### *14.1.1 Input Code Classifier*

The input code classifier is a simple circuit that accepts as its input a 7-bit code represented by the std_logic_vector type input variable code. As the result of this classification, one of two output variables is activated:

- more_ones in the case that the number of ones in the input code is greater than the number of zeros, or
- more_zeros in the case that the number of ones is less than the number of zeros in the code.

The VHDL description of this circuit, presented below, consists of two processes. One process, called counting, counts the number of ones in an input code and exports that number in the form of the signal no_of_ones. Another process, called comparing, compares the number of ones with 3 and determines which output signal will be activated.

```
library ieee;
use ieee.std_logic_1164.all;

entity classifier is
        port
```

```
                (
                code: in std_logic_vector(6 downto 0);
                more_ones: out std_logic;
                more_zeros: out std_logic
                );
end classifier;

architecture beh of classifier is
        signal no_of_ones: integer range 0 to 6;

begin
counting: process(code)
        variable n: integer range 0 to 6;
        begin
                n:=0;
        for i in 0 to 6 loop
                if code(i)='1' then
                        n:=n+1;
                end if;
        end loop;
        no_of_ones <= n;
end process counting;

comparing: process(no_of_ones)
        begin
                if no_of_ones > 3 then
                        more_ones <= '1';
                        more_zeros <= '0';
                else
                        more_ones <= '0';
                        more_zeros <= '1';
                end if;
end process comparing;

end beh;
```

The Max+Plus II Compiler is able to synthesize the circuit from this behavioral description.

### *14.1.2 Sequence Recognizer*

Sequence recognizer checks for two specific sequences in order to start and stop the classification and counting of input codes. For simplicity, we assume that those two sequences consist of four characters. In the beginning of its operation, sequence recognizer starts from initial state and waits until start sequence is recognized. When it is recognized, the start output '1' is produced that enables BCD counters. Otherwise, start signal has the value '0'. When the start sequence is recognized, sequence recognizer starts recognizing a stop sequence. After its recognition it will stop counting process by producing output start equal '0'. This operation of recognizing start and stop sequences one after another is repeated indefinitely. The operation of sequence recognizer is presented with state transition diagram in Figure 14.3. The states in which start sequence is being recognized are labeled as S1, S2, S3, and S4, where S1 means that first character is being recognized, S2 that second character is being recognized, etc. The states in which stop sequence are being recognized are labeled as E1, E2, E3, and E4 with meanings analogue to the preceding example. Output signal, start, has value '0' while the state machine is in the process of recognizing start sequence, and value '1' while it is in process of recognizing the stop sequence. While being in the recognition of either start or stop sequence, state machine is returned to the recognition of first character if any incorrect character is recognized.

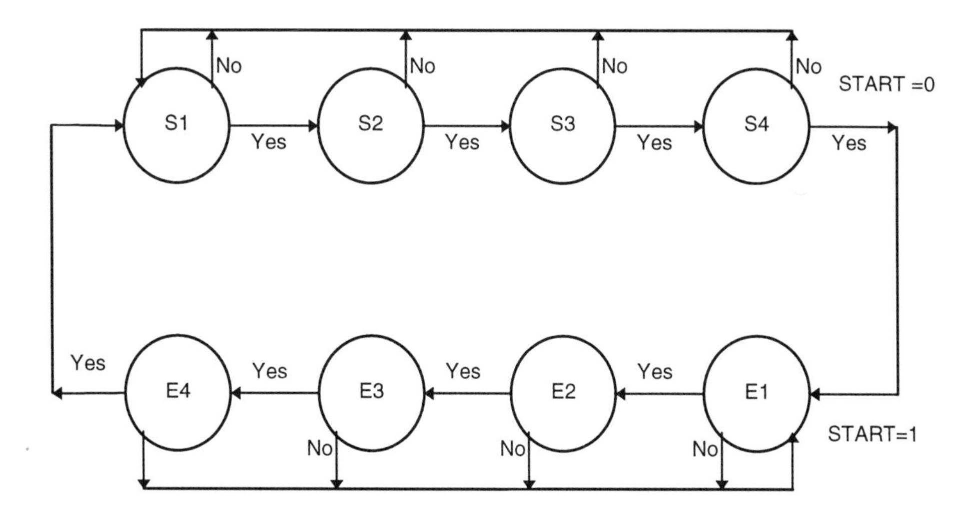


Figure 14.3 Sequence recognizer state transitions

The VHDL description of the sequence recognizer finite state machine (FSM) is derived directly from the state transition diagram in Figure 14.3 and shown below. Characters belonging to the start and stop sequences are declared as the constants in the declaration part of architecture, and can be easily changed.

```
library ieee;
use ieee.std_logic_1164.all;

entity recognizer is
port
(
        inpcode: in std_logic_vector(6 downto 0);
        clk, reset: in std_logic;
        start: out std_logic
);
end recognizer;

architecture start_stop of recognizer is

        type rec_state is (s1, s2, s3, s4, e1, e2, e3, e4);

        signal state: rec_state;

        constant start_seq1: std_logic_vector(6 downto 0):= "0111000";
        constant start_seq2: std_logic_vector(6 downto 0):= "0111001";
        constant start_seq3: std_logic_vector(6 downto 0):= "0111010";
        constant start_seq4: std_logic_vector(6 downto 0):= "0111100";
        constant stop_seq1: std_logic_vector(6 downto 0):= "1111000";
        constant stop_seq2: std_logic_vector(6 downto 0):= "1111001";
        constant stop_seq3: std_logic_vector(6 downto 0):= "1111010";
        constant stop_seq4: std_logic_vector(6 downto 0):= "1111100";

begin

        process(clk)
        begin

        if (clk'event and clk='1') then

                if reset = '1' then
                        state <= s1;
                elseif reset = '0' then
```

```
case state is
    when s1 =>
        if (inpcode = start_seq1) then -- check
            -- for the first character
            state <= s2;
        else
            state <= s1;
        end if;
            start <= '0';


    when s2  =>
        if (inpcode = start_seq2) then
            state <= s3;
        else
            state <= s1;

        end if;
            start <= '0';

    when s3  =>
        if (inpcode = start_seq3) then
            state <= s4;
        else
            state <= s1;

        end if;
            start <= '0';

    when s4  =>
        if (inpcode = start_seq4) then
            state <= e1; -- start sequence
                -- recognized
        else
            state <= s1;

        end if;
            start <= '0';

    when e1  =>
        if (inpcode = stop_seq1) then --check
            -- the first character
```

```
                    state <= e2;
                else
                    state <= e1;

                end if;
                    start <= '1';

            when e2  =>
                if (inpcode = stop_seq2) then
                    state <= e3;
                else
                    state <= e1;

                end if;
                    start <= '1';

            when e3  =>
                if (inpcode = stop_seq3) then
                    state <= e4;
                else
                    state <= e1;

                end if;
                    start <= '1';

            when e4  =>
                if (inpcode = stop_seq4) then
                    state <= s1;  -- stop sequence
                                  -- recognized
                else
                    state <= e1;

                end if;
                    start <= '1';
            end case;
        end if;
    else
            state <=state;
end if;

end process;
end start_stop;
```

The states of the state machine are declared as enumerated type rec_state, and the current state, state, is of that type.

### *14.1.3 BCD Counter*

The BCD counter is a three digit counter that consists of three simple modulo 9 counters connected serially. The individual counters are presented by the processes, bcd0, bcd1, and bcd2, that communicate via internal signals cout and cin which are used to enable the counting process. The ena input is used to enable the counting process of the least significant digit counter, and at the same time is used to enable the entire BCD counter. Each individual counter has to recognize 8 and 9 input changes in order to prepare itself and the next stage for the proper change of state.

```
library ieee;
use ieee.std_logic_1144.all;

entity bcdcounter is
port(
          clk, reset, ena: in std_logic;
          dout0,dout1,dout2: out integer range 0 to 9;
          cout: out std_logic
);
end bcdcounter;

architecture beh of bcdcounter is
          signal cout0,cout1,cin1,cin21, cin20 : std_logic;
begin

bcd0: process(clk)
          variable n: integer range 0 to 9;
begin

          if (clk'event and clk='1') then
                    if reset = '1' then
                              n:=0;
                    else
                              if ena = '1' and n<9 then
                                        if n=8 then
                                                  n := n+1;
                                                  cout0 <= '1';
                                        else
```

```
                                        n:=n+1;
                                        cout0 <= '0';
                              end if;
                    elsif ena = '1' and n=9 then
                              n := 0;
                              cout0 <= '0';
                    end if;
          end if;
     end if;
          dout0 <= n;
end process bcd0;

bcd1: process(clk)
     variable n: integer range 0 to 9;
begin
     cin1 <= cout0;
     if (clk'event and clk='1') then
          if reset = '1' then
               n := 0;
          else
               if cin1 = '1' and n<9 then
                    if n=8 then
                         n := n+1;
                         cout1 <= '1';
                    else
                         n:=n+1;
                         cout1 <= '0';
                    end if;
               elsif cin1 = '1' and n=9 then
                    n := 0;
                    cout1 <= '0';
               end if;
          end if;
     end if;
          dout1 <= n;
end process bcd1;

bcd2: process(clk)
     variable n: integer range 0 to 9;
begin
     cin21 <= cout1;
     cin20 <= cout0;
```

```
        if (clk'event and clk='1') then
                if reset = '1' then
                        n := 0;
                else
                        if cin21 = '1' and cin20 = '1' and n<9 then
                                if n=8 then
                                        n := n+1;
                                        cout <= '1';
                                else
                                        n:=n+1;
                                        cout <= '0';
                                end if;
                        elsif cin21= '1' and cin20 = '1' and n=9 then
                                n := 0;
                                cout <= '0';
                        end if;

                end if;
        end if;
                dout2 <= n;
end process bcd2;

end beh;
```

### *14.1.4 Display Controller*

The display controller receives three binary coded decimal digits on its inputs, passing one digit at time to the output while activating the signal that determines to which 7-segment display the digit will be forwarded. It also performs conversion of binary into 7-segment code. The VHDL description of the display control circuitry is given below. It consists of three processes. The first process, count, implements a modulo 2 counter that selects, in turn, three input digits to be displayed. The counter's output is used to select which digit is passed through the multiplexer, represented by the mux process, and also at the same time selects on which 7-segment display the digit will be displayed. The third process, called converter, performs code conversion from binary to 7-segment code.

```
library ieee;
use ieee.std_logic_1144.all;

entity displcont is
```

```
port (  dig0, dig1, dig2: in integer range 0 to 9;
        clk: in std_logic;
        sevseg: out std_logic_vector(6 downto 0);
        ledsel0, ledsel1, ledsel2: out std_logic
);
end displcont;

architecture displ_beh of displcont is
        signal q, muxsel: integer range 0 to 2;
        signal bcd: integer range 0 to 9;
begin

count: process(clk)
        variable n: integer range 0 to 2;
        begin
                if (clk'event and clk='1') then
                if n < 2 then
                        n:= n+1;
                else
                        n:=0;
                end if;
                if n=0 then
                        ledsel0 <= '1';
                        ledsel1 <= '0';
                        ledsel2 <= '0';
                        q <= n;
                elsif n=1 then
                        ledsel1<='1';
                        ledsel0 <= '0';
                        ledsel2 <= '0';
                        q <=n;
                elsif n=2 then
                        ledsel2 <= '1';
                        ledsel0 <= '0';
                        ledsel1 <= '0';
                        q <=n;
                else
                        ledsel0 <= '0';
                        ledsel1 <= '0';
                        ledsel2 <= '0';
                end if;
        end if;
```

```
end process count;

mux: process(dig0, dig1, dig2, muxsel)
begin
        muxsel <=q;

        case muxsel is
                when 0 =>
                        bcd <= dig0;
                when 1 =>
                        bcd <= dig1;
                when 2 =>
                        bcd <= dig2;
                when others =>
                        bcd <= 0;
        end case;
end process mux;

converter: process(bcd)
begin
        case bcd is
                when 0 => sevseg <= "1111110";
                when 1 => sevseg <= "1100000";
                when 2 => sevseg <= "1011011";
                when 3 => sevseg <= "1110011";
                when 4 => sevseg <= "1100101";
                when 5 => sevseg <= "0110111";
                when 6 => sevseg <= "0111111";
                when 7 => sevseg <= "1100010";
                when 8 => sevseg <= "1111111";
                when 9 => sevseg <= "1110111";
                when others => sevseg <= "1111110";
        end case;
end process converter;

end displ_beh;
```

### *14.1.5 Circuit Integration*

The sequence classifier and recognizer is integrated using already specified components and structural modeling. All components are declared in the architecture declaration part, and then instantiated a required number of times. The interconnections of components are achieved using internal signals declared in the architecture declaration part of the design. As the result of its operation the overall circuit provides two sets of 7-segment codes directed to 7-segment displays, together with the enable signals which select a 7-segment display to which the resulting code is directed.

```
library ieee;
use ieee.std_logic_1164.all;

entity recognizer_classifier is
        port
        (
        code: in std_logic_vector(3 downto 0);
        clk, rst: in std_logic;
        sevsega, sevsegb: out std_logic_vector(6 downto 0);
        leda0, leda1, leda: out std_logic;
        ledb0, ledb1, ledb2: out std_logic;
        overfl0, overfl1: out std_logic
        );
end recognizer_classifier;

architecture structural of recognizer_classifier is
        signal cnt0, cnt1, start: std_logic;
        signal clas0, clas1: std_logic;
        signal succ: std_logic;
        signal d0out0, d0out1, d0out2 : integer range 0 to 9;
        signal d1out0, d1out1, d1out2 : integer range 0 to 9;

        component recognizer
        port (
                inpcode: in std_logic_vector(6 downto 0);
                clk, reset: in std_logic;
                start: out std_logic
        );
        end component;

        component bcdcounter
        port (
                clk, reset, ena: in std_logic;
                dout0,dout1,dout2: out integer range 0 to 9;
                cout: out std_logic
```

```
        );
        end component;

        component displcont
        port (
                dig0, dig1, dig2: in integer range 0 to 9;
                clk: in std_logic;
                sevseg: out std_logic_vector(6 downto 0);
                ledsel0, ledsel1, ledsel2: out std_logic
        );
        end component;

    component classifier
        port (
                code: in std_logic_vector(6 downto 0);
                more_ones: out std_logic;
                more_zeros: out std_logic
        );
        end component;

    begin
        cnt0 <= clas0 and start;
        cnt1 <= clas1 and start;

        recogn: recognizer port map (code, clk, rst, start);
        classif: classifier port map (code, clas1, clas0);
        bcdcnt0: bcdcounter port map (clk, rst, cnt0, d0out0, d0out1,
                d0out2, overfl0);
        bcdcnt1: bcdcounter port map (clk, rst, cnt1, d1out0, d1out1,
                d1out2, overfl1);
        disp0: displcont port map (d0out0, d0out1, d0out2, clk,
                sevsega, leda0, leda1, leda2);
        disp1: displcont port map (d1out0, d1out1, d1out2, clk,
                sevsegb, ledb0, ledb1, ledb2);

    end structural;
```

A modified design of the sequence recognizer and classifier uses only one binary to 7-segment converter as it is illustrated in Figure 14.4. The digits from BCD counters are brought to a common multiplexer from which only one is selected to display, and the corresponding LED selection signal is activated. This design requires less FPLD resources.

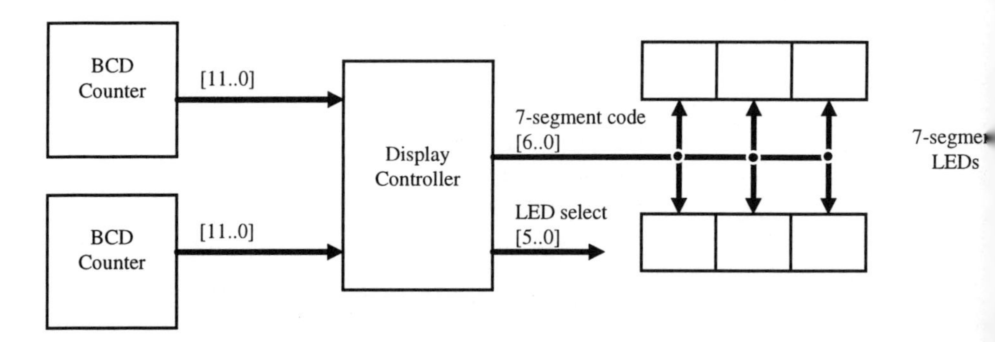


Figure 14.4 The modified display controller

The VHDL description of the modified display controller is given below:

```
library ieee;
use ieee.std_logic_1144.all;

entity displcon1 is
port
(
        adig0, adig1, adig2: in integer range 0 to 9;
        bdig0, bdig1, bdig2: in integer range 0 to 9;
        clk: in std_logic;
        sevseg: out std_logic_vector(6 downto 0);
        aledsel0, aledsel1, aledsel2            : out std_logic;
        bledsel0, bledsel1, bledsel2: out std_logic
);
end displcon1;

architecture displ_beh of displcon1 is
        signal q, muxsel: integer range 0 to 5;
        signal bcd: integer range 0 to 9;

begin

count: process(clk)
        variable n: integer range 0 to 5;
begin
        if (clk'event and clk='1') then
```

```
                        if n < 5 then
                                n:= n+1;
                        else
                                n:=0;
                        end if;
                        if n=0 then
                                aledsel0 <= '1';
                                aledsel1 <= '0';
                                aledsel2 <= '0';
                                q <= n;
                        elsif n=1 then
                                aledsel0 <= '0';
                                aledsel1 <= '1';
                                aledsel2 <= '0';
                                q <=n;
                        elsif n=2 then
                                aledsel0 <= '0';
                                aledsel1 <= '0';
                                aledsel2 <= '0';
                                q <=n;
                        elsif n=3 then
                                bledsel0 <= '1';
                                bledsel1 <= '0';
                                bledsel2 <= '0';
                                q <=n;
                        elsif n=4 then
                                bledsel0 <= '0';
                                bledsel1 <= '1';
                                bledsel2 <= '0';
                                q <=n;
                        else
                                bledsel0 <= '0';
                                bledsel1 <= '0';
                                bledsel2 <= '1';
                        end if;
                end if;
        end process count;

        mux: process(adig0, adig1, adig2, bdig0, bdig1, bdig2, muxsel)
        begin
                muxsel <=q;
                case muxsel is
```

```
                when 0 =>
                        bcd <= adig0;
                when 1 =>
                        bcd <= adig1;
                when 2 =>
                        bcd <= adig2;
                when 3 =>
                        bcd <= bdig0;
                when 4 =>
                        bcd <= bdig1;
                when 5 =>
                        bcd <= bdig2;
                end case;
        end process mux;

        converter: process(bcd)
        begin
                case bcd is
                        when 0 => sevseg <= "1111110";
                        when 1 => sevseg <= "1100000";
                        when 2 => sevseg <= "1011011";
                        when 3 => sevseg <= "1110011";
                        when 4 => sevseg <= "1100101";
                        when 5 => sevseg <= "0110111";
                        when 6 => sevseg <= "0111111";
                        when 7 => sevseg <= "1100010";
                        when 8 => sevseg <= "1111111";
                        when 9 => sevseg <= "1110111";
                        when others => sevseg <= "1111110";
                end case;
        end process converter;

        end displ_beh;
```

## 14.2 Traffic Light Controller

The example demonstrates the use of FPLDs in implementation of a simple yet realistic traffic light controller that controls lights on an intersection of the main road and the side road. We will make the following assumptions:

- Main road traffic is given priority and the traffic light controller enables it as long as there is no side road traffic or no pedestrian requests to cross the roads. The main road state is considered a default state of the traffic light controller.
- Two pairs of sensors on each side of the side road detect the presence of the car near the intersection on the side road and initiate the change of the state of the traffic light controller to let traffic on the side road proceed. One pair of front sensors is used to detect the presence of a car close to the intersection, and another pair of rear sensors detect the presence of a long queue on the side road and allocate an additional time interval for side road traffic.
- If there is a pedestrian input (any of pedestrian buttons pressed) the pedestrian state is entered. In this state all traffic is stopped to enable pedestrians to perform crossings in any desired direction (including diagonally).

For simplicity reasons we will assume that the front sensors on each side of the side road are OR-ed to produce an active signal whenever a car appears on a near end of the side road, and the rear sensors on each side of the side road are also OR-ed to indicate the presence of a long queue of cars on any side of the side road. Similarly, all pedestrian buttons on all corners of the intersection are OR-ed to indicate the presence of a pedestrian request to cross the intersection. These sensor inputs are the only application-specific inputs to the traffic light controller. The outputs are represented by the set of lines that control (logically) turning on and off the traffic lights. Three lines are used to control main road lights, three lines to control side road lights, and two lines to control pedestrian lights.

### *14.2.1 Traffic Light Controller Structure*

The design of the traffic light controller follows basic requirements described in the introduction to this section. It consists of a number of functional parts that manipulate specific information, as is illustrated in Figure 14.5.

The Side Road Subsystem receives the information from two pairs of sensors and provides appropriate information to the main controller FSM. It indicates the presence of a car at a front sensor by activating the front internal signal, or the presence of a long queue by activating the rear internal signal. These signals are deactivated by the action of the main controller FSM by activating clrfront and clrrear signals once the information on the presence of the cars on the side road is processed. Similarly, the Pedestrian Subsystem activates the ped signal if there are pending pedestrian requests, and deactivates that signal by activating clrped when the information on the presence of pedestrians is processed.

The Lights Subsystem is used to decode the current state of the Main Controller FSM and produce appropriate light patterns for all lights: mainlights, sidelights, and pedlights.

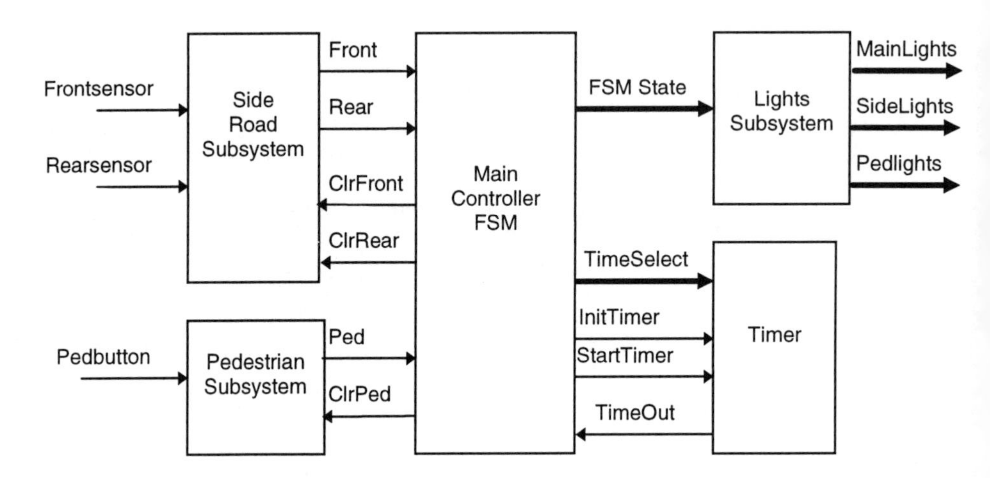


Figure 14.5 Traffic light controller structure

The Timer Subsystem is one of the key components of the traffic light controller. It generates timeout signals after a required number of system clock cycles. The required time-out interval is set by the main controller FSM by selecting an appropriate constant using timeselect lines, and initializing the timer to that constant by inittimer signal. Once the initial value is set, the timer is started (by activating starttimer signal). It counts down to zero, which indicates that the time-out interval has expired. When it reaches zero, the timeout signal is generated informing the Main Controller FSM that the counting process has to be stopped. The Main Controller FSM deactivates the timer by deactivating the starttimer signal. In our case, four different time-out intervals are used to implement various timing situations at the intersection.

### *14.2.2 Main Controller FSM*

Finally, the Main Controller FSM represents the heart of the overall traffic light controller. It provides proper transitions between states in which controller can be, as well as coordination of the operation with other subsystems. The operation of this FSM is illustrated in Figure 14.6 with the state transition diagram. The main FSM passes three main states, maintraffic, pedestrian, and sidetraffic, which are further

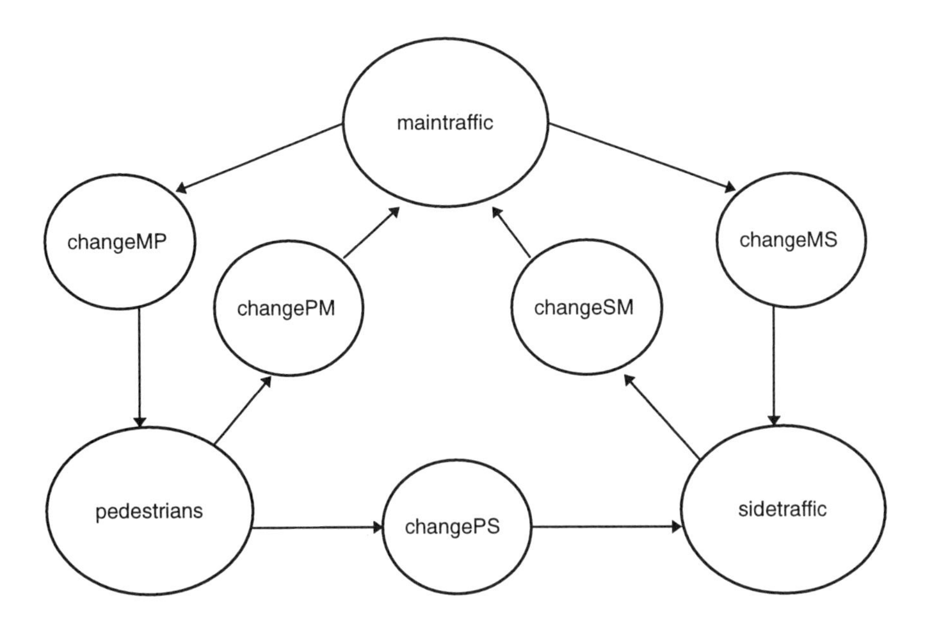


Figure 14.6 Global state transition diagram

decomposed to a number of substates to encounter some specific requirements. A default system state is maintraffic. The controller transfers to either pedestrians or sidetraffic state depending on the conditions provided by Pedestrian and Side Road subsystems, respectively. When in pedestrian state, the controller allows transition to sidetraffic state if there are cars on the side road; otherwise it returns to the maintraffic state. The Main Controller FSM contains a number of intermediate states (change states) that are used to provide safe transition between the main states. The intermediate states are not used for transitions between the sub-states of the main states, because they already provide safe output conditions. The sub-states of the main states together with corresponding transitions are presented in Figure 14.7. The Main Controller FSM spends a limited amount of time in each state, including the sub-states of the main states. The amount of time is specified in the current state for the next state and represents the next time-out interval that will be generated by the timer. The only state which is an exception of this rule is mainM state in which the state

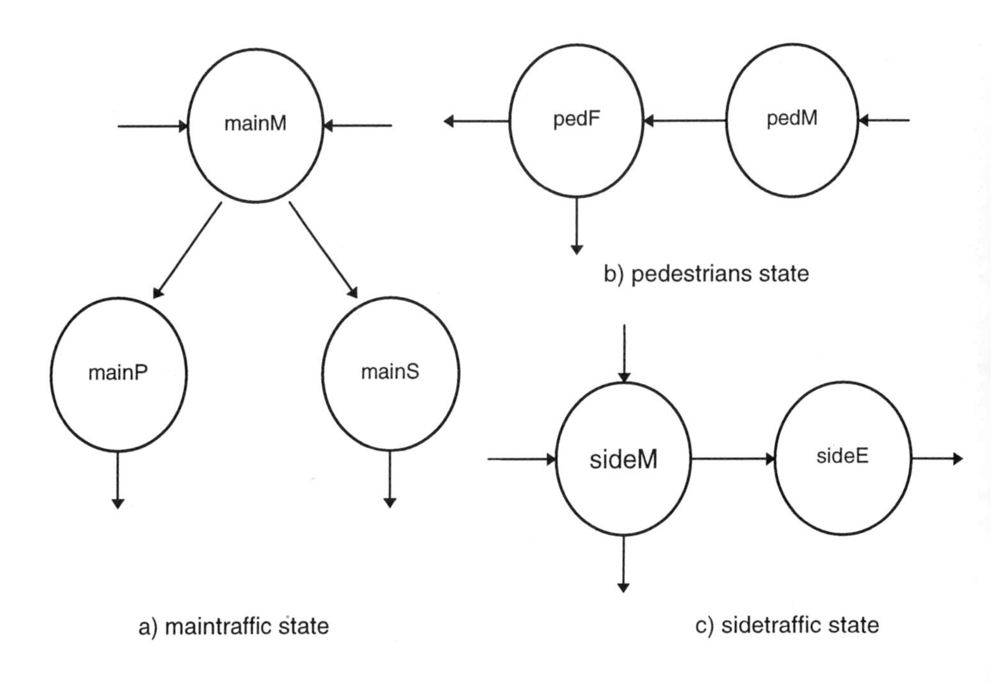


Figure 14.7 Decomposition of main states

machine stays until an event occurs; this event is either a pedestrian request to go across, or the presence of a car at the front sensor of the side road. If these events occur, the FSM will transfer to a sub-state of the maintraffic substate mainP or mainS, which provide the same output conditions as the mainM sub-state, but have predefined termination time. From these states, the main FSM is transferred towards either pedestrians or sidetraffic states driven by the timing conditions. In the case of pedestrian state, the main FSM first enters pedM sub-state and then transfers to pedF sub-state in which it provides transition to other states. The pedF state is usually used to warn pedestrians that the crossing time is expiring (for example, by a flashing lihgt). If the FSM enters sidetraffic state it always first enters sideM sub-state to pass traffic along the side road. If it discovers that there are the cars in the long queue, the sidetraffic state will be extended for an additional time interval (represented by sideE sub-state).

### *14.2.3 VHDL Description and Implementation*

The VHDL description of the traffic light controller is written in the form of single entity, without using hierarchical decomposition. The whole design is presented by a single architecture decomposed into five concurrent and cooperating processes. Processes correspond to the functional units from Figure 14.5. The complete VHDL description is given below. Some comments are given within the code to provide additional explanation.

```
library ieee;
use ieee.std_logic_1164.all;

package my_constants is
          constant t0: integer:= 200; -- timeout ntervals
          constant t1: integer:=100;
          constant t2: integer:=150;
          constant t3: integer:=30;
          constant maxcount: integer:=200
end package;

use work.my_constants.all;

entity traffic is
port
(clk, reset: in std_logic;
frontsensor: in std_logic; -- side road front sensor - or-ed from individual front
sensors
rearsensor: in std_logic; -- side road rear sensors - or-ed from individual rear
sensors
pedbutton: in std_logic; -- or-ed from individual ped buttons
mainlights: out std_logic_vector(2 downto 0); --  red, yellow and green
sidelights: out std_logic_vector(2 downto 0);
pedlights: out std_logic_vector(1 downto 0) --   red and green
);
end traffic;

architecture beh of traffic is

type tlcstates is (main_M, main_P, main_S, ped_M, ped_F, side_M, side_E,
           change_MP, change_PM, change_PS, change_SM, change_MS); -- main
                                                                   -- fsm states
```

```
signal ped, clrped: boolean; -- signals to control ped fsm
signal front, rear, clrfront, clrrear: boolean; -- signals to control side road fsm

signal inittimer, starttimer, timeout: boolean; -- timer control signals
signal timeselect: integer range 0 to 3; -- selection of time constant

signal tlcstate: tlcstates; -- main fsm state variables

begin -- architecture body

-- architecture contains five processes
pedestr: process(clk,reset)
        variable np: boolean; -- indicates pending pedestrians request
begin
if reset ='1' then
        np:=false;
elsif (clk'event and clk = '1') then
        if pedbutton = '1' then
                np:=true;
        elsif clrped then
                np:=false;
        end if;
end if;
        ped<=np; -- indication on waiting pedestrians
end process pedestr;

sidesensors: process(clk,reset)
        variable nf, nr:boolean; -- indicate pending cars on front or rear sensors
begin
if reset = '1' then
        nf:=false;
        nr:=false;
elsif clk'event and clk = '1' then

        if frontsensor = '1' then
                nf:=true;
        elsif clrfront then
                nf:=false;
        end if;

        if rearsensor = '1' then
                nr:=true;
```

```
        elsif clrrear then
                nr:=false;
        end if;

        front <= nf; -- indicates presence of a car at front sensor
        rear <= nr; -- indicates presence of a car at rear sensor

end if;
end process sidesensors;

timer: process(clk)
        variable cnt: integer range 0 to maxcount;
begin
if reset = '1' then
                cnt:=1;
elsif (clk'event and clk ='1') then
                if inittimer then
                        case timeselect is -- timeouts
                        when 0 => cnt:= t0; -- until exiting main state – times
                                -- to wait in states
                        when 1 => cnt:= t1; -- in pedestrian state
                        when 2 => cnt:= t2; -- in side road state
                        when 3 => cnt:= t3; -- change of state
                        end case;
                elsif starttimer then
                        cnt:= cnt-1;
                else
                        cnt:=cnt;
                end if;
                if cnt = 0 then
                        timeout <= true;
                else
                        timeout <= false;
                end if;

end if;

end process timer;

mainfsm: process(clk)
begin
if reset ='1' then
```

```
        tlcstate <= main_M;

elsif (clk'event and clk = '1') then

        case tlcstate is
        when main_M =>
                if ped then -- pedestrians have priority over side road traffic
                        tlcstate <= main_P;
                        inittimer <= true;  -- initialise timer
                        timeselect <= 0;
                elsif front then
                        tlcstate <= main_S;
                        inittimer <= true;
                        timeselect <= 0;
                else
                        inittimer <=false;
                        tlcstate <= tlcstate;
                end if;

        when main_P =>
                if timeout then
                        clrped <=true;
                        starttimer <= false;
                        inittimer <=true;
                        timeselect <= 3;
                        tlcstate <= change_MP;
                else
                        clrped <=false;
                        inittimer <=false;
                        starttimer <=true;
                        tlcstate <= tlcstate;
                end if;

        when main_S =>
                if timeout then
                        clrfront <=true;
                        starttimer <=false;
                        inittimer <=true;
                        timeselect <= 3;
                        tlcstate <= change_MS;
                else
                        clrfront <=false;
                        inittimer<=false;
```

```
                    starttimer <= true;
                    tlcstate <= tlcstate;
            end if;

        when ped_M =>
            if timeout then
                    starttimer <= false;
                    inittimer <=true;
                    timeselect <= 1;
                    tlcstate <= ped_F;
            else
                    inittimer <=false;
                    starttimer <= true;
                    tlcstate <= tlcstate;
            end if;

        when ped_F =>
            if timeout and not front then
                    starttimer <= false;
                    inittimer <=true;
                    timeselect <= 3;
                    tlcstate <= change_PM;
            elsif timeout and front then
                    starttimer <= false;
                    inittimer <=true;
                    timeselect <= 3;
                    tlcstate <= change_PS;
                    clrfront <=true;
            else
                    inittimer<=false;
                    starttimer <= true;
                    tlcstate <= tlcstate;
            end if;

        when side_M =>
            if timeout  and rear then
                    starttimer <= false;
                    inittimer <=true;
                    timeselect <= 2;
                    clrrear<=false;
                    tlcstate <= side_E;
            elsif timeout and not rear then
                    starttimer <= false;
```

```
                    inittimer <=true;
                    timeselect <= 3;
                    tlcstate <= change_SM;

          else
                    inittimer <=false;
                    starttimer <= true;
                    tlcstate <= tlcstate;
          end if;

when side_E =>
          if timeout then
                    starttimer <= false;
                    inittimer <=true;
                    timeselect <= 3;
                    clrrear <= true;
                    tlcstate <= change_SM;
          else
                    inittimer <=false;
                    clrrear <=false;
                    starttimer <= true;
                    tlcstate <= tlcstate;
          end if;

when change_MP =>
          if timeout then
                    starttimer <= false;
                    inittimer <=true;
                    timeselect <= 1;
                    tlcstate <= ped_M;
          else
                    clrped <=false;
                    inittimer <= false;
                    starttimer <= true;
                    tlcstate <= tlcstate;
          end if;

when change_PM =>
          if timeout then
                    starttimer <= false;
                    inittimer <=false;
                    tlcstate <= main_M;
          else
```

```
                    clrped<=false;
                    inittimer <=false;
                    starttimer <= true;
                    tlcstate <= tlcstate;
            end if;

    when change_PS =>
            if timeout then
                    starttimer <= false;
                    inittimer <=true;
                    timeselect <= 1;
                    tlcstate <= side_M;
                    clrfront<=false;
            else
                    clrfront <= false;
                    inittimer <=false;
                    starttimer <= true;
                    tlcstate <= tlcstate;
            end if;

    when change_SM =>
            if timeout then
                    starttimer <= false;
                    inittimer <= true;
                    timeselect <= 3;
                    tlcstate <= main_M;
            else
                    clrrear<=false;
                    clrfront <=false;
                    inittimer <= false;
                    starttimer <= true;
                    tlcstate <= tlcstate;
            end if;

    when change_MS =>
            if timeout then
                    starttimer <= false;
                    inittimer <=true;
                    timeselect <= 1;
                    tlcstate <= side_M;
            else
                    clrfront <=false;
                    inittimer <= false;
```

```
                    starttimer <= true;
                    tlcstate <= tlcstate;
                end if;
        when others =>
                tlcstate <= main_M;

        end case;
end if;
end process mainfsm;

lights: process(tlcstate)
begin
        case tlcstate is

        when main_M =>
                mainlights <= "001"; -- light pattern off, off, green
                sidelights <= "100";        -- light pattern red, off, off
                pedlights <= "10"; -- light pattern red, off

        when main_P =>
                mainlights <= "001";
                sidelights <= "100";
                pedlights <= "10";

        when main_S =>
                mainlights <= "001";
                sidelights <= "100";
                pedlights <= "10";

        when ped_M =>
                mainlights <= "100";
                sidelights <= "100";
                pedlights <= "01";

        when ped_F =>
                mainlights <= "100";
                sidelights <= "100";
                pedlights <= "01"; -- flashing

        when side_M =>
                mainlights <= "100";
                sidelights <= "001";
                pedlights <= "10";
```

```
        when side_E =>
                mainlights <= "100";
                sidelights <= "001";
                pedlights <= "10";

        when change_MP =>
                mainlights <= "010";
                sidelights <= "100";
                pedlights <= "10";

        when change_PM =>
                mainlights <= "100";
                sidelights <= "100";
                pedlights <= "10";

        when change_PS =>
                mainlights <= "100";
                sidelights <= "100";
                pedlights <= "10";

        when change_SM =>
                mainlights <= "100";
                sidelights <= "010";
                pedlights <= "10";

        when change_MS =>
                mainlights <= "010";
                sidelights <= "100";
                pedlights <= "10";

        when others =>
                mainlights <= "001";
                sidelights <= "100";
                pedlights <= "10";


        end case;
end process lights;

end beh;
```

It is interesting to see implementation details of the traffic light controller. The whole design as described by the above description fits into a single MAX7064 device. Actual device utilization depends on the selected clock frequency, which in turn determines number of flip-flops needed for the timer. For a maxcont constant equal 20, the implementation requires only 38 logic cells which represent 59% of capacity of the device. Even much more complex functions of the traffic light controller can be placed into the simplest device.

## 14.3 Digital Frequency Generator/Modulator

The purpose of this example is to demonstrate the use of VHDL and FPLDs in the design of a circuit suitable for the generation of a sine waveform of required frequency. The circuit from this example provides direct digital synthesis of a sinusoidal signal. Circuits of this type are often used in the implementation of digital modulators. As an example, it can be used in Frequency Shift Keying (FSK) modulators/demodulators. The principle of this modulation scheme is that the message signal is constructed by alternating a time signal between two frequencies $f_M$ and $f_S$ according to the two binary values 0 and 1.

A direct digital frequency synthesis technique (DDFS) combined with a digital-to-analog converter (DAC) provides an approach to implementation of a frequency synthesizer. Samples of a sine wave are generated directly from the sine look-up table which is addressed by the output of a digital phase accumulator as it is illustrated in Figure 14.8. We are interested only in the digital part of this circuit which ends with the output of sine the look-up table.

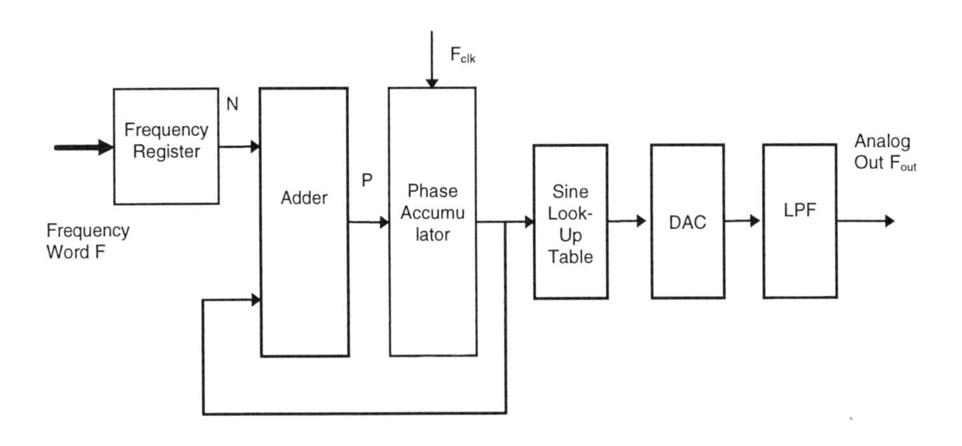


Figure 14.8 Direct Digital Frequency Synthesis

The frequency resolution ($F_{clk}/2^N$) of this synthesizer is determined by its word length (N) and the output frequency is given by

$$F_{out} = (F_{clk}/2^N) \times F$$

where N is the number of bits in the phase accumulator, $F_{clk}$ is the frequency of the system clock, and F is the value of the frequency control register. As an example we can take the system clock frequency of 9600Hz and 4-bit phase accumulator. Table 14.1 shows examples of frequencies of output signals which can be generated by varying frequency control register values.

Table 14.1 Generating output sine signal of frequency $F_{out}$

| F | $F_{out}$(Hz) |
|---|---|
| 2 | 1200 |
| 3 | 1800 |
| 4 | 2400 |

The DDFS approach has several key advantages over conventional analog PLL-based synthesizers, the most important of which is being able to achieve rapid frequency changes with continuous phase. The spectral quality of a DDFS system is related to a number of factors including phase truncation, amplitude quantization, DAC linearity, and the phase noise of the clock source.

The continuous phase FSK signal which codes binary '0' and '1' with two different frequencies of sinusoidal waveform can be achieved using a sequence of discrete sine values stored in a look-up table. The address of each entry to the look-up table is controlled by an address generator. In this case it can be a simple accumulator which forms the address of next value of sine function in the look-up table by adding a constant stored in frequency control register. By changing the value of the frequency control register we effectively change the speed of reading values from the look-up table, and generate output signal with different frequencies. For example, if we code binary '0' with frequency 1800Hz and binary '1' with frequency 1200Hz, we can produce the output frequency corresponding to the input binary signal value by simply using that binary data signal as the select input to a multiplexer which has as its inputs two frequency selection constants. In this way, practically instantaneous change of frequency is achieved by the change of input data signal. This is illustrated by the block diagram in Figure 14.9. Address of the next entry in the look-up table is generated by an accumulator by simply adding frequency constant to previously accumulated sum. The output of the look-up table, which can be implemented in separate memory or in FPLD, has desired number of bits, depending on the accuracy of the signal to be generated. If we assume that we use just 16 samples to represent sine function, then N=4, and the accumulator is a 4-bit accumulator. The look-up table has 16 entries, and the output values are represented by 2's complement 8-bit numbers.

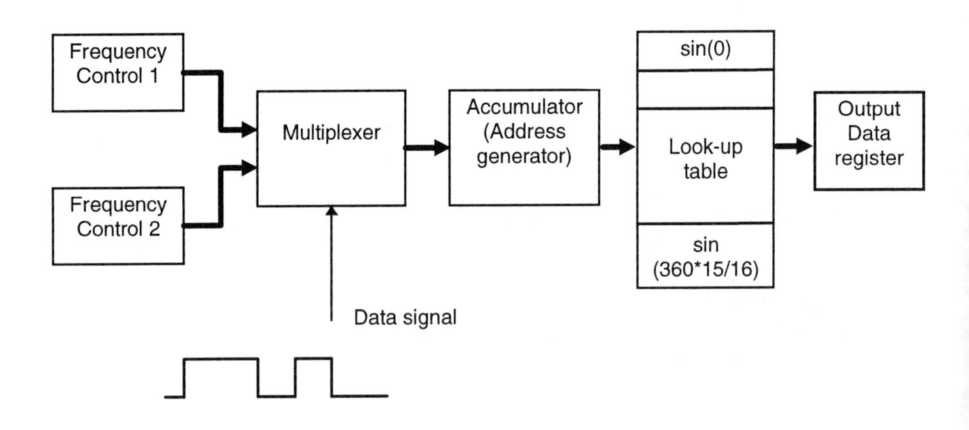


Figure 14.9 Modulating input binary data signal with two frequencies

The VHDL design that describes a simple FSK modulator with DDFS approach is given below. The architecture consists of three processes: accumulator which generates the next address in the look-up table, lookuptable process which implements the table, and output register which captures generated sine value and holds it until the next value is generated.

```
library ieee;
use ieee.std_logic_1164.all;

entity modulator is
        port(clk, reset: in std_logic;
                data: in std_logic;
                dout: out std_logic_vector(7 downto 0));
end modulator;

architecture beh of modulator is
        constant clow: integer:=2; -- frequency constant for 1800Hz
        constant chigh: integer:=3; -- frequency constant for 1200Hz
        signal address: integer range 0 to 15; -- generated address to
                -- look-up table (LUT)
        signal temp: std_logic_vector(7 downto 0); -- output from LUT
begin

accumulator: process(clk)
        variable n: integer range 0 to 15;
begin
        if reset='1' then
                n:=0;

        elsif(clk'event and clk='1') then
                case(data) is  -- implements a multiplexer
                        when '0' =>
                                n:=n+clow;
                        when '1' =>
                                n:=n+chigh;
                        when others =>
                                n:=n;
                end case;
        else
                n:=n;
        end if;
                address <= n;
```

```
        end process accumulator;

        lookuptable: process(address)
        begin
                case(address) is

                        when 0    =>  temp <= "00000000"; -- values
                                      -- of sine function
                        when 1    =>  temp <= "00011000";
                        when 2    =>  temp <= "00101101";
                        when 3    =>  temp <= "00111010";
                        when 4    =>  temp <= "00111111";
                        when 5    =>  temp <= "00111010";
                        when 6    =>  temp <= "00101101";
                        when 7    =>  temp <= "00011000";
                        when 8    =>  temp <= "00000000";
                        when 9    =>  temp <= "11101000";
                        when 10    =>  temp <= "11010011";
                        when 11    =>  temp <= "11000110";
                        when 12    =>  temp <= "11000001";
                        when 13    =>  temp <= "11000110";
                        when 14    =>  temp <= "11010011";
                        when 15    =>  temp <= "11101000";
                        when others      =>  temp <= "00000000";
                end case;

        end process lookuptable;

        outreg: process(clk)
                variable m: std_logic_vector(7 downto 0);
        begin
                if reset='1' then
                        m:="00000000";
                elsif (clk'event and clk='1') then
                        m:= temp;
                else
                        m:=m;
                end if;
                        dout<=m;
        end process outreg;

end beh;
```

Another solution is using two phase accumulators for two frequencies which are used to modulate serial digital stream. The phase accumulator is incremented by a phase step which is different for two frequencies. The correct phase step is selected using a multiplexer with the generated bit value as select input. This scheme is illustrated in Figure 14.10.

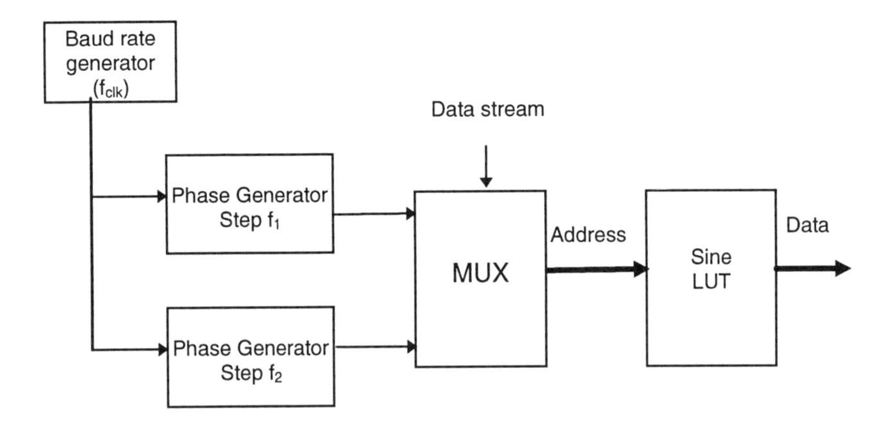


Figure 14.10. FSK generation using two phase accumulators

The VHDL description corresponding to this design is very similar to the preceding one. The major difference is in two parallel phase generators which generate proper accumulated phase (addresses of LUT) of sine functions with both frequencies, and the multiplexer just selects one according to the current value of the data bit.

```
library ieee;
use ieee.std_logic_1164.all;

entity sigg14 is
        port(clk, reset: in std_logic;
                data: in std_logic;
                dout: out std_logic_vector(7 downto 0));
end sigg14;

architecture beh of sigg14 is
        constant clow: integer:=2;
        constant chigh: integer:=14;
        signal out1, out2: integer range 0 to 15;
        signal address: integer range 0 to 15;
        signal temp: std_logic_vector(7 downto 0);
```

```
begin
acc1: process(clk)
        variable n1: integer range 0 to 15;
begin
        if reset='1' then
                n1:=0;
        elsif(clk'event and clk='1') then
                n1:=n1+clow;
        else
                n1:=n1;
        end if;
                out1 <= n1;
end process acc1;

acc2: process(clk)
        variable n2: integer range 0 to 15;
begin
        if reset='1' then
                n2:=0;
        elsif(clk'event and clk='1') then
        n2:=n2+chigh;
        else
                n2:=n2;
        end if;
                out2 <= n2;
        end process acc2;

address <= out1 when data ='0' else
        out2 when data ='1' else
        0;

lookuptable: process(address)
begin
        case(address) is

                when 0   =>  temp <= "00000000";
                when 1   =>  temp <= "00011000";
                when 2   =>  temp <= "00101101";
                when 3   =>  temp <= "00111010";
                when 4   =>  temp <= "00111111";
                when 5   =>  temp <= "00111010";
                when 6   =>  temp <= "00101101";
```

```
                when 7    =>  temp <= "00011000";
                when 8    =>  temp <= "00000000";
                when 9    =>  temp <= "11101000";
                when 10   =>  temp <= "11010011";
                when 11   =>  temp <= "11000110";
                when 12   =>  temp <= "11000001";
                when 13   =>  temp <= "11000110";
                when 14   =>  temp <= "11010011";
                when 15   =>  temp <= "11101000";
                when others    =>  temp <= "00000000";

        end case;
    end process lookuptable;

    outreg: process(clk)
            variable m: std_logic_vector(7 downto 0);
    begin
            if reset='1' then
                    m:="00000000";
            elsif (clk'event and clk='1') then
                    m:= temp;
            else
                    m:=m;
            end if;
                    dout<=m;
    end process outreg;
    end beh;
```

For the addition of the remaining circuitry to implement complete FSK modulator the reader is referred to Problem 14.7 below.

## 14.4 Questions and Problems

14.1. A digital circuit is used to weigh small parcels and classify them into four categories:

- less than 100 grams
- between 100 and 200 grams
- between 200 and 500 grams

- between 500 and 1000 grams

a) The input weight is presented as an 8-bit unsigned binary number, which is output from the weight sensor, with linear dependency between the weight and input binary number in the whole range between 0 and 1023 grams. The weight of the parcel is presented on a four digit seven segment display.

b) Design a circuit that performs the task of classification and weight display and activates a separate output control signal whenever a new parcel is received. The new parcel arrival is indicated by a single bit input signal. The weighing process starts at that point and the weight becomes valid after exactly 100ms.

c) Describe your design using VHDL. Draw a diagram that describes the hierarchy of circuits in the design.

14.2. A temperature control system capable of keeping the temperature inside a group of 8 incubators within any required range between $10^0$ and $35^0$C is to be designed. The required temperature range can be set with a user friendly interface (you are allowed to use either a hexadecimal keypad or any lower number of keys which will provide required functions). The current temperature within an incubator is measured using a temperature sensor. When the current temperature is below the lower limit of the desired range, the incubator must be heated using an AC lamp, and if it is above the upper limit of the desired range, the incubator is cooled using a DC fan. When the current temperature is within the desired range, no control action should have to be undertaken. The current temperature of the selected incubator (with the number displayed at a single digit 7-segment display) is continuously displayed on the 3-digit hexadecimal display to one decimal place significance (for instance, 26.$4^0$C) for a certain time (5-10 sec), and then the temperature in the next incubator is displayed. As an alternative, you are allowed to use an LCD display with capacity of up to 10 digits. Additional LEDs are used to indicate the current state of temperature in each incubator (within the range, below the low limit or above the high limit). The simplified block diagram of the overall temperature control system is shown in Figure 14.11. It is divided into the following subunits:

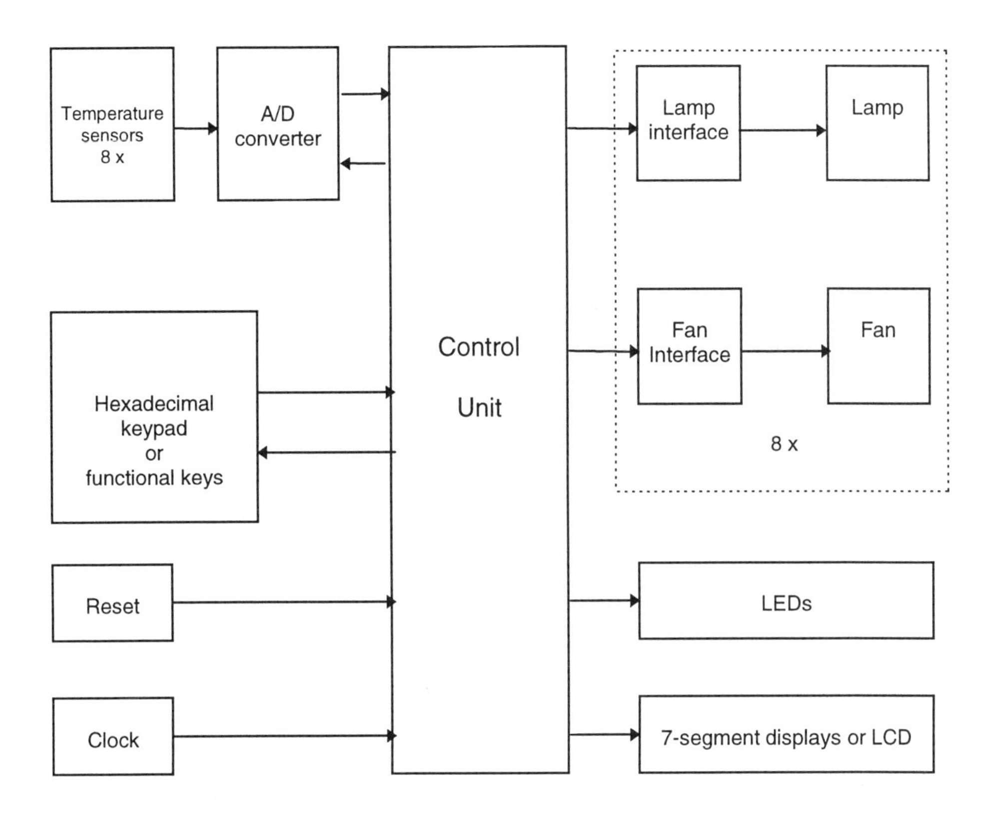


Figure 14.11 Block diagram of the temperature control system

- Temperature sensing circuitry that provides the current temperature in digital form using single 8-channel 8-bit A/D converter

- Keypad circuitry used for setting high and low temperature limits and for operator-controller communication

- Display driving circuitry used to drive both single-digit and 3-digit 7-segment display or a single LCD display

- DC fan control circuitry used to switch on and off the DC fans

- AC lamp control circuitry used to switch on and off AC lamps

- Control unit circuitry which implements the control algorithm and provides synchronization of operations carried out in the controller - since the purpose of the design is not to show any advanced control algorithm, a simple on/off control will be implemented.

Temperature is sensed by sampling it four times in each incubator and averaging it over those four samples.

Consider two cases:

- The temperature sensing, controlling and displaying process are mutually exclusive from the process of communication with a human operator.

- The temperature sensing, controlling and displaying process run simultaneously and independently of the process of communication with a human operator.

The aim is to implement all interface circuits between the analog and digital parts of circuit, including complete control unit, in an Altera CPLD. You are allowed to use a single FLEX10K (up to 10K20) device. Additional circuitry and devices used in your design can be those given in [17], or any other circuits that fulfill design requirements. The A/D converter is assumed to be an 8-channel 8-bit converter with 3 channel select lines. The output from the converter is 8-bit unsigned binary number.

14.3. Consider a more advanced control algorithm for the example from preceding problem. The temperature is maintained in the middle between the upper and lower set. The control action is undertaken only if the temperature changes for a quarter of the range between the higher and the lower set.

14.4. An electronic lock is a circuit that is capable of recognizing a 8 decimal digit input password entered as a sequence of codes from a hexadecimal keypad and indicating that the sequence is recognized using the unlock signal. A sequence of eight 4-bit digits is entered using a hexadecimal keypad. If any of the digits in the sequence is incorrect, the lock resets and indicates that the new sequence should be entered from the beginning. This indication appears at the end of sequence entry in order to increase the number of possible combinations that may be entered, making the task more difficult for a potential intruder.

The electronic lock consists of three major parts, as it is illustrated by the block diagram in Figure 14.12:

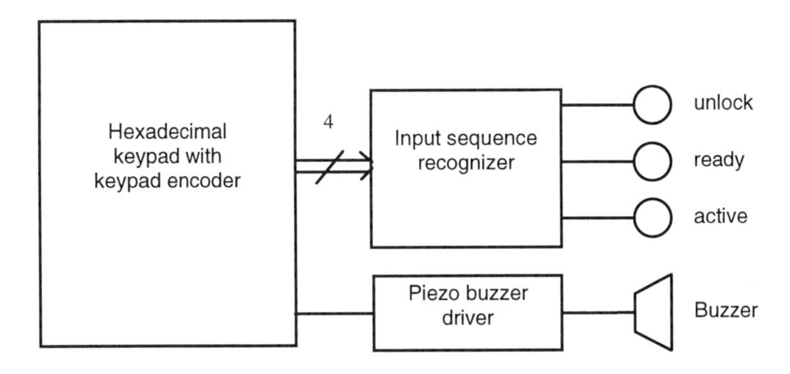


Figure 14.12 The electronic lock block diagram

- Input, represented by a hexadecimal keypad and keypad encoder accepts a key press and produces the corresponding 4-bit binary code.

- Input sequence recognizer, which recognizes the sequence of eight digits entered and produces an unlock signal if the sequence is correct.

- Output, which consists of three LEDs indicating that the lock is ready to accept new sequence, or it is accepting (active) a sequence of five digits, or the correct sequence has been entered (unlock), and a piezzo buzzer output that is activated whenever a key is correctly pressed. When a key is correctly pressed, the buzzer is held high for a short time interval of about ms, corresponding to 5000 clock cycles of a 50kHz clock input.

Non-numeric keys can be used as functional keys with specific functions as assigned by the designer.

Describe the design of the lock as a hierarchy of subdesigns. Use VHDL to describe the whole design.

14.5. Modify the electronic lock from the preceding example by introducing a changeable password that can be entered after entering an 8 decimal digit superpassword. The changeable password is represented by 10 decimal digit sequence. The lock has two modes of operation: a) setting mode in which the password is changed and b) secure mode in which it requires the password to unlock.

14.6. Modify the traffic light controller presented in this Chapter by introducing different priorities for the front and rear sensors on the main and side road. Also, introduce the timers which provide additional green at the main road if the rear sensors indicate long queues.

14.7.Design a complete transmitter that accepts 8-bit parallel data from a microcomputer and transmits it as an FSK modulated signal on its output as it is illustrated in Figure 14.13. The transmitter provides handshaking with the microcomputer in order to synchronize their operation. Express the whole design in VHDL.

Serial binary data stream comes from a buffer in which the data is stored in parallel form (usually byte oriented). Data is shifted serially and a bit counter is used to determine when the buffer is empty. The buffer is loaded with the next (new) data if there are more data to be transmitted, and the process is stopped if there are no more data. Bit counter is incremented at each bit edge as determined by the baud rate generator. The data path implementing required data manipulations is shown in Figure 14.13. Two counters are provided to count the number of remaining data

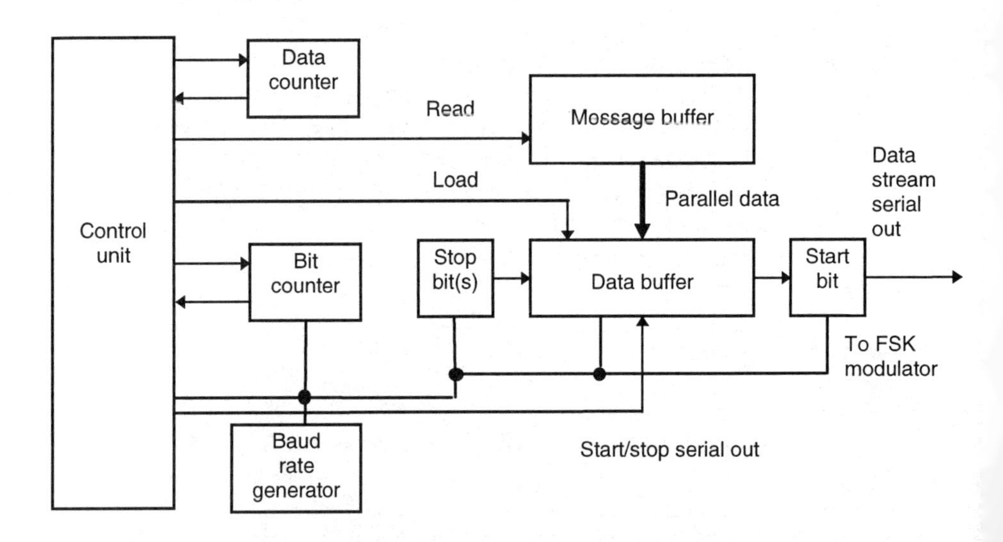


Figure 14.13 Data path supporting data stream generation

bytes in the message to be transferred, and to count the number of remaining bits in the current data byte to be transferred, respectively. Additional storage is provided for start and stop bit concatenation to the data byte. A parallel data path is provided from the message buffer to the data buffer. In this example an EAB implemented message buffer is used. The example is simplified due to the fact that no parity code generation is included.

Coordination and control of all actions in the data path is carried out by the control unit, which represents a part of the overall control unit of the frequency modulator/detector. Control algorithm for this part of FSK modulator is illustrated by the flow control diagram in Figure 14.14. The baud rate generator is used for keeping the track of the phase of the encoded signal and supplying the bit clock. In between the bit boundaries (edges) the modulator behaves as an ordinary "tone generator". At the bit boundary control signals are set according to the bit to be encoded and the encoder phase adjusted depending on the baud rate generator. Starting phase is further adjusted by 0 or 180 degrees to provide phase continuity of the last encoded bit. The design of both data path and control unit is left to be finalized by the reader as an exercise.

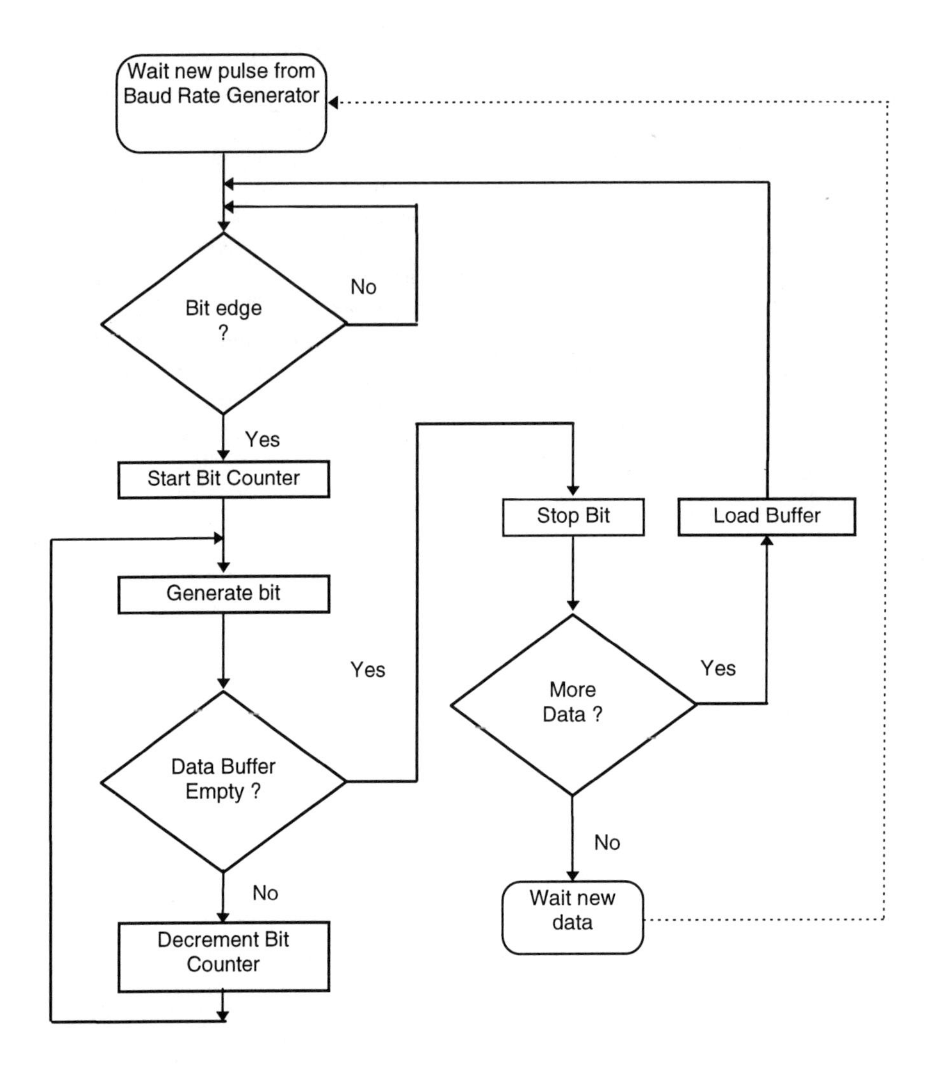


Figure 14.14 Control flow for FSK modulator with phase accumulators

# 15 FLEXSWITCH - A HIGH-SPEED ATM SWITCH

This Chapter presents an Asynchronous Transfer Mode (ATM) Switch, called FlexSwitch, aimed for high-speed applications in ATM-based networks. The initial design supports 8-input/8-output 32-bit port applications, although variations of the number of ports and their widths are possible. The FlexSwitch uses input-queue/output-queue buffering and multi-bit routing to achieve high single-port throughput of 800Mbit/s, or aggregate throughput of up to 6.4 Gbit/s with the system clock of 25MHz, or even higher throughputs at the higher system clock frequency. The implementation is done in two-layer datapath/control layer structure, and is fully described in VHDL. It is implemented with Altera's CPLD FLEX 10K devices, which provide easy reconfigurability and modification of design.

## 15.1 Introduction

The rise in processing power of computers and the increase in exchange of multimedia intensive data such as real time audio and video, and high quality images, enforced the revolution of high-speed communication, which becomes a reality with the maturation of optical communications technology. The urge to transmit different types of data sources through one single network imposed the standardization of B-ISDN/ATM protocol, which invokes the possibility of merging the data communication and telecommunication networks into one single infrastructure in the near future. Generally, two types of communication networks have been used. The traditional telecommunication networks use circuit switching, where voice or data are transmitted across the network through constant length time slots (without a header for each time slot) at a regular time interval called a channel. The system is connection-oriented. The whole channel is occupied once the connection is established, even though free time slots are available within the channel but cannot be used by other channels. On the other hand, the data communication networks use the packet switching where data is transmitted through variable length time slots (with a header attached to each time slot) with a first-come-first-serve contention. The data network is connectionless and is free for contention whenever bandwidth is available. This makes the data network incapable of transmitting real-time audio or video, because there is no guarantee that transmitted data will be received within the required time interval.

ATM is a cell-based switching and multiplexing technology designed to be a general-purpose, connection-oriented transfer mode for a wide range of services. The primary unit in ATM is an ATM *cell.* The ATM standards define a 53 bytes fixed-size cell which consists of a 5-byte header for addressing and controlling information and a 48-byte payload for the carriage of higher level protocols and data transmission. ATM introduces the new concept of packet switching by using the constant length time slot with a header attached during transmission. This modification offers a solution for networks where connection-oriented traffic can be received within a limited time frame (real-time services) whereas the connectionless traffic is taking the benefits of slotting data into free time slots and received by the related terminals through the discrimination of the stamped headers. Hence, ATM handles both connection-oriented traffic and connectionless traffic through a single network which encapsulates different kinds of data sources.

ATM also provides services to the Constant Bit Rate (CBR) traffic which requires a fixed bandwidth during connection, and Variable Bit Rate (VBR) traffic whose bandwidth depens on the client's demand or whose bandwidth can be changed when the network is in heavy demand. This function is particularly important to a share-media network for transmitting different data sources. An increase in the volume of CBR traffic ( for example more people making phone calls), which is normally connection-oriented and time sensitive, requires the bandwidth contribution from VBR traffic (e.g. reduce bandwidth of computer terminals), which normally refers to data terminals that are bandwidth demanding with no time constraint. Hence, ATM provides a better use of available bandwidth.

ATM is particularly well suited to the media-intensive network's needs. It can be used as a backbone for merging the telecommunication and data communication networks together because it has the potential to blur the barriers between telephony, wide-area, local-area, and "desk-area" networks.

The next step in the evolution of media-intensive systems involves allowing multimedia information to reach the application. This means channeling the information to the end users/clients rather than bypassing it with specialized hardware. This will allow a greater variety of multimedia applications that go far beyond teleconferencing, providing services such as on-demand video, home shopping, and digital TV for the home users, and providing distributed services to the office users by connecting the electronic devices such as copy machines, faxes, computers and telephones together to form an "office-area" or "desk-area" network.

The idea behind this is to provide one network with multiple services. One of the key components behind this evolution is the ATM switch, which enables connection of various types of devices with various performances and speed requirements into a single network. This Chapter describes an approach to ATM switch, called *FlexSwitch*. First, we introduce the *FlexSwitch* framework, then describe *FlexSwitch* principles of operation and design, and finally details of its implementation and performance evaluation.

## 15.2 FlexSwitch Framework and Features

Any connection-oriented network requires at least a centralized switch to convey information between end users. This is because each end user has his/her own connection to the network without contention from the others so that data can be received within the limited time. As ATM is a connection-oriented network, which also provides connectionless services, it requires switches to enable routing of the traffic within the network. An ATM switch consists of a set of input ports and output ports which are interconnected through a cell switch fabric. The switch is assumed to provide only a cell relay service, but may be used to interconnect end users, other switches and other network elements into various topologies.

Our approach emerged from an initial idea to only put the most basic functionality in the network hardware infrastructure regarded as the ATM switch. The other objective was to design a simple high speed yet low cost ATM switch that is flexible and suitable to operate in different network environments such as the media-intensive Desk Area Network (DAN), the traffic-intensive LAN or WAN, and the image (and video) intensive medical networks. By selecting and implementing the basic switching functions into the core design, leaving the higher level control and management functions to be conducted by the switch interfaces, a low cost, high speed ATM switch may be possible to achieve. The next goal was to use FPLDs as the main implementation technology. There were two major reasons for this. First, the hardware prototype is easily implemented as the FPLD chips are already available in the market. Second, our decision was to use VHDL to specify switch design. As such, design specification is ready to be transported to another target technology with practically no modifications. An additional advantage of using FPLDs was to use the reconfigurability of FPLD chips for the purpose of modification of design or providing design for different specifications (especially, a variable number of input and output ports and a variable width of the ports) without changing the PCB design.

As illustrated in Figure 15.1, the *FlexSwitch* is an 8 input ports/8 output ports switching unit with each port being capable of transferring data through a standard 32-bit uni-directional data path. Each I/O port consists of a Switch Interface (SI), which is outside of the *FlexSwitch*, and a Port Interface (PI). The SI is optional to *FlexSwitch*, and it is only present when a client is required to transmit information through the network. The SI is sometimes referred to as host interface in other systems, and provides supports for data streaming, packet reassembly, and other specific functions for different end users or clients.

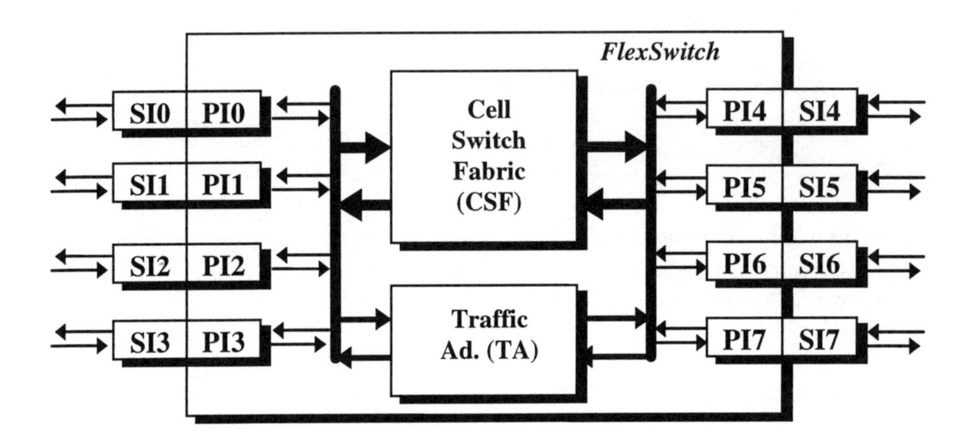


Figure 15.1 The FlexSwitch Design

Serial bit stream from the cable line is received by SI, converted to parallel data stream and stored in a buffer. Packet reassembly is normally done through the buffering on a per-VCI basis, and only interrupts the Cell Switch Fabric (CSF) through input port for transmission when a full packet (an ATM cell) is received. Other network functions, such as bandwidth policing, cell header conversion, error checking etc. are required on demand basis which may be optional to certain network environments.

The PI, which is part of the *FlexSwitch*, accounts for the simple interface topology that has been adopted in the initial *FlexSwitch* proposal. By taking advantage of the on-chip reconfiguration function offered by FPLD device, PI is capable of providing 8-bit, 16-bit, 32-bit, 64-bit or 128-bit data bus to SI, depending on the number of I/O ports that *FlexSwitch* has been configured to.

For an 8 I/O ports configuration, PI can offer 8-bit, 16-bit and 32-bit interface to SI. However, regardless of the data bus interface between SI and PI, the internal switching is done through 32-bit bus for 8 I/O ports, 64-bit bus for 4 I/O ports, and 128-bit for 2 I/O ports.

The core component of *FlexSwitch* system is the Cell Switch Fabric (CSF) which is described in more details in the following sections. The ATM switch should have a Traffic Administrator (TA), as shown in Figure 15.1, which provides bandwidth reservation/release functions in order to prevent bandwidth contention between connection-oriented and connectionless traffics. However, the current version of *FlexSwitch* is concentrating on the standard 8 I/O ports 32-bit bus design. In this Chapter we concentrate on the design without the TA. The bandwidth policing is supposed to be provided by a more sophisticated SI design.

## 15.3 FlexSwitch Operation and Design

A key goal in designing the *FlexSwitch* hardware has been simplicity, with an objective of building network hardware from the off-the-shelf VLSI components. An early decision to use FPLD devices was made to simplify functionality and provide flexibility of the network hardware.

### *15.3.1 Cell Format and Selection of Switching Strategy*

The cell format is based on a modified ATM standard 53-byte cell: 11 bytes are added to the 53-byte ATM cell (7 bytes to the beginning and 4 bytes to the end) in order to make a cell 64 bytes in length, which we call a *Flexcell.* This makes a cell eight byte-aligned permitting a direct interface to 8-, 16-, 32- up to 128-bit bus architectures that are more suitable for computers, data terminals or telecommunications transmission. Such an extended cell also provides provisions for future extensions and additional functions.

Figure 15.2 illustrates format of the *Flexcell.* The 53-byte ATM cell, which will be transferred across the switch without modification, is encapsulated from the 8$^{th}$ byte to the 60$^{th}$ byte of the entire *Flexcell.* The extra 4 bytes are attached to the end of the ATM cell for Cell Parity Check (CPC), which will be explained in a later section. Seven bytes are added to the beginning of the ATM cell, with the first 4 bytes being unused and the next three bytes carrying the routing tag. Each bit of the routing tag represents the destination output port for the transmitting cell. Because there are only 8 I/O ports in the current design, only the 5$^{th}$ byte is being used to represent the routing tag, leaving the others for future expansion.

The CSF performs two basic functions : buffering and routing. Due to the statistical nature of network traffic, buffering in any packet switch network (ATM is a type of packet switching) is unavoidable. There are 3 distinct types of queuing methods, respectively, the input-queue buffering, output-queue buffering and shared-memory buffering.

| | | |
|---|---|---|
| Bytes1 to 4 | Reserved | |
| Bytes5 to 8 | **Routing tag** | Reserved |
| Byte 9 to Byte 60 | **53-byte ATM cell** **(from byte 8 to byte 60)** | |
| Bytes 61 to 64 | **CPC (Cell Parity Check)** | |

Figure 15.2 The cell format of FlexSwitch

The shared-memory buffering strategy is very similar to the design concept adapted in today's conventional packet switches, where the switch memory is shared by all input and output ports. Arriving packets on all N input ports are multiplexed and stored in common memory and then organized into N separate output queues, one for each output port. Thus, the central controller must have the capability to sequentially process N incoming packets from N input ports, and select N outgoing packets for N output ports, in each time slot. The bandwidth of a shared memory switch is fundamentally limited by the access time of the RAM. While the required bandwidth is 2NV (where V is the speed of each I/O port) for a shared-memory switch, the actual bandwidth that can be provided by RAM modules is much lower, due to the slow access time of current technology. This allows the shared-memory switch to be a medium-speed ATM switch and provides a moderate solution.

An output-queued switch is also limited by the memory bandwidth. Output-queue buffering, which refers to buffering the transmitting cell at the related output port after the cell has been routed or switched, requires a buffer with a memory bandwidth equal to the aggregate throughput. This may be solved by implementing N queuing buffers for N input ports at each output port and the memory bandwidth can be reduced to the arrival rate of a single port. However, this presents an inappropriate use of memory, cost ineffective, and physically non-practical design.

The input-queue buffering, in which incoming cells are stored in buffers before switching to the appropriate output ports, requires the least memory bandwidth, and is well suited to meet the requirements of ATM switching. Each I/O port is only required to buffer cells at the arrival rate of a single port, rather than at a multiple of the arrival rate or the aggregate arrival rate. Hence, the *FlexSwitch* adapts the input-queue buffering topology, but also uses output buffers, as it will be shown below.

The routing is carried out by introducing a routing tag (in our design, it is called destination vector) at the beginning of each cell which is used to identify the output port the related cell is destined for. There are several kinds of switching methods available for routing but the multiple bus approach is adapted to maximize the total throughput. The multiple bus approach, which requires a dedicated bus for each I/O port, allows cells to be transmitted to all output ports simultaneously, providing the best solution for cell broadcasting and multicasting. However, port contention must be solved in this case.

The solution for port contention is based on two simple rules : first-come-first-serve and priority-queue discipline if more than one input port is required to transfer data to the same output port in the same time slot. Output ports are always available to the input port which made the request first. If more than one input port is contending for the same output port, service is given to the input port with higher priority. The current version of the *FlexSwitch* uses the simplest design solution, giving higher priority to the input port with the lower index.

### *15.3.2 Switch Organisation*

Figure 15.3 shows the overall *FlexSwitch* organisation. The *FlexSwitch* consists of two operational layers, the data flow layer and the control layer. The data flow layer provides the data paths between the input ports and output ports. The control layer synchronizes *FlexSwitch*, resolves the contention for an output port, and provides appropriate control signals for the data flow layer. A hierarchical view of switch functions is presented in Figure 15.4.

Each input and output port consists of a pair of First-In-First-Out (FIFO) buffers that feed through a Switching Unit called DPSU. Clients connect to the *FlexSwitch* via the ports, through which the reading, writing, and routing of ATM cells are controlled by the individual control unit of each port. During the operation of the switch, a client wishing to transmit a cell first writes the incoming cell to the input port buffer through which the routing tag of that cell will be extracted and passed

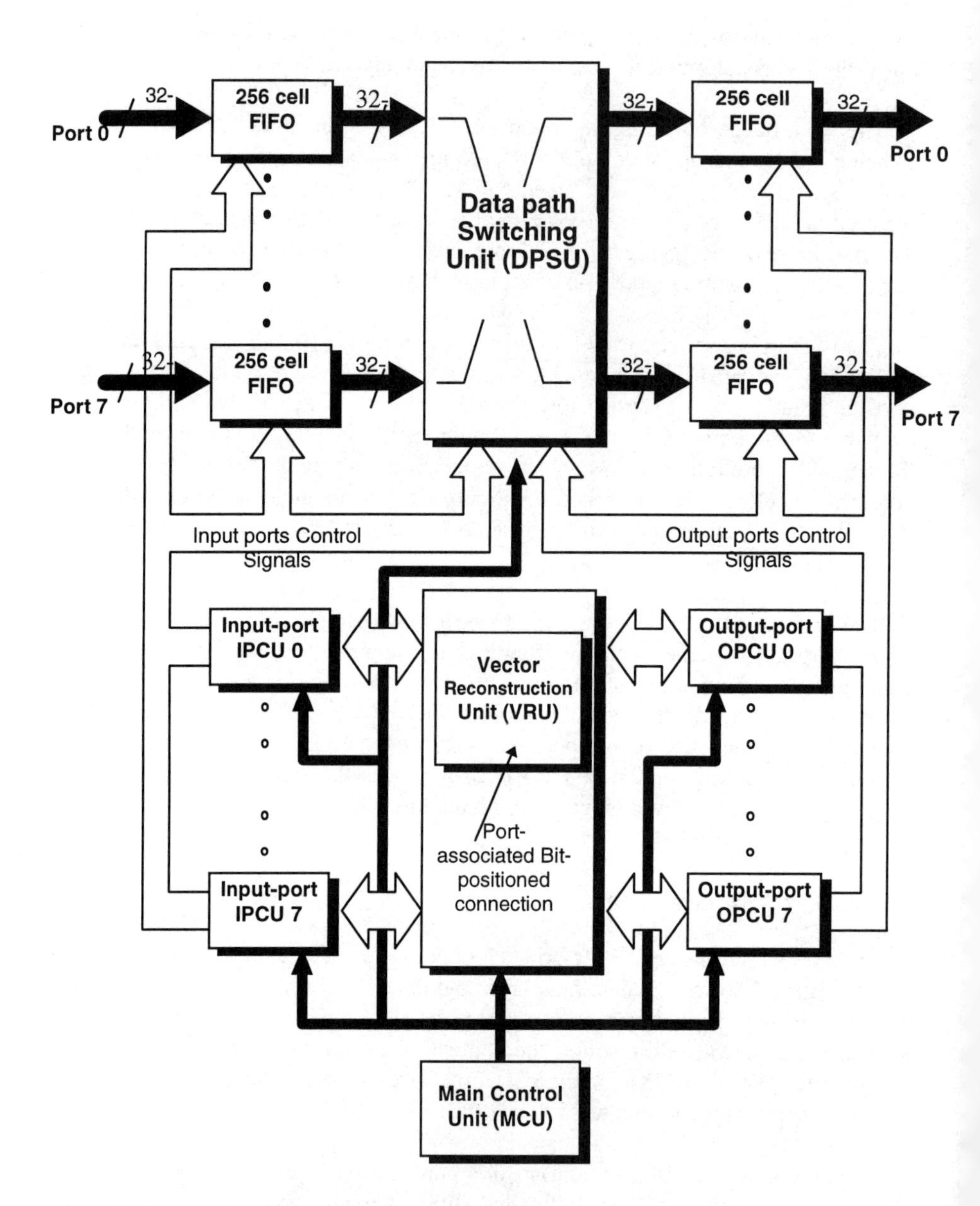


Figure 15.3 The Overall block diagram of FlexSwitch.

on to the control layer for resolving the output port contention and setting up the data path. Once the appropriate data path has been set up, the transmit cell is routed through DPSU to the appropriate output port buffer. In the case of multicast or broadcast messages, appropriate data paths must be set up to enable the routing of the transmit cell to the appropriate output ports' buffers.

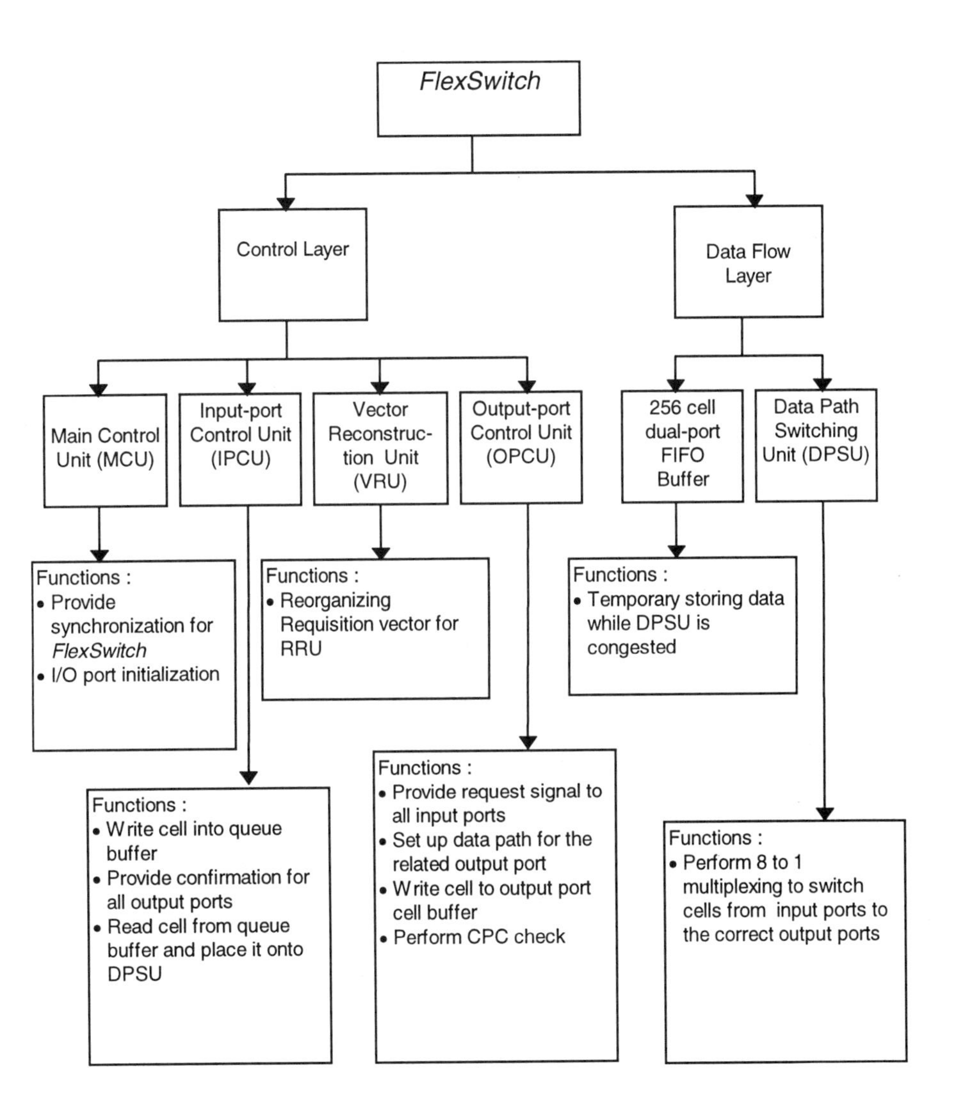


Figure 15.4 An overview of the *FlexSwitch* functions

The buffers used in FlexSwitch are external static RAM memory units which can store 256 cells in total. The memory units are selected to be dual-port FIFOs which allow the incoming cell to be written to from one end and the transmit cell to be read out from the other end. This has significantly increased the memory bandwidth and uplift the system's performance.

The DPSU performs a switching and routing function for *FlexSwitch.* It is designed using a number of a 8-to-1 multiplexers. The concept behind DPSU is simple. By connecting the same positional-bit of all input ports to the input of 8-to-1 multiplexer, and letting the control unit of the related output port select the agreed input port, the incoming cell can be routed and also broadcast or multicast to the correct output ports. The input port buffers, output port buffers, and DPSU form the data path layer.

The control layer consists of four functional units, the Main Control Unit (MCU), Input-Port Control Unit (IPCU), Output-Port Control Unit (OPCU), and Vector Reconstruction Unit (VRU).

*FlexSwitch* adapts the distributed control approach in which each port has a dedicated control unit in charge of its own operation. The IPCU is responsible for writing the incoming cell into the input port buffer, providing confirmation hand-shake with OPCU, and reading the transmit cell from the input port buffer to DPSU. The OPCU, on thc other hand, provides the appropriate control signals for the output port which is responsible for initializing the transmission by sending a request signal to IPCU for confirmation, setting up the output port data path, and reading the transmitted cell from DPSU to the output port buffer while performing Cell Parity Check (CPC). Each I/O port requires a pair of control units (IPCU to control input port and OPCU for output port). Hence, totals of 8 pairs of IPCU/OPCU are required in *FlexSwitch.*

Although FlexSwitch adapts the distributed control, from the global point of view it is designed to be a synchronous switch by introducing a high-speed system clock. Therefore, the MCU is also used to synchronize all functional units. The presence of the MCU also enables another important feature of *FlexSwitch,* to initialize each port during set-up period and monitor the presence of Switch Interface. Finally, the VRU is used to convert the routing tags, called Destination Vectors, to receiving tags called Requisition Vectors for OPCU. The Destination Vectors from all input ports are merged to VRU which will then be repositioned as Requisition Vectors for the output ports. The Requisition Vector represents the request for transmission signal from the input ports in a time instance. If more than 1 input port is requesting transmission at the same time, the requests will be granted in Priority-Queue order according to the bit-position in Requisition Vector.

## 15.4 Design Packages

This section provides an overview of the design process of the *FlexSwitch* system so that some VHDL description files can be introduced as examples in a later section. In order to simplify the *FlexSwitch* VHDL description and allow easy modifications of some of its parameters, three VHDL design packages are declared first. The first package various_constants, declares a number of constants, primarily values used for declaring the width and length for various data types as found in the latter package.

```
library ieee;
use ieee.std_logic_1164.all;

package various_constants is

constant data_width        : integer := 32;   -- Data Path is 32-bit wide
constant coder8_width      : integer := 3;        -- decoder is 3-bit wide
constant cell_ptr_width    : integer := 8;        -- cell pointer is 8-bit wide
constant address_width     : integer := 12;   -- address pointer is 12-bit wide
constant cell_cycle_width : integer := address_width - -cell_ptr_width;
         -- cell cycle is 4-bit width for counting the address within a cell.
constant cell_cycle_range : integer := 16;   -- 16 clock cycles for 1 cell
constant max_no_of_cell   : integer := 256;  -- 256 cells in 1 buffer
constant des_vector_width: integer := 8;        -- routing tag is 8-bit width
constant number_of_port   : integer := 8;        -- 8 I/O ports are required
constant max_port_fitted  : integer := number_of_port/2;

         -- design is partitioned into half due to lack of LABs. So, Max no. of I/O
         -- port that can be fitted in one CPLD chip is 4.

end various_constants;
```

The second package various_types, defines a number of types that represent standard logic vectors of different lengths.

```
use work.various_constants.all;

package various_types is

-- The following declares new data types

subtype cell_pointer is std_logic_vector (cell_ptr_width-1 downto 0);
subtype  address_pointer is std_logic_vector (address_width-1 downto 0);
```

```
subtype destination_code is std_logic_vector (number_of_port-1 downto 0);
subtype handshake_inp is std_logic_vector (max_port_fitted-1 downto 0);
subtype handshake_out is std_logic_vector (number_of_port-max_port_fitted-1
downto 0);
subtype quadrant_cell_width is std_logic_vector (data_width/4-1 downto 0);
subtype half_cell_width is std_logic_vector (data_width/2-1 downto 0);
subtype cell_width is std_logic_vector (data_width-1 downto 0);
subtype mux_sel3bit is std_logic_vector (coder8_width-1 downto 0);
subtype port_request is std_logic_vector (number_of_port-1 downto 0);

-- The following declares a state machine and vectors of new data types

type state_machine is (initial_mode, test_mode, operating_mode, locked_mode);
type cell_pointer_vector is array(natural range<>) of cell_pointer;
type address_pointer_vector is array(natural range<>) of address_pointer;
type  destination_code_vector is array(natural range<>) of destination_code;
type  handshake_inp_vector is array(natural range<>) of handshake_inp;
type handshake_out_vector is array(natural range<>) of handshake_out;
type quadrant_cell_width_vector is array(natural range<>) of quadrant_cell_width;
type half_cell_width_vector is array(natural range<>) of half_cell_width;
type cell_width_vector is array(natural range<>) of cell_width;
type mux_sel3bit_vector is array(natural range<>) of mux_sel3bit;

end various_types;
```

The third package my_components, contains components declaration for the use of structural design . Structural design provides a better way of designing a high speed ATM switch. Each function is designed separately and then they group together as the components of the switch.

```
library IEEE;
use IEEE.std_logic_1164.all;
use work.various_constants.all;
use work.various_types.all;

package my_components is

component XOR3byte      -- 3-inputs 8-bit XOR logic
port ( inputa, inputb, inputc: in std_logic_vector(7 downto 0);
       q: out std_logic_vector(7 downto 0));
end component;
```

```
component dlatch          -- 1-bit data latch logic
port ( d, clr, pre, load: in std_logic;
          q: out std_logic);
end component;

component  dfflop         -- 1-bit D-flip-flop
port ( d, clk, clr, pre, load: in std_logic;
          q: out std_logic);
end component;

component tffena          -- 1-bit T-flip-flop
port ( t, clk, clr, pre, ena, load, d: in std_logic;
          q: out std_logic);
end component;

component reg3bit         -- 3-bit register
port ( clk, clr, pre, load    : in std_logic;
          data_in             : in std_logic_vector (2 downto 0);
          data_out            : out std_logic_vector (2 downto 0));
end component;

component reg8bit         -- 8-bit register
port ( clk, clr, pre, load    : in std_logic;
          data_in   : in std_logic_vector (number_of_port-1 downto 0);
          data_out : out std_logic_vector (number_of_port-1 downto 0));
end component;

component reg32bit        -- 32-bit register, combination of four 8-bit registers
port ( data_in: in cell_width;
          clk, clr, pre, load  : in std_logic;
          data_out : out cell_width);
end component;

component reg8aclr-- 8-bit register with each bit can be asynchronously reset
port ( clk, pre, load: in std_logic;
          data_in, clr: in std_logic_vector (number_of_port-1 downto 0);
          data_out : out std_logic_vector (number_of_port-1 downto 0));
end component;

component opstate         -- The Port Control Logic  State Machine
port ( clk, enable, cell_error, reset, tested_OK: in std_logic;
          control: in std_logic_vector(cell_cycle_range-1 downto 0);
```

```
        rt_ena, clr_CSU, clr_RRU, port_fail: out std_logic);
end component;

component Ccontrol        -- The Main Control Logic
port ( clr, clk: in std_logic;
        hold: out std_logic;
        control: out std_logic_vector(cell_cycle_range-1 downto 0));
end component;
component rwfifo          -- Basic components to access FIFO
port ( data: in destination_code;
        addressA, addressB: in std_logic_vector (cell_ptr_width-1 downto 0);
        add_sel  : in std_logic;
        clr_wreq: out std_logic;
        q: out destination_code);
end component;

component fifocont        -- FIFO controller
port ( write_clk, read_clk, clr, direction, wcnt_en, rcnt_en: in std_logic;
        write_counter, read_counter: out cell_pointer;
        full, empty: out std_logic);
end component;

component asfifo -- Asynchronous Read/Write FIFO, Combination of
                          -- rwfifo and fifocont
port ( data: in destination_code;
        wreq, rreq, direction, clr: in std_logic;
        write_ctr, read_ctr: OUT std_logic_vector (cell_ptr_width-1 downto 0);
        FIFO_full, FIFO_empty: out std_logic;
        q: out destination_code);
end component;

component search          -- The Search control Logic
port ( clk, clr, q_empt: in std_logic;
        data_in, request, confirm: in destination_code;
        ld_reg, ld_clr, ld_req, ld_confirm: in std_logic;
        match, ld_data: out std_logic;
        accept: out destination_code);
end component;

component writecel        -- The process to write a cell into buffer
port ( clr, write_clk, new_cell, end_cell, wreq_clk: in std_logic;
        address_in: in cell_pointer;
```

```
        control: in std_logic_vector(cell_cycle_range-1 downto 0);
        address_out: out address_pointer;
        write_ack, write, wena, wreq: out std_logic);
end component;

component readcell        -- The process to read a cell out from buffer
port ( clr, clk, match, new_cell, end_cell: in std_logic;
        address_in: in cell_pointer;
        address_out: out address_pointer;
        read, oena: out std_logic);
end component;

component inports         -- The Input Port Control Unit (IPCU)
port ( data, request, confirm, oport_full: in destination_code;
        clr, clk, write_clk, new_cell, end_cell, port_ena: in std_logic;
        control: in std_logic_vector(cell_cycle_range-1 downto 0);
        write_add, read_add: out address_pointer;
        FIFO_full, write_ack, read, write, oena, wena: out std_logic;
        wreq: out std_logic;
        q, accept: out destination_code);
end component;

component requests        -- The Request Control Logic
port ( clk, clr, q_empty    : in std_logic;
        data_in, accept, clr_input: in destination_code;
        control: in std_logic_vector(cell_cycle_range-1 downto 0);
        request, confirm: out destination_code;
        match, req_empty: out std_logic;
        mux_sel : out std_logic_vector(coder8_width-1 downto 0));
end component;

component outports        -- The Output Port Control Unit (OPCU)
port ( data_in, accept, clr_CSU: in destination_code;
        clr, clk, wreq, port_ena: in std_logic;
        control   : in std_logic_vector(cell_cycle_range-1 downto 0);
        mux_ena, FIFO_full: out std_logic;
        request, confirm: out destination_code;
        mux_sel : out std_logic_vector(coder8_width-1 downto 0));
end component;

component errcheck        -- The CPC error checking process
port ( data_in: in cell_width;
```

```
        clr, clk, port_ena: in std_logic;
        data_out : out cell_width);
end component;

component bitlatch          -- 1-bit Requisition vector
port ( clr, wreq, hold, clr_wreq: in std_logic;
        d: in std_logic;
        latch_d: out std_logic;
        q: out std_logic);
end component;

component grplatch          -- 8-bit Requisition vector
port ( clr, hold: in std_logic;
        wclk: in destination_code;
        control: in std_logic_vector(cell_cycle_range-1 downto 0);
        d: in destination_code;
        wreq: out std_logic;
        q: out destination_code);
end component;

component reqdata           -- The Vector Reconstruction Unit (VRU)
port ( clr, hold: in std_logic;
        wclk: in std_logic_vector(number_of_port-1 downto 0);
        control: in std_logic_vector(cell_cycle_range-1 downto 0);
        d: in destination_code_vector(number_of_port-1 downto 0);
        wreq: out std_logic_vector(number_of_port-1 downto 0);
        q: out destination_code_vector(number_of_port-1 downto 0));
end component;

component port_mux          -- 8 ports, 16-bit multiplexer with clock input for mux
                            -- selection signals
port ( data_in      : in std_logic_2D(number_of_port-1 downto 0, data_width/4-1
        downto 0);
        mux_sel : in mux_sel3bit;
        mux_clk : in std_logic;
        data_out : out quadrant_cell_width);
end component;

component inp_data          -- Flip-flops that latch the input data and provide data
                            -- type conversion
port ( data_in   : in quadrant_cell_width_vector(number_of_port-1 downto 0);
        clock_in : in std_logic_vector(number_of_port-1 downto 0);
```

```
        data_out : out std_logic_2D(number_of_port-1 downto 0, data_width/4-1
        downto 0));
end component;

end my_components;
```

## 15.5 Operation of the Control Layer

The following sections describe the control layer in more details. Some VHDL design files are included as examples for describing simple operations.

### 15.5.1 Main Control Unit

The *FlexSwitch* is designed to be a synchronous switch by introducing a high-speed system clock. The MCU provides synchronous operation of the *FlexSwitch*. It provides two main operations; synchronization of functional units to the centralized system clock, and initialization and monitoring of the operation of each I/O port. Therefore, all the functional units can be operated synchronously by referring to the MCU. Although the *FlexSwitch* is self-synchronizing, it accepts asynchronous read and write processes prior to the add-on Switch Interface. This allows the Switch Interface to operate at a variable transmission rate up to the maximum speed of the *FlexSwitch*. To synchronize *FlexSwitch*, the MCU provides 16 logic states that are synchronized to the system clock. This is because the 64-byte cell takes 16 clock cycles (32-bit per clock cycle) to transmit. Therefore, all control signals and decision making has to be done within 16 logic states. The following VHDL file illustrates the design of 16 logic states which we called Main Control Logic :

```
library ieee;
library lpm;
use ieee.std_logic_1164.all;
use work.various_constants.all;
use work.various_types.all;
use work.my_components.all;
use lpm.lpm_components.all;

entity mcontrol is
port (    clr, clk   : in std_logic;       -- provide clock and clear input signals

-- Hold signal is produced to present both writing and reading destination vector
-- into EAB at the same time.
          hold       : out std_logic;
```

```
-- the control signal represents the output of 16 logic states.
        control   : out std_logic_vector (cell_cycle_range-1 downto 0));
end mcontrol;

architecture arch1 of mcontrol is

-- Internal bus and signals
signal control_sig: std_logic_vector (cell_cycle_range-1 downto 0);
signal invert, inv_start, start, clr_hold: std_logic;

begin
        generate_control_sig : for i in control'range generate

-- generating the first logic state.
        generate_start_bit :        if i = 0 generate
                FF_1st_logic        : Dfflop
                        port map ( clk    => clk,
                                   clr          => clr,
                                   pre          => GND,
                                   load         => VCC,
                                   d            => inv_start,
                                   q            => control_sig(i));
                end generate;

-- generating the rest of the logic states.
        generate_others   :         if i > 0 generate
                FF_control          : Dfflop
                        port map (        clk          => clk,
                                          clr          => clr,
                                          pre          => GND,
                                          load         => VCC,
                                          d            => control_sig(i-1),
                                          q            => control_sig(i));
                end generate;

-- generating the hold signal.
        generate_hold     :         if  i = 14 generate
                FF_hold             : Dfflop
                        port map (        clk          => clk,
                                          clr          => clr_hold,
                                          pre          => GND,
                                          load         => control_sig(i-1),
```

```
                                        d       => control_sig(i-1),
                                        q       => hold);
                end generate;
        end generate;

-- The start signal to generate the 1st logic state.
                FF_start        : Dfflop
                        port map (      clk     => clk,
                                        clr     => clr,
                                        pre     => GND,
                                        load    => VCC,
                                        d       => invert,
                                        q       => start);

        VCC <= '1';
        GND <= '0';
        clr_hold <= clr or control_sig(1);
        control  <= control_sig;
        invert   <= not control_sig(cell_cycle_range-2);
        inv_start <= not start;

end arch1;
```

On the other hand, MCU is also responsible for initializing and monitoring each I/O port. Because there are 8 I/O ports, 8 state machines are required for monitoring the presence of Switch Interface before the individual port control units can be activated. Figure 15.5 illustrates the state transition diagram of the finite state machine that monitors an I/O port.

Initially, the state machine is in *Initialize State* and the *enable* signal from Switch Interface is required to be active high for a period of time (port_initialize = 1) in order to advance to *Test State*. In *Test State*, Switch Interface sends a set of test cells that are destined to test the operation of the remaining functional units. If the received cells match the test pattern, a *Tested_OK* signal is generated which releases the state machine to *Operating State*. If the error rate is too high or the *Tested_OK* signal has not been acknowledged within the limited time period, the system has failed and is locked in *Locked State* until it is reset by Switch Interface. Even though the I/O port is in *Operating State*, the *enable* signal from SI and the cell error rate will be closely monitored by the state machine. If the error rate is getting too high, the system is re-initialized by returning to *Initialize State*. In both *Test State* and *Operating State*, the state machine can be reset to *Initialize State* if the *enable* signal is active low for a period of time.

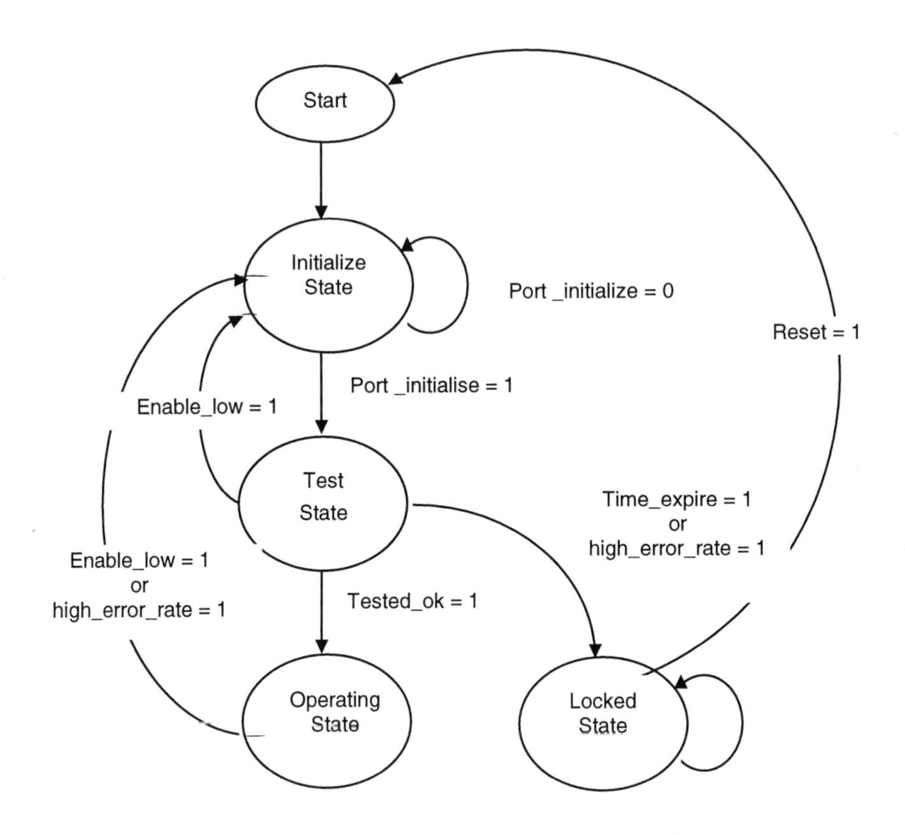


Figure 15.5 The State transition diagram of monitoring each I/O port.

The following VHDL file illustrates the design of the state machine that implements Port Control Logic:

```
library ieee;
library lpm;
use ieee.std_logic_1164.all;
use work.various_constants.all;
use work.various_types.all;
use work.my_components.all;
use lpm.lpm_components.all;

entity opstate is
port (clk, reset: in std_logic;
```

```
        control: in std_logic_vector(cell_cycle_range-1 downto 0);

-- input signals monitored by the state machine.
        enable, cell_error, tested_OK: in std_logic;

-- output signals from state machine.
        port_ena, clr_CSU, clr_RRU, port_fail: out std_logic);

end opstate;

architecture arch1 of opstate is

-- Internal bus and signals
signal state: state_machine;
signal count_ena : std_logic_vector(7 downto 0);
signal Timer, count_error: std_logic_vector(15 downto 0);
signal S0, cnt_ena1, cnt_ena0, reset_cnt, clear, ena_sig, error_sig: std_logic;
signal increment1, increment0, reset_cnt_error, high_error: std_logic;
signal ena_cnt_error1, ena_cnt_error0, reset_ena, initialise_sig: std_logic;
signal increment_error1, increment_error0: std_logic;
signal port_initialise, high_error_rate, enable_low, inv_ena: std_logic;

begin

process (clk, reset)
begin
        if reset = '1' then
                state       <=      initial_mode;
        elsif (clk'event and clk = '1') then

                case state is

                        when    initial_mode        =>

                                clr_CSU             <=      '1';
                                clr_RRU             <=      '1';
                                port_ena            <=      '0';
                                S0                  <=      '1';
                                port_fail           <=      '0';

                                if port_initialise = '1' then
                                        state       <=      test_mode;
                                else
                                        state       <=      initial_mode;
```

```
            end if;

    when    test_mode           =>

            clr_CSU             <=      '0';
            clr_RRU             <=      '0';
            port_ena            <=      '1';
            S0                  <=      '0';

            if tested_OK = '1' then
                    port_fail   <=      '0';
                    state       <=      operating_mode;
            elsif enable_low = '1' then
                    port_fail   <=      '0';
                    state       <=      initial_mode;
            elsif port_initialise = '1' or high_error_rate =
                                                            '1'
            then
                    port_fail   <=      '1';
                    state       <=      locked_mode;
            else
                    port_fail   <=      '0';
                    state       <=      test_mode;
            end if;

    when    operating_mode =>

            clr_CSU             <=      '0';
            clr_RRU             <=      '0';
            port_ena            <=      '1';
            S0                  <=      '0';
            port_fail           <=      '0';

            if enable_low = '1' or high_error_rate = '1'

                                                    then
                    state       <=      initial_mode;
            else
                    state       <=      operating_mode;
            end if;

    when    locked_mode =>

            clr_CSU             <=      '1';
            clr_RRU                     <=      '1';
            port_ena            <=      '0';
            S0                  <=      '0';
```

```
                        port_fail       <=      '1';
                        state           <=      locked_mode;

                when    others  =>

                        clr_CSU         <=      '0';
                        clr_RRU         <=      '0';
                        port_ena        <=      '1';
                        S0              <=      '0';
                        port_fail       <=      '0';
                        state           <=      operating_mode;
            end case;
        end if;
end process;

-- 16-bit counter that counts the enable pin from Switch Interface in
-- initialise_mode or serves as the timer to reset the error counter outside the
-- initialise_mode.

        upperbyte: lpm_counter
            generic map (   LPM_WIDTH       =>      8)
            port map (      clock   =>      control(7),
                            cnt_en  =>      cnt_ena1,
                            updown  =>      GND,
                            aclr    =>      GND,
                            aset    =>      reset_cnt,
                            q       =>      Timer(15 downto 8));

        lowerbyte           :       lpm_counter
            generic map (   LPM_WIDTH       =>      8)
            port map (      clock   =>      control(7),
                            cnt_en  =>      cnt_ena0,
                            updown  =>      GND,
                            aclr    =>      GND,
                            aset    =>      reset_cnt,
                            q       =>      Timer(7 downto 0));

        cnt_ena1  <= cnt_ena0 and increment0;
        cnt_ena0  <= enable or not S0;
        reset_cnt  <= reset or (S0 and not enable) or(control(15)
                        and(high_error_rate or enable_low) and not S0);

        with timer (7 downto 0) select
            increment0 <= '1' when "00000000",
```

```
                '0' when others;
        with timer(15 downto 8) select
                increment1 <= '1' when "00000000",
                '0' when others;
        initialise_sig <= increment0 and increment1;

        FF_port_init: Dfflop
                port map ( clk          =>      control(15),
                           clr          =>      clear,
                           pre          =>      GND,
                           load         =>      VCC,
                           d            =>      initialise_sig,
                           q            =>      port_initialise);

-- 8-bit counter which counts enable pin and reset if it is
-- not active for more than 163.84us.

        cnt_enable: lpm_counter
                generic map (  LPM_WIDTH     =>      8)
                port map ( clock        =>      control(7),
                           cnt_en       =>      inv_ena,
                           updown       =>      GND,
                           aclr         =>      GND,
                           aset         =>      reset_ena,
                           q            =>      count_ena);

        inv_ena  <= not enable and not S0;
        reset_ena <= reset or enable or S0;

        with count_ena select
                ena_sig  <= '1' when "00000000",
                        '0' when others;

        FF_enable_low: Dfflop
                port map ( clk          =>      control(15),
                           clr          =>      clear,
                           pre          =>      GND,
                           load         =>      VCC,
                           d            =>      ena_sig,
                           q            =>      enable_low);

-- 12-bit counter which counts the error rate of the receiving cell. If the error rate is -
- more than 6.25% (256 error cell out of 64k cell), port will be reset.
```

```
cnt_error_byte1: lpm_counter
        generic map (  LPM_WIDTH      =>      8)
        port map ( clock        =>      control(7),
                   cnt_en       =>      ena_cnt_error1,
                   updown       =>      GND,
                   aclr         =>      GND,
                   aset         =>      reset_cnt_error,
                   q            =>      count_error(15 downto 8));

cnt_error_byte0: lpm_counter
        generic map (  LPM_WIDTH      =>      8)
        port map ( clock        =>      control(7),
                   cnt_en       =>      ena_cnt_error0,
                   updown       =>      GND,
                   aclr         =>      GND,
                   aset         =>      reset_cnt_error,
                   q            =>      count_error(7 downto 0));

ena_cnt_error1 <= cell_error and not S0 and increment_error0;
ena_cnt_error0 <= cell_error and not S0;
reset_cnt_error <= reset or (port_initialise and not S0) or S0;

with count_error(15 downto 8) select
        increment_error1 <= '1' when "11110000",
                            '0' when others;
with count_error(7 downto 0) select
        increment_error0 <=     '1' when "00000000",
                                '0' when others;

error_sig <= increment_error1 and increment_error0;

FF_high_error: Dfflop
        port map (  clk         =>      control(15),
                    clr         =>      clear,
                    pre         =>      GND,
                    load        =>      VCC,
                    d           =>      error_sig,
                    q           =>      high_error_rate);

VCC <= '1';
GND <= '0';
clear <= reset or control(7);
```

```
end arch1;
```

The VHDL design files that shown above describe the operation of Main Control Logic and Port Control Logic. The MCU consists of a Main Control Logic and 8 Port Control Logic blocks, one for each of the 8 I/O ports. This is described in the following VHDL design file:

```
library ieee;
library lpm;
use ieee.std_logic_1164.all;
use work.various_constants.all;
use work.various_types.all;
use work.my_components.all;
use lpm.lpm_components.all;

entity maincontrol is
port ( clk, reset:  in std_logic;
        enable, cell_error, clear: in std_logic_vector (number_of_port-1 downto 0);
        hold:  out std_logic;
       control: out std_logic_vector (cell_cycle_range- downto 0);
       clr_CSU, clr_RRU: out std_logic_vector (number_of_port-1 downto 0);
       port_ena,  port_fail: out std_logic_vector (number_of_port-1 downto 0));

end maincontrol;

architecture arch1 of maincontrol is
signal control_sig:  std_logic_vector(cell_cycle_range-1 downto0);

begin

-- replication of the 8 Port Control Logic circuits

        generate_port: for i in port_ena'RANGE generate

        Port_control        : opstate
                port map (  clk             =>      clk,
                            enable          =>      enable(I),
                            cell_error      =>      cell_error(I),
                            reset           =>      reset,
                            control         =>      control_sig,
                            port_ena        =>      port_ena(i),
                            clr_CSU         =>      clr_CSU(i),
```

```
                              clr_RRU      =>      clr_RRU(i));
        end generate;

-- generation of the Main Control Logic.

        Main_control: Ccontrol
                port map (  clr        =>      reset,
                            clk        =>      clk,
                            hold       =>      hold,
                            control    =>      control_sig);

        control <= control_sig;

end arch1;
```

### 15.5.2 Input-Port Control Unit

The IPCU provides three main operations to the related input port. First, it writes data into the input port buffer when a new cell is arriving from Switch Interface. Second, it negotiates with OPCU to determine the destination output port. If the destination output port(s) is/are not available in the next coming time slot, transmission is held and data is kept in the buffer. Third, it reads the transmit cell from buffer and places it on DATA bus. The DATA bus is connected to DPSU in which the transmit cell will be routed to the destination output port(s). Two VHDL design files will be described here, in order to illustrate the write and read operation of IPCU.

The following design file describes the write operation of IPCU :

```
entity writecel is
port ( clr: in std_logic;      -- Clear signal
        control: in std_logic_vector(cell_cycle_range-1 downto 0);

-- system clock from Switch Interface for writing data into buffer.
        write_clk: in std_logic;

-- control signal from Switch Interface to mark the beginning and end of cell.
        new_cell, end_cell: in std_logic;

-- system clock from MCU to latch the destination vector and store it in EAB.
        wreq_clk: in std_logic;
```

```
-- cell pointer from internal EAB controller.
        address_in: in cell_pointer;

-- address bus that output to input port buffer for writing incoming data.
        address_out: out address_pointer;

-- acknowledgement signal to Switch Interface for confirming the end of cell.
        write_ack: out std_logicl;

-- write and write enable signals for input port buffer.
        write, wena: out std_logic;

-- write request signal to latch destination vector into EAB.
        wreq: out std_logic);

end writecel;

architecture arch1 of writecel is
signal  wreq_in, wreq_ena, wreq_sig, wreq_out: std_logic;
signal write_sig, write_ack_sig: std_logic;
signal wcnt_ena, ld_ack, clr_wcnt_ena: std_logic;
signal wcnt_add  : std_logic_vector (cell_cycle_width-1 downto 0);

begin

-- latch the write address for one cell cycle (or time slot).
        register_address_out: reg8bit
                port map(  clk          =>      write_clk,
                           clr          =>      clr,
                           pre          =>      GND,
                           load         =>      new_cell,
                           data_in      =>      address_in,
                           data_out     =>      address_out(address_width-1
                                                downto cell_cycle_width));

-- this counter provides the last 4-bit of address to input port buffer.

        write_cnt          :     lpm_counter
                generic map      (LPM_WIDTH  =>    cell_cycle_width)
                port map         ( clock     =>    write_sig,
                                   cnt_ena   =>    wcnt_ena,
                                   updown    =>    VCC,
```

```
                                aclr            =>      clr_wcnt_ena,
                                aset            =>      GND,
                                q               =>      wcnt_add);

        address_out(cell_cycle_width-1 downto 0) <=     wcnt_add;

-- the FF_wcnt_ena is used to enable the write counter which will be active high
-- for the whole cell cycle.
        FF_wcnt_ena: Dfflop
                port map        (  clk          =>      write_clk,
                                   clr          =>      clr_wcnt_ena,
                                   pre          =>      GND,
                                   load         =>      new_cell,
                                   d            =>      VCC,
                                   q            =>      wcnt_ena);

        write_sig <= not wcnt_ena or write_clk;
        write <= write_sig;
        wena <= wcnt_ena;

-- the FF_clr_wcnt_ena is used to wcnt_ena at the end of cell cycle.
        FF_clr_wcnt_ena: Dfflop
                port map        (  clk          =>      write_clk,
                                   clr          =>      clr_wcnt_ena,
                                   pre          =>      GND,
                                   load         =>      VCC,
                                   d            =>      ld_ack,
                                   q            =>      clr_wcnt_ena);

-- This is the flip flop which produces an acknowledgement signal after the whole
-- cell has been written into memory.

        FF_write_ack    :       Dfflop
                port map        (  clk          =>      write_clk,
                                   clr          =>      GND,
                                   pre          =>      GND,
                                   load         =>      VCC,
                                   d            =>      ld_ack,
                                   q            =>      write_ack_sig);

        ld_ack <= '1' when wcnt_add = "1111" else
                        '0';
        write_ack <= write_ack_sig;
```

```
-- the following flip-flops are used to produce the write request (wreq) signal for
-- ASfifo after the ATM cell has been confirmed written into memory. When wreq –
is active high, the destination vector for the writen cell is stored in asfifo
-- for further comparison.

        FF_wreq_in: Dfflop
                port map        ( clk       =>      wreq_ena,
                                  clr       =>      wreq_out,
                                  pre       =>      GND,
                                  load      =>      VCC,
                                  d         =>      VCC,
                                  q         =>      wreq_sig);

        wreq_ena <= end_cell and write_ack_sig;
        wreq_in <= wreq_sig and not control (cell_cycle_range-2);

        FF_wreq: Dfflop
                port map        ( clk       =>      wreq_clk,
                                  clr       =>      GND,
                                  pre       =>      GND,
                                  load      =>      VCC,
                                  d         =>      wreq_in,
                                  q         =>      wreq_out);

        wreq <= wreq_out;
        VCC <='1';
        GND <= '0';
end arch1;
```

The following design file describes the read operation of the IPCU :

```
entity readcell is
port ( clr, clk: in std_logic; -- clear and clock signals

-- match signal is activated if cell is required to be read in the next time slot.
        match   : in std_logic;

-- mark the beginning and end of read cycle. new_cell is connected to logic state
-- Control(0) and end_cell is connected to logic state Control(15).
        new_cell, end_cell: in std_logic;

-- cell pointer from internal EAB controller.
```

```
            address_in: in cell_pointer;

-- address bus that output to input port buffer for reading data.
            address_out: out address_pointer;

-- read and output enable signals for input port buffer.
            read, oena: out std_logic);

end readcell;

architecture arch1 of readcell is
signal read_sig, inv_read_sig, rcnt_ena, inv_rcnt_ena, inv_clk: std_logic;
signal oena_sig, ld_oena, ld_add, add_clk, clr_cnt, clr_cnt_sig: std_logic;

begin

-- latch the read address for one time slot.
            register_address_out: reg8bit
                        port map (      clk         =>      add_clk,
                                        clr         =>      clr,
                                        pre         =>      GND,
                                        load        =>      ld_add,
                                        data_in     =>      address_in,
                                        data_out    =>      address_out(address_width-1
                                                            downto cell_cycle_width));

            ld_add   <=      new_cell and oena_sig;
            add_clk  <=      new_cell and not (oena_sig and clk);

-- this counter provides the last 4-bit address to read the whole cell from memory.

            read_cnt: lpm_counter
                        generic map     ( LPM_WIDTH => cell_cycle_width)
                        port map (  clock       =>      inv_read_sig,
                                    cnt_en      =>      rcnt_ena,
                                    updown      =>      VCC,
                                    aclr        =>      clr_cnt,
                                    aset        =>      GND,
                                    q           =>      address_out(cell_cycle_width
                                                        - 1 downto 0));

            clr_cnt <= clr_cnt_sig or clr;
```

```
        read_sig  <= rcnt_ena and clk;
        inv_read_sig <= not read_sig;
        read <=  read_sig;

        FF_rcnt_ena: Dfflop  -- enable the above counter.
                port map        ( clk           =>      inv_clk,
                                  clr           =>      clr,
                                  pre           =>      GND,
                                  load          =>      new_cell,
                                  d             =>      oena_sig,
                                  q             =>      rcnt_ena);

-- FF_oena is used to produce the output enable signal to buffer.
        FF_oena: Dfflop
                port map        ( clk           =>      inv_clk,
                                  clr           =>      clr,
                                  pre           =>      GND,
                                  load          =>      end_cell,
                                  d             =>      ld_oena,
                                  q             =>      oena_sig);

        ld_oena  <= match and end_cell;
        oena <= oena_sig or rcnt_ena;

-- Clear signal to reset counter at the end of each time slot.
        FF_clr_oena: Dfflop
```

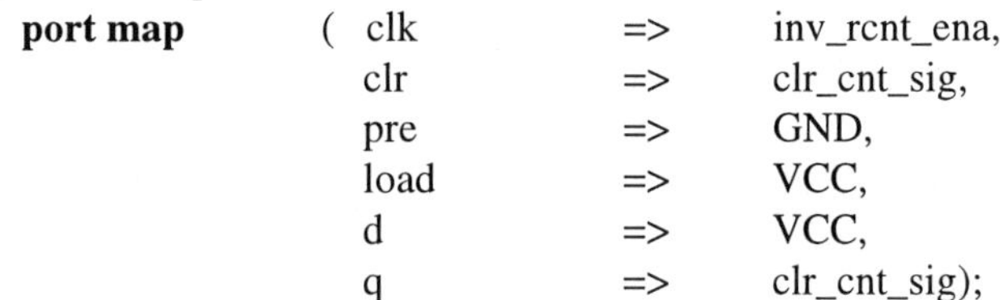

```
                port map        ( clk           =>      inv_rcnt_ena,
                                  clr           =>      clr_cnt_sig,
                                  pre           =>      GND,
                                  load          =>      VCC,
                                  d             =>      VCC,
                                  q             =>      clr_cnt_sig);

        inv_rcnt_ena <= not rcnt_ena;
        inv_clk <= not clk;
        VCC <= '1';
        GND <= '0';
end arch1;
```

### *15.5.3 Output-Port Control Unit*

The OPCU provides four main operations to the related output port. First, it analyses the Requisition Vector according to the Priority-Queue order and allows transmission for one input port in every time slot. Second, it provides the control signals for DPSU to select the correct data path. Third, it reads the transmit cell from DPSU and performs Cell Parity Check (CPC). Finally, the received cell is written to output port buffer so that it can be read by Switch Interface at a speed slower than that of the *FlexSwitch*. Further details of OPCU will not be discussed here.

### *15.5.4 Destination Vector and Requisition Vector*

During operation, the Destination Vectors of each cell will be extracted and stored in a look-up table in IPCU for transmission confirmation. The Destination Vector indicates to which output ports the current ATM cell is destined to. It is a signal from one input port to all output ports and, therefore, all the Destination Vectors from 8 input ports must be repositioned and converted to the request signal for each output port. This results in the creation of the Requisition Vector.

The Requisition Vector is the transmission request from all input ports to a particular output port in the same time slot. When there is only 1 input port request for the service from the output port in one time slot, transmission will be available according to its arrival (i.e. first-come-first-served). But, when there are more than 1 input ports requiring transmission from the same output port in the same time slot, priority is given to the lower number of input port according to the bit position of Requisition Vector.

The Destination Vectors of all the incoming cells are extracted and stored in an internal FIFO, called destination look-up table in IPCU. The look-up table is implemented using the Embedded Array Block (EAB) featured in FLEX10K CPLD devices. The look-up table is configured in 8-bit width which can store up to 256 destination vectors. In other words, a totals of 256 cells can be stored in the input port buffer with each destination vector representing the routing tag of each cell.

On the other hand, there is a Requisition look-up table presented in OPCU which keeps track of the requisition vectors queued in the output port. Both tables are required to provide the hand-shake operation that is discussed later in section 15.5.6.

The following VHDL design file describes the implementation of the look-up table :

```
entity asfifo is
port (clr: in std_logic;
-- 8-bit input data path.
        data: in destination_code;

-- write_request and read_request signals for writing and reading data to FIFO.
        wreq, rreq: in std_logic;

-- write pointer and read pointer. Write pointer is pointing to the end of FIFO
-- while read pointer is pointing to the top of FIFO.
        write_ptr, read_ptr: out std_logic_vector (cell_ptr_width-1 downto 0);

-- empty and full signal for the FIFO.
        FIFO_full, FIFO_empty: out std_logic;

-- 8-bit output data path.
        q: out destination_code);

end asfifo;

architecture arch1 of asfifo is
signal write_ctr, read_ctr: std_logic_vector (cell_ptr_width-1 downto 0);
signal status_ctr, address: std_logic_vector (cell_ptr_width-1 downto 0);
signal full_sig, empty_sig, inv_full, inv_empty: std_logic;
signal wreq_flag, inv_wreg_flag, write_pulse: std_logic;
signal delay_write, extend_pulse_width, read_clk: std_logic;
signal data_in, q_sig: destination_code;

begin

-- set the write request flag when a write is required..

        FF_wreq_flag: Dfflop
                port map (      clk             =>      wreq,
                                clr             =>      write_pulse,
                                pre             =>      GND,
                                load            =>      VCC,
                                d               =>      VCC,
                                q               =>      wreq_flag);

-- latch the input data into register.
        register_data_in: reg8bit
```

```
                port map (      clk         =>      wreq_flag,
                                clr         =>      clr,
                                pre         =>      GND,
                                load        =>      wreq,
                                data_in     =>      data,
                                data_out    =>      data_in);

-- latch the output data into register.
        register_q_out: reg8bit
                port map (      clk         =>      read_clk,
                                clr         =>      clr,
                                pre         =>      GND,
                                load        =>      inv_empty,
                                data_in     =>      q_sig,
                                data_out    =>      q);
-- the write pointer.
                X: lpm_counter
                generic map (   LPM_WIDTH   =>      cell_ptr_width)
                port map    (   clock       =>      inv_wreg_flag,
                                cnt_en      =>      inv_full,
                                updown      =>      VCC,
                                aclr        =>      clr,
                                aset        =>      GND,
                                q           =>      write_ctr);

        inv_wreg_flag <= not wreq_flag;
        inv_full  <= not full_sig;

-- the read pointer.
                Y: lpm_counter
                generic map (   LPM_WIDTH   =>      cell_ptr_width)
                port map    (   clock       =>      read_clk,
                                cnt_en      =>      inv_empty,
                                updown      =>      VCC,
                                aclr        =>      clr,
                                aset        =>      GND,
                                q           =>      read_ctr);

        read_clk  <= rreq and not wreq;
        inv_empty <= not empty_sig;

-- status counter is one less than read counter. Hence if write counter
```

```
-- equal to status counter, FIFO is full.

            Z: lpm_counter
            generic map (  LPM_WIDTH   =>   cell_ptr_width)
            port map    (  clock       =>   read_clk,
                           clk_en      =>   inv_empty,
                           updown      =>   VCC,
                           aclr        =>   GND,
                           aset        =>   clr,
                           q           =>   status_ctr);

      full_sig <= '1' when write_ctr = status_ctr else
                           '0';
      empty_sig <= '1' when read_ctr = write_ctr else
                           '0';

-- if write request flag has been set, the write pointer is selected as the address bus
-- for EAB.

            with wreq_flag select
                  address <= write_ctr when '1',
                  read_ctr when others;

-- flip-flop X is used to introduce delay for setting up write_address.

            X: Dfflop
            port map (     clk    =>   wreq_flag,
                           clr    =>   extend_pulse_width,
                           pre    =>   GND,
                           load   =>   VCC,
                           d      =>   VCC,
                           q      =>   delay_write);

-- flip-flop Y is used to produce the write request signal.

            Y: Dfflop
            port map (     clk    =>   delay_write,
                           clr    =>   extend_pulse_width,
                           pre    =>   GND,
                           load   =>   VCC,
                           d      =>   VCC,
                           q      =>   write_pulse);
```

```
-- finally, flip-flop Z is used to extend the width of the write pulse signal so that
-- data did write into memory even though memory setup/hold time is slightly
-- large.

            Z: Dfflop
            port map (      clk     =>      write_pulse,
                            clr     =>      extend_pulse_width,
                            pre     =>      GND,
                            load    =>      VCC,
                            d       =>      VCC,
                            q       =>      extend_pulse_width);

-- lpm megafunction to access EAB.

            W: lpm_ram_dq
            generic map ( lpm_width              =>      number_of_port,
                          lpm_widthad            =>      cell_ptr_width,
                          lpm_indata             =>      unregistered,
                          lpm_address_control    =>      unregistered,
                          lpm_outdata            =>      unregistered)
            port map (      data                 =>      data_in,
                            address              =>      address,
                            we                   =>      write_pulse,
                            q                    =>      q_sig);

      VCC <= '1';
      GND <= '0';
      write_ptr <= write_ctr;
      read_ptr  <= read_ctr;
      full <= full_sig;
      empty <= empty_sig;

end arch1;
```

### *15.5.5 Vector Reconstruction Unit*

The purpose of VRU is to perform vector conversion for the Destination Vectors from all input ports to the Requisition Vectors for all output ports. Figure 15.6 illustrates the relationship between Destination vector and Requisition vector. The Destination vector that is received by the IPCU of each input port represents the output port(s) that the received cell is(are) destined for. A ‘1’ in the first bit of the destination vector means the related cell is destined for the output port 1, a '1' in the second bit means it is destined for output port 2 and so on.

The Matrix on the left side of Figure 15.6 shows the destination vector list from IPCUs of all input ports at an time instance (slot). The first bits of all the destination vectors are repositioned as bit 1 to bit 8 of the requisition vector of OPCU 0 for the next available time slot. Vice versa, the second, third and fourth bits of all the destination vectors are repositioned as the requisition vectors for OPCU 1, OPCU 2, OPCU 3, and so on. Each of the requisition vectors is kept in the look-up table of each OPCU until the related cell is received by output port. The details of the vector conversion will not be discussed further.

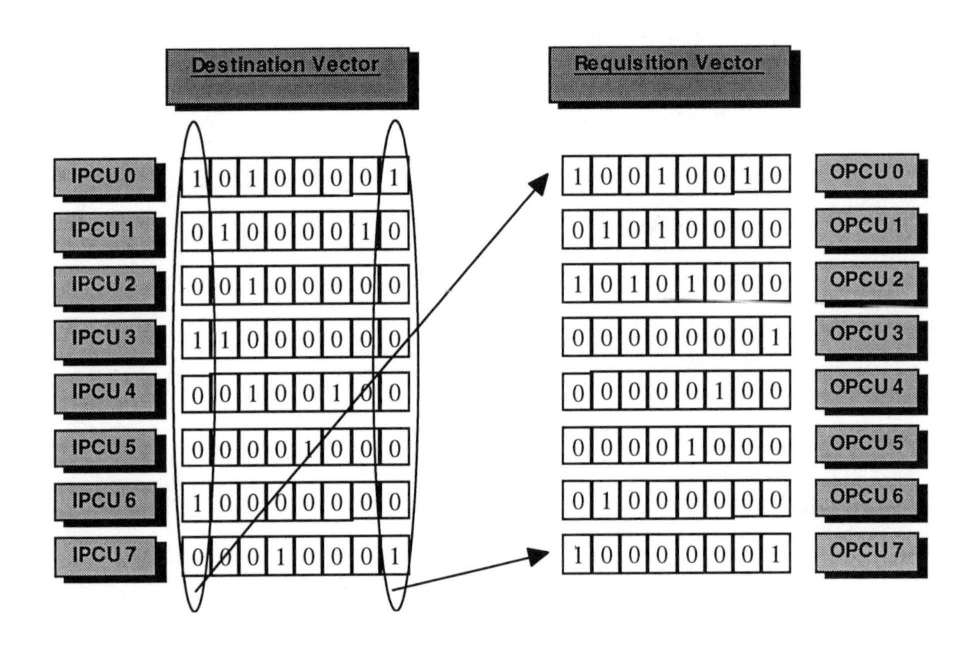


Figure 15.6 The Relationship between Destination vector and Requisition vector

### *15.5.6 Hand-shake Between IPCU and OPCU*

The head-of-queue contention problem, that is created by applying the input-queue buffering, generates a difficulty of designing the distributed control units (IPCU for input port and OPCU for output port). This is because "hand-shake" confirmation is required between input ports and output ports for every transmission. Head-of-queue contention occurs when more than 1 input ports is destined for the same output port at the same time. This is solved by applying the priority-queue order according to the bit-position of requisition vector so that only one input port is available for transmission in any one time slot. While the cell blockage in the input port that is put on hold is caused by arrival of new cells, a mismatch may occur if another output port requires the transmission of the blocking input port whose cell is behind the head-of-queue cell. Hence, hand-shake confirmation is required.

To perform hand-shake confirmation, three types of signals are required: the *request* signals, *confirm* signals and *accept* signals. First, the requisition vectors from all output ports are decoded and broadcast as *request* signals to all input ports. For example, the requisition vector for OPCU 0 in the current time slot is "01010010" (this means input ports 1, 3, and 6 have cells for transmission in the same time slot), then "01000000" is broadcast to all input ports for the current time slot, "00010000" is broadcast in the next time slot and so on. Hence, only one input port is available for transmission in each time slot and for the current time slot, input port 1 is given the first priority for cell transmission.

Upon receiving the *request* signals, the input ports match the *request* signal with the head-of-queue destination vectors to provide *confirm* signals back to all output ports. For instance, the IPCU 1 has received a *request* signal of "10010001" (this means that output ports 0, 3, and 7 are available for transmission in the next time slot). However, if the head-of-queue destination vector of IPCU 1 is "10000001" (this means that the ATM cell that is waiting at the head of the queue is destined for output port 0 and output port 7), the *confirm* signal is "10000001" (this means that only output ports 0 and 7 will be available for transmission in the next time slot and output port 3 is neglected).

Upon receiving the *confirm* signals, the output ports will match the *confirm* signals with *request* signals and broadcast the *accept* signals back to all input ports. For example, the *request* signal that was sent by OPCU 0 is "01000000" and the *confirm* signal is "01000000", then the *accept* signal is also "01000000". This means that a match is found and the data path will be set up for input port 1 in the next time slot. In the meantime, the requisition vector of OPCU 0 will be updated from "01010010" to "00010010" (this means that input port 1 has been served and will be removed from the list). However, if no match is found or an error has occurred, the same *request* signal will be broadcast again in the next time slot. A good example is the output port 3. OPCU 3 has sent a *request* signal "01000000" which is expecting transmission to input port 1 in the next time slot. However, since the cell that is at the head of the queue is not destined for output port 3, the request will be rejected by input port 1 and the *confirm* signal of "00000000" will be received by OPCU 3. This means that no match was found and the *accept* signal of "00000000" will be broadcast to all input ports.

Upon receiving the *accept* signals, the input ports will match the *accept* signals with the *confirm* signals. If a match is found, the transmit cell will be read out from the input port buffer and the head-of-queue destination vector will be updated. For instance, the IPCU 3 sends out *confirm* signal "10000001" which will receive the *accept* signal "10000001". Therefore, the destination vector will be updated from "10010001" to "00010000". This means that the transmit cell will be sent to output ports 0 and 7 in the next time slot. Hence, the same cell is destined only for output port 3 for the upcoming time slot until it has been served.

Figure 15.7 illustrates the basic concept behind the hand-shaking mode. Once a match is found between input port and output port, the input port will read the cell from the input port buffer and send it to DPSU in the next time slot. On the other hand, the related output port will also respond by setting up the data path between

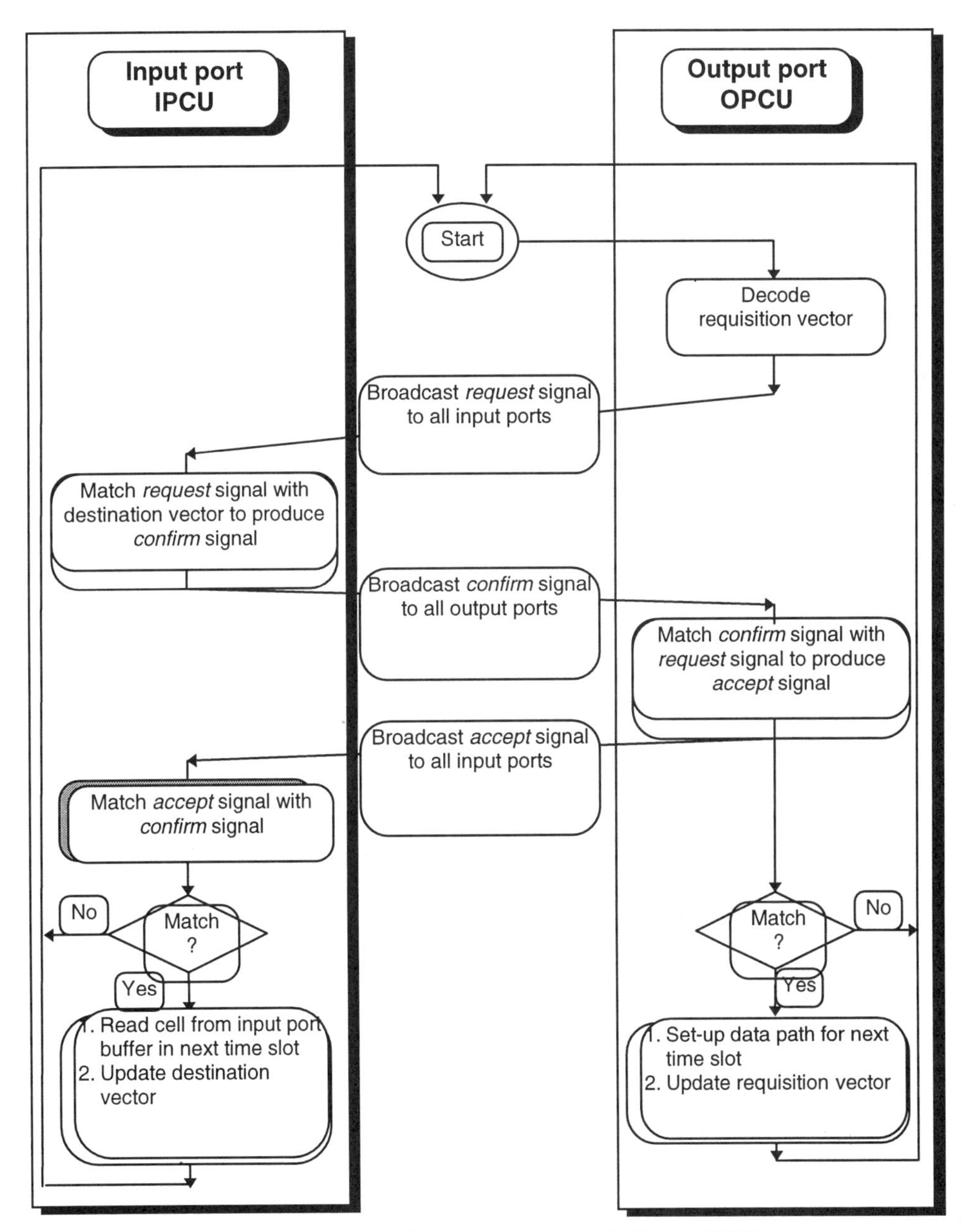


Figure 15.7 The State diagram of hand-shaking between IPCU and OPCU

them. It will read the transmit cell from DPSU and write it to the output port buffer in the next time slot.

The following VHDL design file describes the Search (or Confirm) operation performed in IPCU:

```
entity search is
port (clk, clr: in std_logic;

-- empty signal from destination look-up table. If table is empty, no search is
required
        q_empty : in std_logic;

-- head-of-queue destination vector from the look-up table.
        data_in : in destination_code;

-- request and confirm signals from OPCU.
        request, confirm: in destination_code;

-- control signal to carry out operations in different logic states.
        control: in std_logic;

-- match signal to "readcell.vhd" design file to inform that a match has been found.
        match: out std_logic;

-- load data signal to look-up table so that HOQ destination vector can be latched
-- into register for search operation.
        ld_data: out std_logic;

-- accept signal to OPCU.
        accept: out destination_code);
end search;

architecture arch1 of search is
signal match_reg, request_reg, compare_reg: destination_code;
signal clr_compare, empty_code, accept_sig, match_sig: destination_code;
signal status, match_bit, clr_match, clr_request: std_logic;
signal reg_empty, ld_compare: std_logic;

begin

-- check if compare register is empty. If yes, HOQ destination vector can be read-in
```

```
-- for comparison.
        reg_empty <= '1' when compare_reg = "00000000" else
                             '0';
        ld_data <= reg_empty;
-- the following is the Compare Register. Each bit of the Compare Register can be
-- asynchronously clear, after the request from the output port has been confirmed.

                register_compare: reg8aclr
                port map (   clk       =>     clk,
                             clr       =>     clr_compare,
                             pre       =>     GND,
                             load      =>     ld_compare,
                             data_in   =>     data_in,
                             data_out=>       compare_reg);

-- if compare register is empty and look-up table is not empty, HOQ destination
-- vector is allowed to read into compare register at control(0) logic state.

        ld_compare <= reg_empty and control(0) and (not q_empty);


-- the following is the Request Register. The request signal from all output ports
-- will be latched into this register at control(1) logic state.

                register_request   : reg8bit
                port map (   clk       =>     clk,
                             clr       =>     clr_request,
                             pre       =>     GND,
                             load      =>     control(1),
                             data_in   =>     request,
                             data_out=>       request_reg);

        clr_request <= clr or control(7);

-- output the accept signal to OPCU.

        accept_sig <= request_reg and compare_reg;
        accept <= accept_sig;

-- the following is the Match Register. Confirm signal will be latched into this
-- register at control(5) logic state.

                register_match: reg8bit
```

```
                    port map (      clk        =>      clk,
                                    clr        =>      clr_match,
                                    pre        =>      GND,
                                    load       =>      control(5),
                                    data_in    =>      match_sig,
                                    data_out=>         match_reg);

        match_sig <= confirm and accept_sig;
        clr_match <= clr or control(0);

-- check if a match has been found.

        match_bit <= '1' when match_reg /= "00000000" else
                                    '0';
-- the appropriate bit(s) in compare register will be cleared if match has been
-- found.

        generate_clr: for i in clr_compare'range generate
                clr_compare(i) <= clr or (match_reg(i) and control(7));
        end generate;

-- output the match signal to read control logic (readcell.vhd).

                    FF_match: Dfflop
                    port map (      clk        =>      clk,
                                    clr        =>      clr_match,
                                    pre        =>      GND,
                                    load       =>      VCC,
                                    d          =>      match_bit,
                                    q          =>      match);

        VCC <= '1';
        GND <= '0';

end arch1;
```

The following VHDL design file describes the Request operation performs in OPCU.

```
entity requests is
port ( clk, clr: in std_logic;
```

```
-- empty signal from requisition look-up table. If table is empty, no request is sent
        q_empty : in std_logic;

-- head-of-queue requisition vector from the look-up table.
        data_in: in destination_code;

-- accept signal from IPCU.
        accept: in destination_code;

-- control signal to carry out operations in different logic states.
        control: in std_logic;

-- request and confirm signals to IPCU.
        request, confirm: out destination_code;

-- match signal to inform the DPSU that a match has been found.
        match : out std_logic;

-- load data signal to look-up table so that HOQ requisition vector can be
-- latched into register for request operation.
        ld_data: out std_logic;

-- 3-bit multiplexer control signal to select the correct input for multiplexer.
        mux_sel: out std_logic_vector (coder8_width-1 downto 0));

end requests;

architecture arch1 of requests is
signal request_reg, confirm_reg, request_sig, confirm_in: destination_code;
signal clr_select, empty_code: destination_code;
signal reg_empty, ld_select, clr_confirm, NULL0: std_logic;

begin

-- check if compare register is empty. If yes, HOQ destination vector can be read-in
-- for comparison.
        reg_empty <= '1' when request_reg = "00000000" else
                                    '0';
        ld_data <= reg_empty;

-- The following is the Request Register. Each bit of the Request Register can be
-- asynchronously cleared, after the "hand-shake" has been confirmed.
```

```
        register_select: reg8aclr
        port map (      clk             =>      clk,
                        clr             =>      clr_select,
                        pre             =>      GND,
                        load            =>      ld_select,
                        data_in         =>      data_in,
                        data_out        =>      request_reg);

-- if request register is empty and look-up table is not empty, HOQ requisition
-- vector is allowed to read into request register at control(15) logic state.

    ld_select <= reg_empty and control(15) and (not q_empty);

-- the appropriate bit(s) in request register will be cleared at control(7) logic state if
-- request has been confirmed by IPCU.

        generate_clr: for i in clr_select'range generate
        clr_select(i) <= clr or (confirm_reg(i) and control(7));
        end generate;

-- the following is the Confirm Register. Confirm signal will be latched into this
-- register at control(3) logic state.

        register_confirm: reg8bit
        port map (      clk             =>      clk,
                        clr             =>      clr,
                        pre             =>      GND,
                        load            =>      control(3),
                        data_in         =>      confirm_in,
                        data_out        =>      confirm_reg);

    confirm_in <= request_sig and accept;
    confirm <= confirm_reg;

-- priority encoder to find the position of the selected input port.
-- For example "01000000" → "001" (input port 1).

        mux_control: pcoder8
        port map (      enable          =>      VCC,
                        status          =>      VCC,
                        d               =>      confirm_reg,
                        Oena            =>      match,
                        GS              =>      NULL0,
```

```
                    q                =>        mux_sel);

-- the following process is used to find the priority request signal.
-- for example "00111011" → "00100000".

output_request: process
        variable temp: std_logic;

        begin
                temp := '0';
                for i in 0 to request_sig'length-1 loop
                        if temp = '1' then
                                request_sig(i) <= '0';
                        else
                                temp:= select_reg(i);
                                request_sig(i) <= select_reg(i);
                        end if;
                end loop;
        end process output_request;

        VCC <= '1';
        GND <= '0';
        request <= request_sig;
        ld_data <= reg_empty;

end arch1;
```

## 15.6 *FlexSwitch* Implementation

Figure 15.8 shows the assignment of the *FlexSwitch* functional units to CPLD devices. The *FlexSwitch* uses 16 IDT7M1015 RAM Modules manufactured by GEC as dual-port FIFO buffers. There are 8 I/O ports in *FlexSwitch* and each I/O port has 2 buffers, one serves as the input port buffer and the other as the output port buffer. The IDT7M1015 is a 4K x 36bit BiCMOS Dual-Port Static RAM Module constructed on a confired ceramic substrate using IDT7015 (4K x 9) asynchronous Dual-Port RAMs. The RAM module is fully utilized and allows to store up to 256 ATM cells, each cell being 64 bytes long (256 x 64 = 16Kbytes). However, *FlexSwitch* has 32-bit wide Data Bus, and therefore 4 bits from each address are left unused.

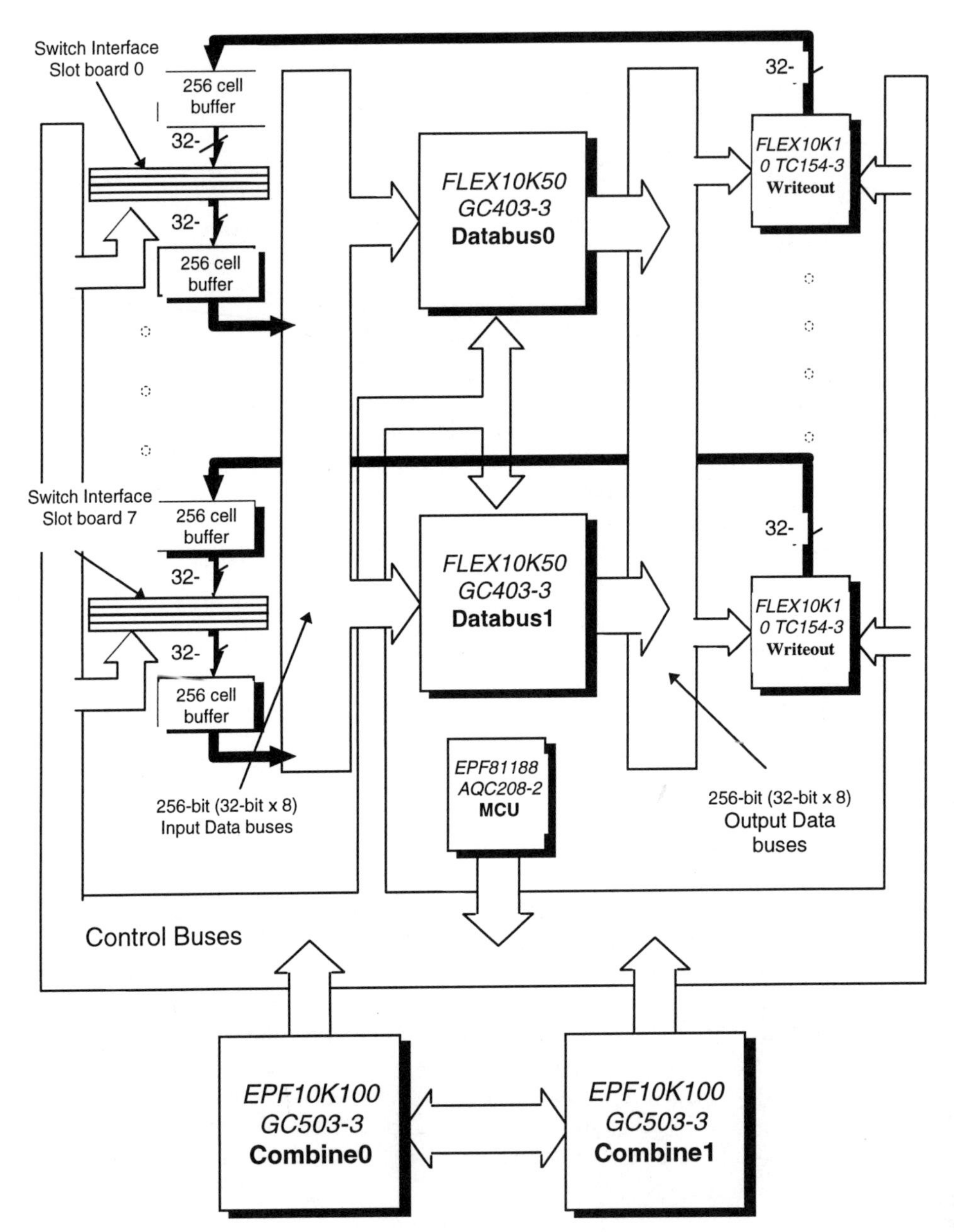


Figure 15.8Allocation of FlexSwitch functions to hardware devices

All FlexSwitch core functions are implemented in Altera FLEX 10K and 8000 Complex Programmable Logic Devices (CPLDs). Two FLEX10K devices are used for the core of the switching function of *FlexSwitch.* As the DPSU design is solarge that it cannot be fitted into a single device, it is split into 2 identical designs (called *Databus0* and *Databus1*) and fitted into 2 FLEX10K50 CPLD devices, with each design accepting 16 input data bus from 8 Input ports, and 16 output data bus to 8 output ports. This gives 32-bit data bus for each I/O port. A total of 300 I/O pins from each EPF10K50GC403-3 chip have been used and 73% of the Logic Array Blocks (LABs) are utilized.

The IPCUs, VRU and OPCUs are fitted into 2 FLEX10K100 devices (use EPF10K100GC503-3 chips). There are 8 pairs of IPCU/OPCU control units. The first 4 pairs, as well as the VRU are fitted into the first chip called *Combine0*, while the rest (the other 4 pairs) are fitted into the second chip called *Combine1*.

A total of 382 I/O pins from *Combine0* have been used which utilizes 43% of LABs and 66% of Embedded Array Blocks (EABs). Similarly, 380 I/O pins from *Combine1* are used which utilizes 35% of LABs and 66% of EABs. The purpose of combining the IPCUs, VRU, and OPCUs together is to reduce the complexity of the PCB.

The MCU is fitted into a separate FLEX8000 CPLD device in order to provide more cleaner and accurate synchronous signals to other functional units. The EPF81188AQC208-2 chip has been used to implement the MCU which requires 83 I/O pins and utilizes 81% of LABs.

Finally, 8 FLEX10K10 CPLD devices are used as the interfaces between DPSU and output ports buffers. They are used for signal filtration and the CPC (cell parity check) process. Noise is introduced to the output data from DPSU which is not stable enough to be written to the output port buffer. Therefore, the 32-bit data (for each output port) output from DPSU is passed through a Flex10K10 CPLD device for signal filtration before storing it in the output port buffer. The CPC process of OPCU is also done in this chip before writing data to the output port buffer. The CPLD devices have been used in this part of the design. The implemented devices are named *Writeout*, and require 87 I/O pins and each utilizes 41% of LABs.

## 15.7 Performance Evaluation

There are 3 parameters that need to be analyzed to evaluate the overall performance. These parameters are port throughput, cell throughput and overall time delay.

### *15.7.1 Port Throughput Evaluation*

As mentioned earlier in this paper, the transmission rate is dependent on the memory access time, which decides the operating frequency of the system. The *IDT7M1015* RAM modules have a memory access time of 15ns which yields a clock cycle of 30ns (15ns clock pulse for memory access). Therefore, 33.33 MHz is the maximum operating frequency. In this design, a 20ns clock pulse has been used for memory access which results in 40ns per clock cycle. Hence, the system is simulated at 25 MHz.

32-bit data bus is adapted for each I/O port. With an operating frequency of 25 MHz, data is transmitted at 800 Mbit/sec per port. This is called the I/O throughput or port throughput which is the minimum throughput of *FlexSwitch* if all the input ports are contending for transmission to the same output port at the same time. However, an aggregate throughput of 6.4 Gbit/sec can be achieved if there is no head-on blockage during transmission. This means that all 8 input ports are destined for different output ports at the same time. Full switching can be achieved under this circumstance.

Theoretically, FlexSwitch should be able to operate at 33.33 MHz (30 ns clock rate) because the IDT7M1015 RAM Module has access time as fast as 15ns. This yields an I/O throughput of 1.066 Gbit/sec and a maximum throughput of 8.533 Gbit/sec. This is still a conceptual result as the system has not been simulated and tested at 30 ns clock rate.

### *15.7.2 Cell Throughput Evaluation*

Although the I/O throughput is 800 Mbits/sec, the performance of the *FlexSwitch* is evaluated upon the number of cells that can be transmitted (cell throughput) within a second. This is because although an ATM cell only requires a minimum of 16 clock cycles (16 x 4-byte = 64-byte), it may take 18 clock cycles to be transmitted if necessary, with the extra clock cycles used for cell boundary confirmation.

The Switch Interface is able to operate at a frequency lower than the *FlexSwitch.* Therefore, boundary confirmation is a must while writing cells into the input port buffer and reading cells from the output port buffer; 18 clock cycles are required to transfer a single ATM cell and total of 1.3889 Mcells can be transferred every second. Once the cells are stored in buffers, they are switched from the input port buffer through DPSU to the output port buffer at a maximum speed of 16 clock cycles per cell. In such case, ATM cells are transferred at a rate of 1.5625 Mcells/sec.

### *15.7.3 Time Delay Evaluation*

The port-to-port delay is measured from the first row of ATM cells stored in the input port buffer through to the first row of data retrieved from the output port buffer. During this period, data is stored in input port buffer, then the data path is required to be set-up, and finally ATM cells are transmitted to the related output port buffer. The time delay is calculated to be a minimum of 1.84 μs with a maximum delay of 2.16 μs. The time variation is due to the asynchronous arrival of the ATM cell from Switch Interface which will be synchronized to the operation of the *FlexSwitch.*

It should be noted that the time delay has been calculated under assumption that there is no output port contention (In other words no head-of-queue congestion during switching). In the worst case, when 8 input ports are contending for the same output port at the same time, a maximum delay of 7 time slots is required for the least priority input port which raises the maximum delay to 6.64 μs.

## 15.8 Future of the FlexSwitch

The current *FlexSwitch* project is based on the standard 8 I/O ports, 32-bit bus design. This design has been simulated and tested thoroughly at 25 MHz system clock frequency. Our next step is to design a PCB and implement it using FLEX 10K devices which were only announced at the time of design simulation.

Once the PCB has been built, a true test can be carried out on the current design. Beyond that, the next phase of design shall concentrate on different I/O configuration files and implementing the Traffic Administrator into the *FlexSwitch.* This requires no further modification to the designed PCB, making the *FlexSwitch* a true high speed and flexible ATM switch by taking full advantage of the use of FPLD devices.

The expand to the 16 or more I/O ports, *FlexSwitch* requires extra components and therefore it must be re-designed. However, by replacing the FLEX10K100 device with the newly announced FLEX10K130, FLEX10K150 or FLEX10K250 CPLD devices, a slight modification in PCB design may expand the *FlexSwitch* system to 16 I/O ports or more.

The philosophy behind the *FlexSwitch* system is to design a high speed ATM switch that is flexible and able to operate in different network environments, and suits networks for distributed multimedia applications. The *FlexSwitch* project is exploring the hardware-intensive design using FPLD devices, and yet it is offering a network environment which is suitable for both the hardware-intensive or software-intensive processing of media, including audio and video, through the modularity of Switch Interface that interacts with end users. The *FlexSwitch* has created an excellent environment for research in multimedia applications and adds a new dimension to the development of ATM networks.

## 15.9 Questions and Problems

15.1. Analyze a modified design of the *FlexSwitch* in which input and output external buffers are replaced with internal EABs of the FLEX10K FPLD family. Discuss the consequences of using internal memories on the overall performance of the *FlexSwitch*. State the advantages and disadvantages of the modified design.

# 16 FLIX - A CUSTOM-CONFIGURABLE MICROCOMPUTER

In this chapter we present a more complex example of a microcomputer that is using benefits of both technologies: FPLDs to make custom-configurability reality at the designer's desk, and VHDL to achieve it in a very short design time. The basic ideas and inspiration for this design come from our early works on SimP and CCSimP custom-configurable microprocessors [14, 15, 17], but extend beyond their objectives in the direction of more complex and yet flexible and customisable microcomputer cores. The approach is uses an open concept leaving the designers option to change design in a practically unlimited number of directions. Taking the benefits of embedded memory arrays, the whole microcomputer, called a FLexible Instruction eXecution unit, or FLIX, is implemented in a single FPLD. The FLIX can be considered as a core for various application-specific computing machines, although it can be used for many applications without any modifications. The FLIX contains a set of basic microcomputer features that can be used directly without any change for some simple applications, or can be extended by the user in many application-specific directions. The FLIX architecture consists of a data path that uses two internal buses to interconnect microprocessor resources and a flexible control unit able to accommodate new control tasks. The data path can be modified by removing or adding other microprocessor resources (such as registers, or processing elements which perform simple data transformations, or even changing some of the fundamental concepts introducing separate program and data memory). Furthermore, the control unit may support those changes by providing control of the microoperations at the register transfer level. If more complex processing tasks that are computationally demanding are required, then the extensions of the FLIX can be achieved in two ways:

- By adding new instructions or other features to the Flix's core preserving the core architecture and basic principles of its operation. This approach becomes obvious once the operation of the FLIX is fully understood.
- By attaching functional units to the core without any change to the core. This approach requires no or just minor changes of the FLIX core, but the main task becomes the design of functional units. The FLIX architecture supports the

addition of functional units with minimum effort to the interfacing, and also leaves the designer with the possibility to use more advanced options.

The FLIX's design is completely specified in VHDL. Using a structured approach to hardware design together with the power of VHDL, the methodology FPLD-based microcomputer is developed which provides easy modification of the FLIX core or the addition of new features. In this chapter, we first represent FLIX's basic architectural principles and rationale behind the most important design decisions. In the second part, the whole design is presented and explained. The FLIX architecture represents a nucleus for many application-specific computers. As an example, Chapter 17 illustrates an approach to functional units used for the performance improvement of image enhancement algorithms.

## 16.1 Basic Architecture

There are a few design objectives that were set before the development of the FLIX architecture started. Those objectives can be summarized as follows:

- Simplicity of programming model. The programming model is represented by microprocessor resources that include registers, addressing modes and an instruction set. The FLIX operation is based on the load/store type model, assuming relatively frequent transfers of operands and results of operation between microprocessor and memory. The whole programming model is explained in the latter sections.
- Single address space architecture. This approach provides an unique treatment of memory and registers of input/output devices. It is adopted because it is expected that many of the input/output device interfaces will be implemented within the single chip together with the FLIX core. Also, a part of physical memory is implemented in the same chip. This memory, although implemented in the SRAM embedded blocks, can be treated as either RAM or ROM.
- Simple and uniform instruction execution cycle. The instruction execution takes always exactly four machine cycles. This uniformity requires a slightly more complex data path and registers for temporary storage of the current values of other registers or predictive calculation of the values which will never be used or used very rarely. Despite the additional internal data path resources, this approach leads to a fairly simple control unit, and enables further performance improvement using pipelining.
- Allocation of complex calculations to external functional units. All more complex operations or whole computational tasks which exceed the capabilities

of simple register transfers are detached to functional units. The control is transferred from the FLIX core to a functional unit in a sort of cooperating finite state machine which implements both the FLIX control unit and a functional unit control unit. A variety of operations can be performed under the control of functional units.

- Algorithm control from the FLIX program. The assumption is that the FLIX core executes the main program thread. When required, functional units are activated by simple FLIX instructions. Truly parallel operation of the FLIX core and functional unit(s) is possible, but with the programmer's responsibility for proper use of synchronization mechanisms provided as specialized instructions within the FLIX core. This mechanism will be briefly described, but not shown in this implementation.

### *16.1.1 FLIX Basic Features*

The FLIX core is a microcomputer with the following basic features implemented in the current version presented in this book:

- 16-bit processing unit that performs operations on 16-bit operands. It performs a number of simple arithmetic and logic operations, which can be further extended as required.
- 16-bit external data bus and 16-bit address bus that enable direct access to up to 64K 16-bit memory locations. A part of the low address memory locations is implemented in embedded memory blocks. The basic core has 256 16-bit RAM locations starting with address H"0000". A larger memory can be implemented using external chips.
- There are two programmer visible 16-bit working registers (accumulators), called A and B registers, which are used to store operands and results of data transformations. The data transformations can be performed within the logic implemented around these registers, or within the arithmetic-logic unit (ALU).
- Communication with the input and output devices is carried out using memory-mapped input/output. This presumes that all interfacing aspects are left for application-specific consideration.
- There are three registers supporting addressing modes. They are a 16-bit address register X used in index indirect addressing mode, a 16-bit stack pointer SP used for stack addressing mode, and a 6-bit page register PR used to support direct (absolute) addressing mode. Due to the limitation of instruction size to a single 16-bit word, FLIX supports direct addressing with 10 addressing

bits allowing direct access to 1K lowest addresses. This is considered as the access to the page 0 of the address space. Actually, direct address bits are concatenated with 6 bits of page register to enable direct access to any location, paying the price of additional instructions needed to control the value of the PR register.

- Addition of application-specific instructions is performed easy with the instructions similar to existing ones and reserved a number of operation codes, or by using a huge space of undefined instruction codes that are dedicated to functional units that execute them.
- It is implemented in a single Altera FLEX 10K20 CPLD with the spare resources for application-specific extensions, or can be targeted to any larger FLEX10K device.

Other FLIX features will be introduced in the following sections.

### *16.1.2 Instruction Formats and Addressing Modes*

The FLIX instructions have very simple formats. All instructions are 16-bits long requiring one memory word. These formats are shown in Figure 16.1. Addressing modes are coded

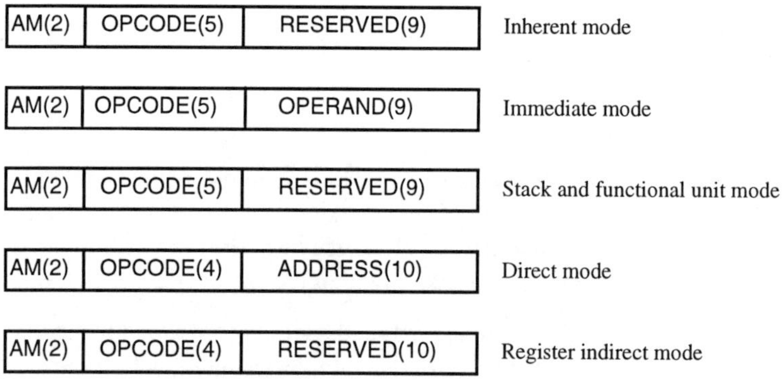

| | | | |
|---|---|---|---|
| AM(2) | OPCODE(5) | RESERVED(9) | Inherent mode |
| AM(2) | OPCODE(5) | OPERAND(9) | Immediate mode |
| AM(2) | OPCODE(5) | RESERVED(9) | Stack and functional unit mode |
| AM(2) | OPCODE(4) | ADDRESS(10) | Direct mode |
| AM(2) | OPCODE(4) | RESERVED(10) | Register indirect mode |

Figure 16.1 Instruction formats

with two most significant bits. The following addressing modes are available:

- Inherent addressing mode, which is used in most instructions, provides an access to FLIX registers to carry out operations on the contents of these registers.
- Immediate addressing mode, which is currently used in only one instruction to initialize the value of page register PR. The operand is contained in 6 least significant bits of instruction. This mode can be used for other specific purposes, such as initialization of the value of the other registers, provided that the value fits into nine bits.
- Stack addressing mode, which is used in a number of instructions to that use stack implicitly or explicitly.
- Functional unit addressing mode, which is used to initiate operations within functional units or to transfer operation codes of functional unit specific operations. These operations are fully dependent on the FLIX customization.
- Direct addressing mode, which is used to directly specify addresses of memory locations. Due to the limited instruction word length, it provides addressing capabilities within the single 1K word page defined by the contents of the PR register. Programming model requires that the programmer keeps a track of the current contents of the PR register, or, in small applications, use just page zero as default page.
- Register indirect addressing mode, which is used in to access any memory location or groups of adjacent locations using the X register as an address register. The contents of X register can be manipulated by a number of simple instruction with inherent addressing mode.

Further modifications of addressing modes are possible within customization concept as required with a specific application.

### *16.1.3 Register Set, Memory and I/O Registers*

FLIX contains a number of registers which are used as data storage to provide the operands when performing data transformations or various types of data transfers regarded as FLIX microoperations or register transfers, respectively. Some of the registers belong to the programming model and are accessible by the user, and the other ones are used to support internal microoperations of the processing unit. All user visible registers are presented in Figure 16.2.

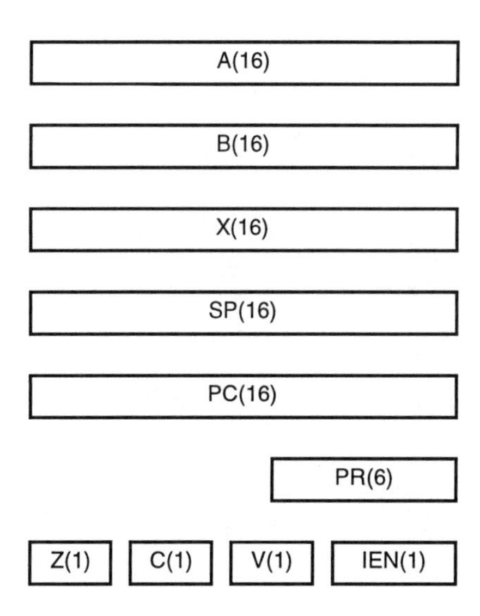


Figure 16.2 User-visible registers

Two further registers, the program counter temporary register TEMPC, and the auxiliary stack pointer register ST, are neither directly nor indirectly accessible by the user. They are used in some internal operations to save values (copies) of the program counter and stack pointer and provide a very simple and uniform instruction execution cycle. As we shall see, all instructions execute in exactly four machine (clock) cycles, thus providing eventual performance improvement using pipelining in a more advanced version of the FLIX microcomputer. Finally, two other registers are used to support internal operations: the instruction register IR is used to store the most recently fetched instruction for decoding, and the address register AR, which is used to store an effective address in direct addressing mode, or to support other addressing modes when needed, as we shall see in the latter sections. All user invisible registers are illustrated in Figure 16.3.

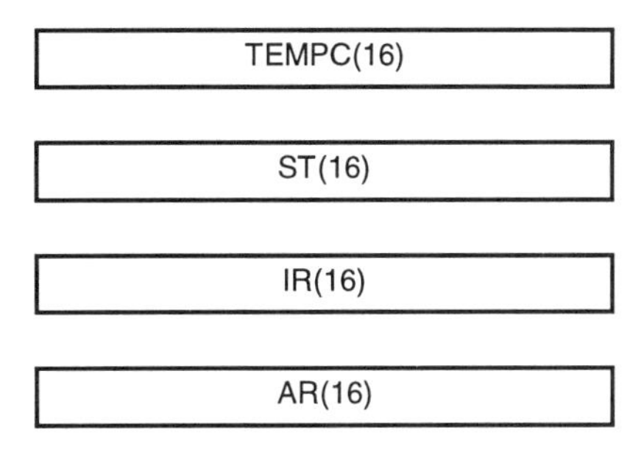


Figure 16.3 User invisible registers

The FLIX register set contains a number of additional single-bit registers used for control purposes during system reset and interrupt processing. They will be explained when discussing the FLIX control unit.

### *16.1.4 Instruction Set*

The FLIX core instruction set is illustrated in Table 16.1. It consists of the following instruction groups:

- Memory reference instructions, which use direct or register indirect addressing mode to transfer data between the FLIX core and memory or I/O registers.
- Register reference instruction, which operate on the contents of the FLIX registers.
- Program flow control instructions, which enable conditional and unconditional change of program flow.

Besides these, there is a group of functional unit related instructions that will be explained in a latter section that deals with the FLIX extensions to functional units.

Table 16.1 FLIX Instruction Set

| Instruction mnemonic | Function |
|---|---|
| Memory reference | |
| LDA or LDAX | $A \leftarrow M[address]^{*}$ or $A \leftarrow M[X]$ |
| LDB or LDBX | $B \leftarrow M[address]$ or $B \leftarrow M[X]$ |

| | |
|---|---|
| STA or STAX | M[address] ← A or M[X] ← A |
| STB or STBX | M[address] ← B or M[X] ← B |
| LDX | X ← M[address] |
| LDSP | SP ← M[address] |
| STOX | M[address] ← X |
| PSHA | stack ← A, SP ← SP-1 |
| PULA | SP ← SP+1, A ← stack |
| Register reference | |
| ADAB | A ← A + B |
| SBAB | A ← A - B |
| ANAB | A ← A and B |
| ORAB | A ← A or B |
| LDPR | PR ← address |
| CLA | A ← 0 |
| CLB | B ← 0 |
| CMB | B ← B' |
| INCB | B ← B + 1 |
| DECB | B ← B - 1 |
| INCX | X ← X + 1 |
| DECX | X ← X - 1 |
| INCS | SP ← SP + 1 |
| DECS | SP ← SP - 1 |
| CLflag | flag ← 0 (flag can be C, Z, V) |
| ION | IEN ← 1, Enable interrupts |
| IOF | IEN ← 0, Disable interrupts |
| Program flow control | |
| SZ | If Z=1, PC ← PC + 1 |
| SC | If C=1, PC ← PC + 1 |
| SV | If V=1, PC ← PC + 1 |
| JMP | PC ← address |
| JSR | stack ← PC, PC ← address, SP ← SP-1 |
| RET | SP ← SP+1, PC ← stack |
| NOOP | No Operation |

* M[address] means the contents of memory location specified with address

## 16.2 Processor Data Path

The FLIX internal structure is presented by the simplified data path in Figure 16.4. The processor contains two 16-bit internal buses: a data bus, often referred to as DBUS, and an address bus, referred to as ABUS, enabling two simultaneous register transfers to take place within the data path. The DBUS is connected to the external pins and enables easy connection with external memory up to 64K 16-bit words or to the registers of input and output devices in a memory-mapped scheme. In the FLIX version presented in this chapter, external data bus is implemented with separated input and output data lines enabling the user to make it bi-directional if required. The ABUS is available only for internal register transfers.

In the internal structure of Figure 16.4, describing the FLIX data path, all user visible and non-visible registers, except condition code and interrupt-related registers, are shown. The 16-bit instruction register, IR, is connected to the instruction decoder which is a part of the control unit. Actually, the instruction decoder decodes only upper 8 bits, while the remaining 8 lower bits are available as external signals to the application-specific functional units. The details of the use of all registers will be explained in the following sections.

The Arithmetic-Logic Unit (ALU) performs simple arithmetic and logic operations on 16-bit operands as specified in the instruction set. In its first version, the ALU performs only four operations, “addition”, “subtraction”, “logical and”, and “logical or”. It can easily be extended to perform additional operations as required. Some data transformations, such as incrementing and decrementing contents of registers, and initialization of working registers, are carried out by registers’ surrounding logic.

The FLIX memory consists of two parts: internal memory block of the size of 256 16-bit words (which can be extended as the capacity of CPLD used allows), and an external memory up to the maximum capacity of the FLIX address space. Access to the external memory and input output devices is provided through a number of multiplexers that are used to form the buses. An external address is generated on the address lines, A[15..0], as the result of selection performed on selection lines of memory multiplexer, ADRMUX. Usually, the effective address is contained in address register AR, but in some cases it will be taken from another source, stack pointer SP, auxiliary stack pointer ST, or X register when register indirect addressing is used.

Two other multiplexers which are used to form the ABUS and DBUS are not shown in Figure 16.4, but it is obvious where they are placed. It is also obvious from Figure 16.4 that several registers can be the source of data on both these buses. Two multiplexers, called the address bus multiplexer, ABMUX, and data

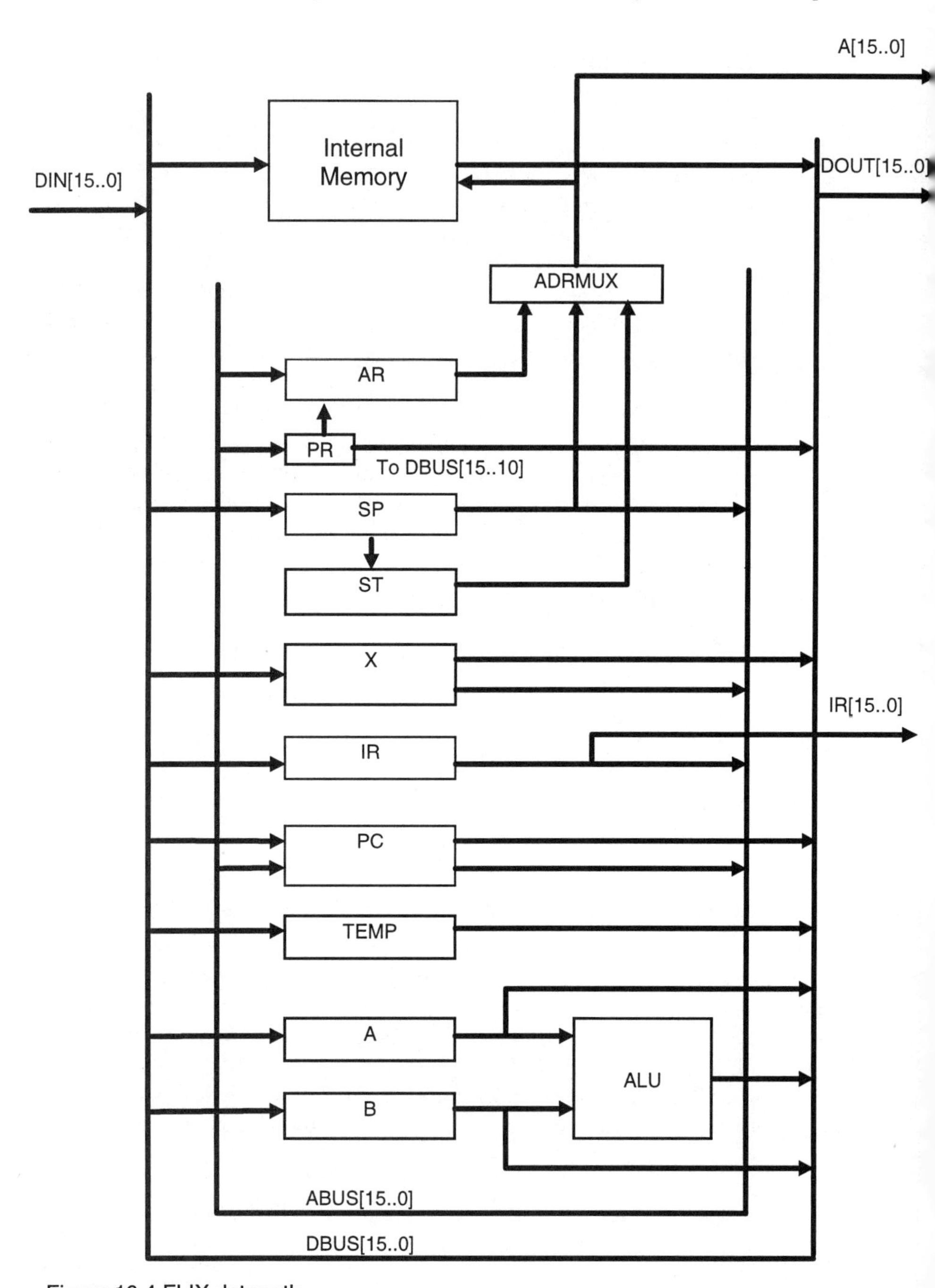


Figure 16.4 FLIX datapath

bus multiplexer, DBMUX, are used to enable access to the ABUS and DBUS, respectively. The only register that can be the source of data for both these buses is the program counter, PC. The details of register transfers that can take place will be discussed together with implementation details.

Memory can be both the source and destination in data transfers determined by the memory control lines that specify either the memory read or memory write operation. Memory location that takes place in data transfer is specified by the value of the output of ADRMUX which in turn specifies the effective address.

All register transfers are initiated and controlled by the FLIX control unit. Control unit carries out selection of the source of data for each of internal buses, destination of data transfer, as well as operations local to individual resources. For instance, it activates memory read or write operation, initialization of individual registers, performs such operations as the incrementing or decrementing of the contents of a register, and selects the operation of the arithmetic-logic unit. All register transfers are synchronized by the system clock and take place at the next clock tick.

## 16.3 Instruction Execution

The FLIX core instructions are executed as sequences of microoperations represented by register transfers or simple operations within the individual registers and completely follow execution of its predecessor [ ]. The basic instruction cycle is divided into three major steps that take place in the four machine clock cycles denoted by T0, T1, T2, and T3:

1. Instruction fetch, when a new instruction is fetched from external memory location pointed to by program counter. It is performed in two machine cycles. The first cycle, T0, is used to transfer the address of the next instruction from the program counter to the address register, and the second cycle T1 is used to actually read the instruction from the memory location into instruction register, IR.
2. Instruction decode, when the recognition of the operation that has to be carried out is done, and preparation of effective memory address is performed. This is done in the third machine cycle T2 of instruction cycle.
3. Instruction execution, when the actual operation specified by operation code is carried out. This is done in the fourth machine cycle T3 of instruction cycle.

Besides these fundamental operations in each machine cycle, various auxiliary operations are also performed that enable each instruction to be executed in exactly four machine cycles. They also provide the consistency of contents of all processor registers at the beginning of each new instruction cycle.

Instructions are executed in the sequence in which they are stored in memory. The only programmer controlled exception is if program flow change instructions are used. Besides this, the FLIX provides a very basic single-level interrupt facility that enables the change of the program flow based on the occurrence of external events represented by hardware interrupts. A hardware interrupt can occur at any instance of time as it is controlled by an external device. However, the FLIX control unit checks for existence of the hardware interrupt at the end of each instruction execution, and, in the case that the interrupt has been required, it sets an internal flip-flop called interrupt flip-flop IFF. At the beginning of each instruction execution, the FLIX control unit checks if IFF is set. If it is not set, the normal instruction execution takes place. If the IFF is set, FLIX enters an interrupt cycle in which the current contents of the program counter is saved on the stack, and the execution is continued with the instruction specified by the contents of memory location called interrupt vector, INTVEC. The interrupt vector represents the address of the memory location which contains the first instruction of Interrupt Service Routine, which further executes as any other program sequence. At the end of this sequence a return to the interrupted sequence, represented by the returning address saved on the stack at the moment of the interrupt acknowledgment, is performed using the "ret" instruction.

The overall instruction execution and control flow of the FLIX control unit, including normal execution and interrupt cycle, is represented by the state flowchart of Figure 16.5. This flowchart is used as the basis for the state machine that is implemented as the part of the control unit.

Some other 1-bit registers (flip-flops) appear in the flowchart of Figure 16.5. The first is the interrupt request flip-flop, IRQ, which is used to record the active transition on the interrupt request input line of the microprocessor. When the external device generates an interrupt request, the IRQ flip-flop will be set, and, under the condition that interrupts are enabled, it will cause the IFF flip-flop to be set. Consequently, the interrupt cycle will be initiated instead of normal instruction execution cycle. Control of the interrupt enable, IEN, flip-flop is carried out by programmer using instructions to enable or disable interrupts. Initially, all interrupts are enabled automatically. After recognition of an interrupt, further interrupts are disabled automatically. All other interrupt control is under programmer responsibility. During the interrupt cycle, the IRQ flip-flop will be cleared enabling

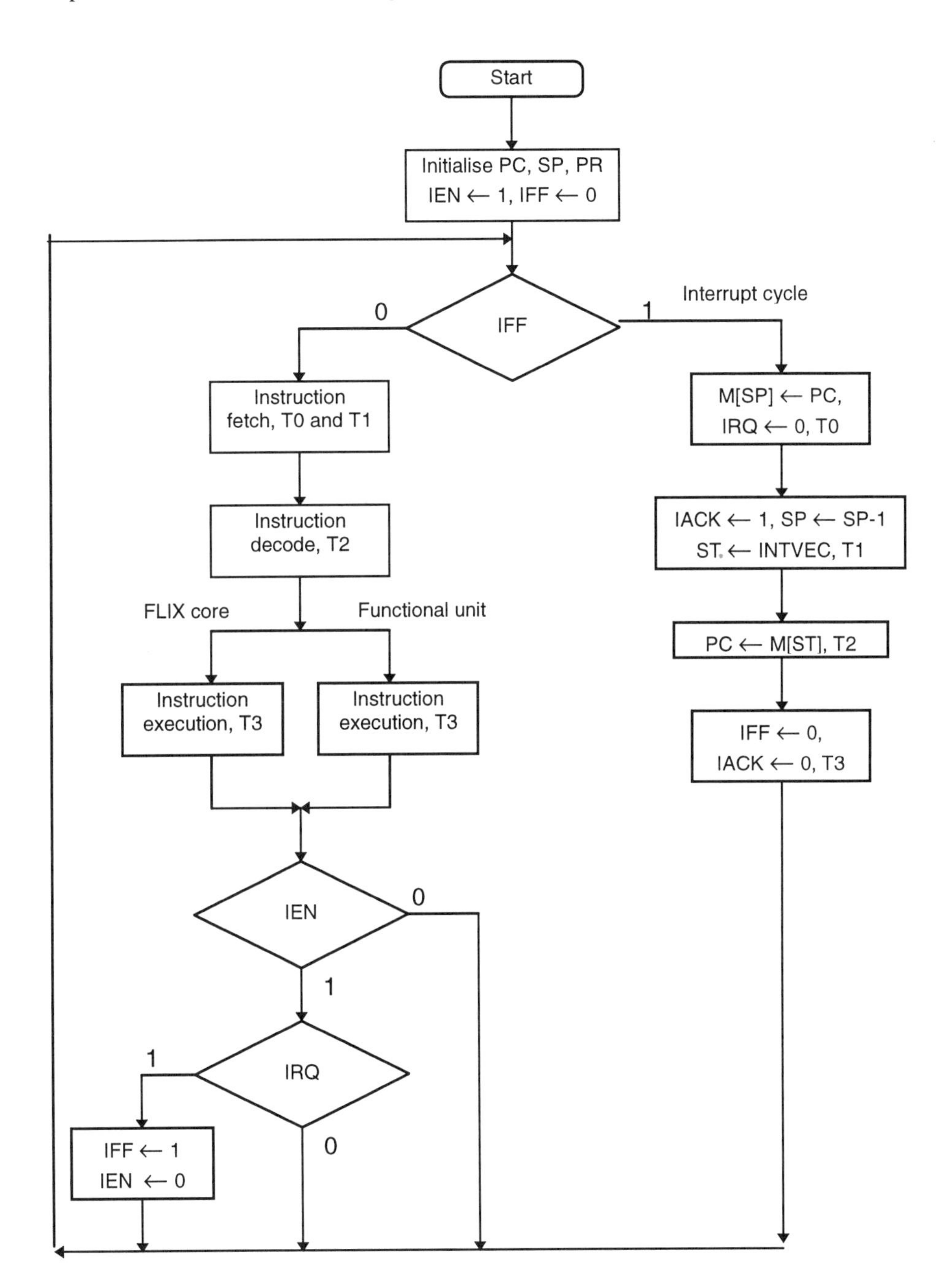


Figure 16.5 Control flow diagram

new interrupt requests to be recorded. Also, interrupt acknowledgment information will be transferred to the interrupting device in the form of the pulse that lasts two clock cycles (IACK flip-flop is set in the machine cycle T1, and cleared in the cycle T3 of the interrupt cycle).

Now, we can describe the normal instruction execution cycle. It is illustrated by the flowchart in Figure 16.6. In the first machine cycle, T0, the contents of the program counter is transferred to the address register preparing the address of the memory location where the next program instruction is stored. The next machine cycle, T1, is firstly used to fetch and transfer an instruction to the instruction register to enable further decoding. In the same cycle, two other microoperations are performed.

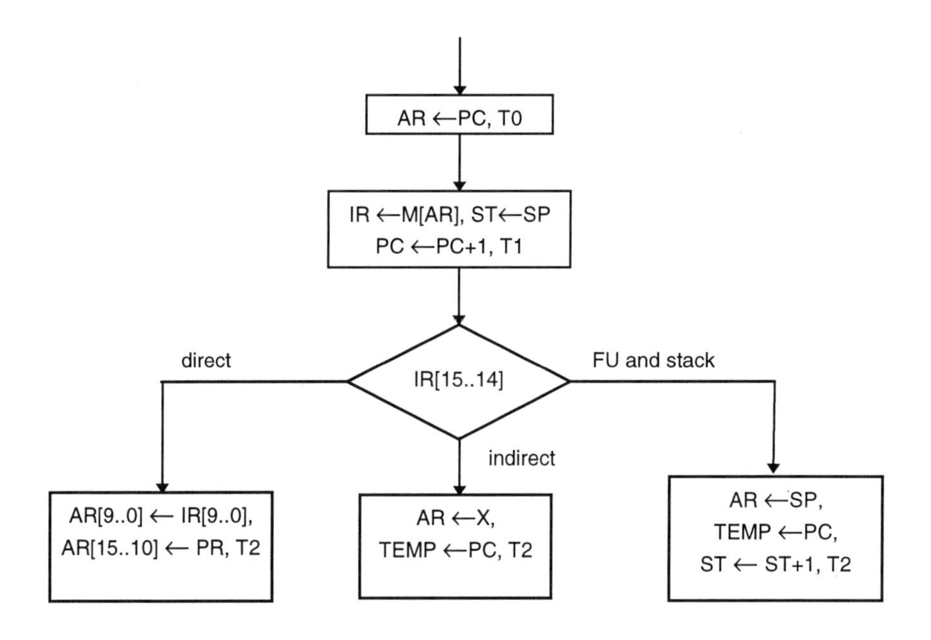


Figure 16.6 Instruction execution flowchart

The program counter is incremented to point to the next instruction which should be executed if there is no change in program flow. Also, the stack pointer SP is copied into the ST. This actually represents preparation for the case the control unit finds that the instruction uses stack addressing mode in the next machine cycle.

Register transfers that will take place in the next machine cycle, T3, depend on the addressing mode used in the instruction, represented by a value of the two most significant bits. These bits are decoded as bits IR[15..14] of the instruction register. In the case of direct addressing mode, the lower 10 instruction bits, IR[9..0] represent the effective address which will be used during the instruction execution step. Therefore, they are transferred to the address register preparing the effective address for the last machine cycle if needed. The other address bits are fetched from the PR register. In the case of indirect addressing mode, the the value contained in the X register will represent the effective address. Therefore, it is transferred from the X to AR register. If none the former cases apply then it is the case of functional unit or stack addressing mode. If the functional unit addressing is used, it is instruction which executes custom, application-specific instruction in a functional unit. Actions undertaken by the control unit for this case will be explained latter. Otherwise, the instruction belongs to one using the stack addressing mode. In order to execute these instructions efficiently, preparation for all possible directions of program flow in which instruction execution can continue are done. Firstly, the stack pointer is copied into the address register preparing for instructions which are pushing data onto the stack ("push" and "jump to subroutine" instruction) during the execution step. Secondly, the program counter is copied into the TEMP register in order to prepare for instructions that must save the contents of the program counter onto the stack and change the value of the program counter ("jump to subroutine" instruction). Finally, the ST register is incremented to prepare for the instructions that are pulling data from the stack ("pull" and "ret" instructions). These steps also enable proper updating (incrementing or decrementing) of the SP register in the T3 machine cycle, while the stack is accessed using the AR or ST register as the source of the effective address.

The instruction execution step is performed in the T3 machine cycle for all FLIX core instructions. It is presented for all instructions in Table 16.2.

Table 16.2 Instruction execution step

| Instruction mnemonic | Function |
|---|---|
| Memory reference | |
| LDA or LDAX | A ← M[AR] |
| LDB or LDBX | B ← M[AR] |
| STA or STAX | M[AR] ← A |
| STB or STBX | M[AR] ← B |
| LDX | X ← M[AR] |
| LDSP | SP ← M[AR] |
| STOX | M[AR] ← X |

| | |
|---|---|
| PSHA | M[AR] ← A, SP ← SP-1 |
| PULA | SP ← SP+1, A ← M[AR] |
| Register reference | |
| ADAB | A ← A + B |
| SBAB | A ← A - B |
| ANAB | A ← A and B |
| ORAB | A ← A or B |
| LDPR | PR ← IR[5..0] |
| CLA | A ← 0 |
| CLB | B ← 0 |
| CMB | B ← B' |
| INCB | B ← B + 1 |
| DECB | B ← B - 1 |
| INCX | X ← X + 1 |
| DECX | X ← X - 1 |
| INCS | SP ← SP + 1 |
| DECS | SP ← SP - 1 |
| CLflag | flag ← 0 (flag can be C, Z, V) |
| ION | IEN ← 1, Enable interrupts |
| IOF | IEN ← 0, Disable interrupts |
| Program flow control | |
| SZ | If Z=1, PC ← PC + 1 |
| SC | If C=1, PC ← PC + 1 |
| SV | If V=1, PC ← PC + 1 |
| JMP | PC[15..10] ← PR, PC[9..0] ← address or PC ← M[AR] |
| JSR | M[AR] ← TEMP, PC[15..10]← PR, PC[9..0]← IR[9..0], SP ← SP-1 or M[AR] ← TEMP, PC← M[AR], SP ← SP-1 |
| RET | PC ← M[ST], SP ← SP+1 |
| NOOP | No Operation |

## 16.4 Customising FLIX with Functional Units

FLIX allows two types of extensions: a) to add simple instructions that perform operations on the existing data path (simple register transfers and operations) not requiring any serious modification of either data path or control unit, and/or b) to add instructions that invoke more complex operations without limitation on the execution time; these operations are carried out in functional units added to existing core and controlled and synchronized by user-specific instructions. Number and the meaning of these instructions is practically unlimited. Generally, they support two modes of operations of functional units with respect to the core:

- synchronous operation, which presumes that the core halts program execution (and waits) until the functional unit completes the requested operation (instruction)
- asynchronous operation, which presumes that the core, after starting an instruction in a functional unit continues execution of program and synchronizes with the functional unit by executing wait instruction.

This enables writing of very flexible programs to implement and control the entire application. The designer of the functional unit only has to assign operation codes to the instructions which will be executed in the functional unit, and concentrate on the design of the functional unit itself. The framework for FLIX customization with functional units is illustrated in Figure 16.7.

FLIX executes a program which consists of instructions from the FLIX core instruction set, which are executed using the FLIX data path under control of the FLIX control unit, and new instructions that represent commands to a functional unit. It should be noted that several functional units may be attached to the FLIX core simultaneously. The functional unit can execute one or more instructions from its own instruction set. Instruction execution in the functional unit is under the full control of its own control unit. Overall program execution control is transferred from the FLIX control unit to the functional control unit and vice versa, as required by a program. The FLIX control unit transfers control to the functional unit (FUCU) control unit which, in turn, controls both its own and the core's data path as required. Upon completion of the required operations, the FUCU returns control to the FLIX control unit. In order to communicate with functional units, the instruction set required by the specific application has to be specified. A number of generic instructions is specified which may be appropriate for most of the functional units. They are presented in Table 16.3.

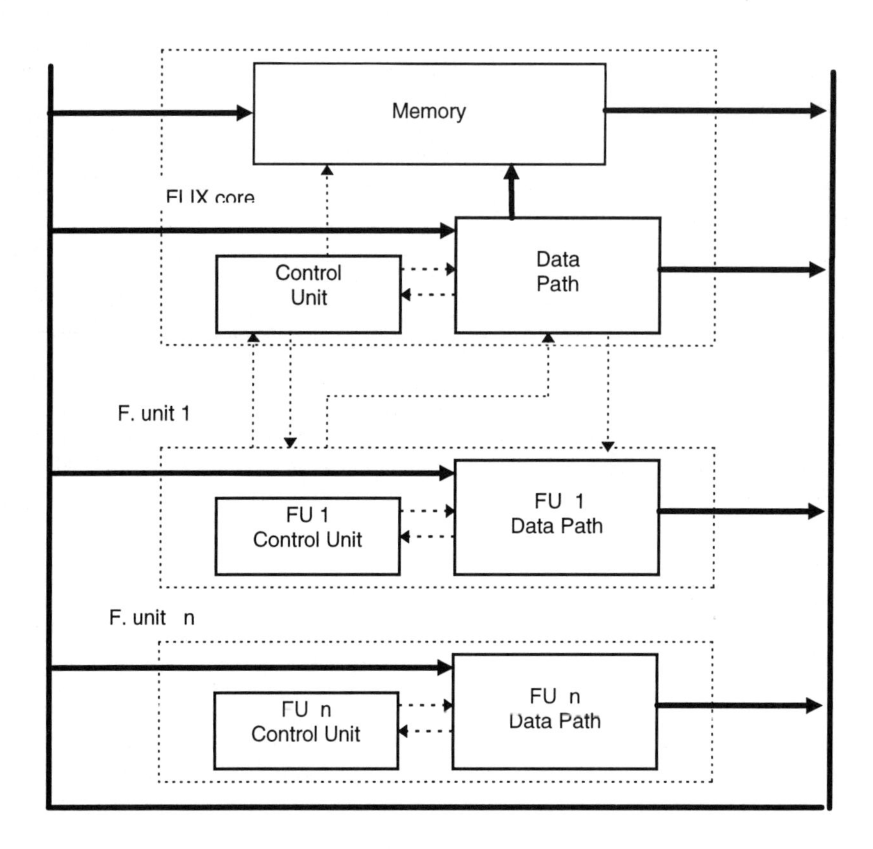


Figure 16.7 Extending FLIX with functional units

Table 16.3 Functional unit generic instructions and their meaning

| Instruction Mnemonic | Function |
|---|---|
| FTAF n | FIN[n] ← A, Transfer accumulator A to FIN[n] , n is FU ID |
| FTFA n | A ← FOUT[n], Transfer FOUT[n] to accumulator A, n is FU ID |
| FIOP n | Execute instruction in functional unit n specified by OP(code), FLIX to wait state until the specified operation completed |
| FSOP n | Execute instruction in functional unit n specified by OP(code), and continue with execution of instructions from common memory |
| FWOP n | FLIX to wait state until the current operation in functional unit n completed |

They comply with the general model of the functional units as shown in Figure 16.8. This model assumes the existence of functional unit control unit and data path. The FUCU can generally be in one of four major states:

- initialization, when the functional unit circuitry is initialized
- waiting on a new instruction (command) from the FLIX core
- executing instruction synchronously with FLIX operation, and
- executing instruction asynchronously with FLIX operation or
- waiting to synchronize with the FLIX core.

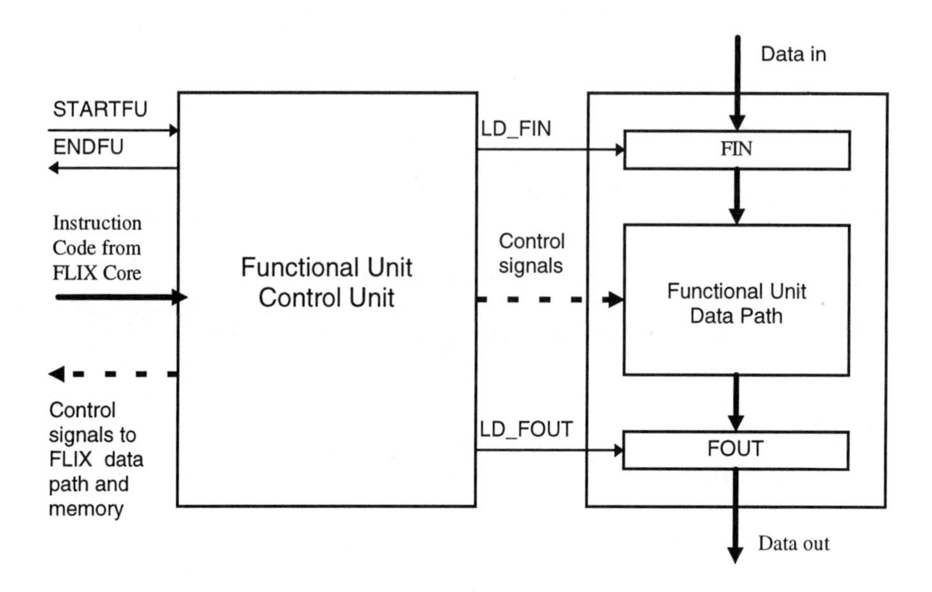


Figure 16.8 General model of the FLIX functional unit

An example of extending the FLIX core with functional units for image enhancement is presented in details in Chapter 16.

## 16.5 FLIX Implementation

The FLIX design that follows represents a traditional approach of digital system design partitioned into the data path and control unit parts as it is illustrated in Figure 16.9. The data path consist of all registers, interconnect structures including various multiplexers, and data processing resources. It enables register transfers under the control of multiplexer selection signals and control signals of the registers, local operations on the contents of the registers, data transformations in the arithmetic-logic unit, and data exchange with the outside world (memory and input/output devices). From an external point of view, it provides the 16-bit address bus, and 16-bit data bus. The control unit provides proper timing, sequencing and synchronization of microoperations, and activation of control signals at various points in the data path as required by the microoperations. It also provides control signals which are used to control external devices such as memory operations and the interrupt acknowledgment signal. The operation of the control unit is based on information provided by the program (instructions fetched from memory), results of previous operations, as well as the signals received from the outside world. In our case the only such signal is the interrupt request received **from the interrupting device.**

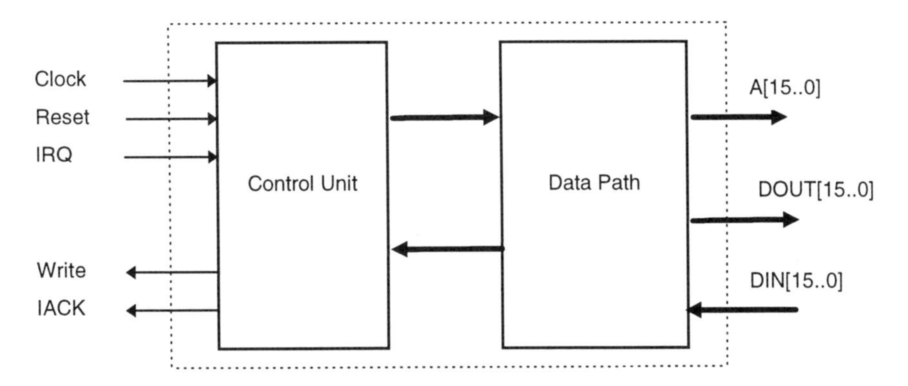


Figure 16.9 Basic Partition of FLIX Design

### *16.5.1 Design Packages*

In order to simplify the FLIX VHDL description and allow easy modifications of some of its parameters for customization purposes, three VHDL design packages are declared first. The first package, flix_types, defines a number of types that represent standard logic vectors of different lengths, as found in the latter design.

```
library ieee;
use ieee.std_logic_1164.all;
package flix_types is
        subtype bit_16 is std_logic_vector(15 downto 0);
        subtype bit_10 is std_logic_vector(9 downto 0);
        subtype bit_8 is std_logic_vector(7 downto 0);
        subtype bit_6 is std_logic_vector(5 downto 0);
        subtype bit_5 is std_logic_vector(4 downto 0);
        subtype bit_4 is std_logic_vector(3 downto 0);
        subtype bit_3 is std_logic_vector(2 downto 0);
        subtype bit_2 is std_logic_vector(1 downto 0);

end flix_types;
```

The second package various_constants, declares a number of constants, primarily values used for selecting data sources at the multiplexers inputs, and selection of the operations of the arithmetic-logic unit.

```
library ieee;
use ieee.std_logic_1164.all;
use work.flix_types.all;

package various_constants is

-- dbus selection  dbsel
        constant selpc2db: bit_3:="000";
        constant seltemp2db: bit_3:="001";
        constant selx2db: bit_3:="010";
        constant seldata2db: bit_3:="011";
        constant selpr2db: bit_3:="101";
        constant selalu2db: bit_3:="100";
        constant passa2db: bit_3:="110";
        constant passb2db: bit_3:="111";

-- abus selection absel
        constant selar2ab: bit_2:="00";
        constant selpc2ab: bit_2:="01";
        constant selsp2ab: bit_2:="10";
        constant selir2ab: bit_2:="11";

-- address selection constants adsel

        constant selar2adr: bit_2:="00";
        constant selsp2adr: bit_2:="10";
        constant selst2adr: bit_2:="11";

-- alu operation selection alusel
        constant addd: bit_2:="01";
        constant subb: bit_2:= "00";
        constant andd: bit_2:="10";
        constant orr: bit_2:="11";

end various_constants;
```

The third package contains codes for different addressing modes and operation codes for all instructions.

```
library ieee;
use ieee.std_logic_1164.all;
use work.flix_types.all;

package opcodes is

-- addressing modes
        constant impl: bit_2:="00";
        constant funit: bit_2:="01";
        constant direct: bit_2:="10";
        constant indirect: bit_2:="11";

-- operations with direct and indirect a.m.
        constant lda: bit_4:="0000";
        constant ldb: bit_4:="0001";
        constant sta: bit_4:="0010";
        constant stb: bit_4:="0011";
        constant ldx: bit_4:="0100";
        constant stox: bit_4:="0101";
        constant ldsp: bit_4:="0110";
        constant jmp: bit_4:="1000";

-- operations with implicit and immediate a.m.
        constant adab: bit_5:="00000";
        constant sbab: bit_5:="00001";
        constant anab: bit_5:="00010";
        constant orab: bit_5:="00011";
        constant clra: bit_5:="00100";
        constant clrb: bit_5:="00101";
        constant comb: bit_5:="00110";
        constant incb: bit_5:="00111";
        constant decb: bit_5:="01000";
        constant incx: bit_5:="01001";
        constant decx: bit_5:="01010";
        constant incs: bit_5:="01011";
        constant decs: bit_5:="01100";
        constant clrc: bit_5:="01101";
        constant clrz: bit_5:="01110";
        constant clrv: bit_5:="01111";
        constant ion: bit_5:="10000";
        constant ioff: bit_5:="10001";
        constant skc: bit_5:="10010";
```

```
        constant skz: bit_5:="10011";
        constant skv: bit_5:="10100";
        constant ldpr: bit_5:="11000";
        constant noop: bit_5:="11111";

-- operations with stack a.m. and functional units operations
-- stack related opcodes start with 0 and functional units have opcodes
-- beginning with '1'

        constant pula: bit_4:="0000"; -- opcodes begin with '0'
        constant psha: bit_4:="0001";
        constant ret: bit_4:="0010";
        constant jsr: bit_4:="0011";

end opcodes;
```

### *16.5.2 Data Path Implementation*

In order to describe all resources of the data path represented by the block diagram of Figure 16.4, we must first identify data inputs and data outputs of each of the resources in the data path, the operations that can be carried out and the control signals that initiate those operations.

The whole data path is presented as a single entity that contains a number of cooperating processes. Processes are used to describe all individual resources: registers and processing elements. Besides them, a number of library parametrized components have been used, as well as the other features of VHDL that enable full synthesis of the description. Instead of giving any further description, we will comment features of these resources in the VHDL description below.

```
-- this package specifies characteristics of FLIX's internal memory

package ram_constants is
        constant addr_width: integer:= 8;
        constant data_width: integer:= 16;
end ram_constants;

library ieee;
library lpm;
use ieee.std_logic_1164.all;
use ieee.std_logic_unsigned.all;
use ieee.std_logic_arith.all;
```

```
use work.flix_types.all;
use lpm.lpm_components.all;
use work.ram_constants.all;
use work.various_constants.all;
entity datapath is
port(clk: in std_logic;

-- specify the input and output data lines, output address lines, and output
-- instruction register lines (lower 8 bits are available to the user)

        datain: in bit_16;
        dataout: inout bit_16; -- used as both output data lines and internal
                                  -- dbus lines
        addressout: out bit_16;
        irout: out bit_8;

-- select lines to three multiplexers and arithmetic-logic unit
        dbsel: in bit_3;
        absel: in bit_2;
        adsel: in bit_2;
        alusel: in bit_2;

-- inputs from the control unit that control individual resources
        init_pc, lda_pc, ldd_pc, inc_pc: in std_logic; -- program counter
        ld_x, inc_x, dec_x: in std_logic; -- address (indirect) register
        ld_arl, lda_arh, ldp_arh, init_ar: in std_logic; -- memory address
                                                          -- register
        ld_ir: in std_logic; -- instruction register
        ld_temp: in std_logic; -- temporary storage for program counter
        ld_a, clr_a: in std_logic; -- accumulator a
        ld_b, inc_b, dec_b, clr_b, com_b: in std_logic; -- accumulator b
        ld_pr: in std_logic;  -- page register
        init_st, inc_st, ld_st: in std_logic; --auxiliary stack pointer
        init_sp, inc_sp, dec_sp, ld_sp: in std_logic; -- stack  pointer

-- control inputs to condition code bits
        clr_c, clr_z, clr_v: in std_logic;
        ld_c, ld_z, ld_v: in std_logic;

-- write control signal
        write: in std_logic;
```

```
-- outputs from condition code bits
            zout, cout, vout: out std_logic;
            );

end datapath;

architecture combined of datapath is

-- internal buses and signals

                        signal intdbus: bit_16;
                        signal intabus: bit_16;
                        signal dmemout: bit_16;
                        signal maddress: bit_16;
                        signal pc_hold, ar_hold, x_hold, acca_hold, accb_hold,
                                temp_hold, sp_hold, st_hold, ir_hold: bit_16;
                        signal pr_hold: bit_6;
                        signal aluout: bit_16;
                        signal data: bit_16;
                        signal result: bit_16;
                        signal z1, c1, v1: std_logic;

begin

-- instance of internal RAM

  int_ram: lpm_ram_dq
        generic map (lpm_widthad => addr_width,
                        lpm_width => data_width  )
        port map (data => intdbus, address => maddress(7 downto 0), inclock=>
        clk, outclock=> clk, we => write, q=> dmemout);

-- instance of adder/subtractor - alusel(0) selects addition and subtraction
        ad_sub: lpm_add_sub
        generic map (lpm_width=> data_width, lpm_representation =>
        "UNSIGNED", lpm_direction => "DEFAULT" )
        port map (dataa => acca_hold, datab => accb_hold, add_sub => alusel(0),
        result => result, cout=>c1, overflow=> v1);

-- program counter has two data inputs, one from dbus and the other from abus
-- and page register, load control is split into upper six and lower 10 bits, initial
-- value can easily be changed within design or declared as constant in the
```

```
-- topmost packages

        pc: process(clk)
                variable mpc: bit_16;
        begin
        if(clk'event and clk='1') then
                if init_pc='1' then
                        mpc:= "0000000000000000";  -- initial value
                elsif lda_pc= '1' then
                        mpc(9 downto 0):= intabus(9 downto 0);
                        mpc(15 downto 10):= pr_hold;
                elsif ldd_pc = '1' then
                        mpc:= intdbus;
                elsif inc_pc = '1' then
                        mpc:= mpc + 1;
                else
                        mpc:=mpc;
                end if;
        end if;
                pc_hold <=mpc;

        end process pc;

-- address (indirect) register; has only one data input from dbus; can be
-- incremented and decremented

        xreg: process(clk)
                variable mx: bit_16;
        begin
        if(clk'event and clk='1') then
                if ld_x= '1' then
                        mx:= intdbus;
                elsif dec_x = '1' then
                        mx:= mx-1;
                elsif inc_x = '1' then
                        mx:= mx + 1;
                else
                        mx:=mx;
                end if;
        end if;
                x_hold <=mx;
        end process xreg;
```

```
-- memory address register; holds effective address; divided into the upper 6 and
-- lower 10 bits, which are controlled separately

        areglow: process(clk)  -- low part
                variable marl: std_logic_vector(9 downto 0);
        begin
        if(clk'event and clk='1') then
                if ld_arl= '1' then
                        marl:= intabus(9 downto 0);
                else
                        marl:=marl;
                end if;
        end if;
                ar_hold(9 downto 0) <=marl;

        end process areglow;

        areghigh: process(clk) – high part
                variable marh: std_logic_vector(5 downto 0);
        begin
        if(clk'event and clk='1') then
                if lda_arh='1' then
                        marh:=intabus(15 downto 10);
                elsif ldp_arh ='1' then
                        marh:=pr_hold;
                else
                        marh:=marh;
                end if;
        end if;
                ar_hold(15 downto 10) <=marh;

        end process areghigh;

-- page register; can only be loaded from abus; initially  zero

        preg: process(clk)
                variable mpr: std_logic_vector(5 downto 0);
        begin
        if(clk'event and clk='1') then
                if init_ar='1' then
                        mpr:="000000";
                elsif ld_pr='1' then
```

```
                        mpr:=intabus(5 downto 0);
                else
                        mpr:=mpr;
                end if;
        end if;
                pr_hold <=mpr;

        end process preg;

-- instruction register; can only be loaded from abus

        ireg: process(clk)
                variable mir: bit_16;
        begin
        if(clk'event and clk='1') then
                if ld_ir= '1' b
                        mir:= intdbus;
                else
                        mir:=mir;
                end if;
        end if;
                ir_hold <=mir;

        end process ireg;

        irout<= ir_hold(7 downto 0);

-- temporary storage for program counter value; loaded from dbus

        tempreg: process(clk)
                variable mtemp: bit_16;
        begin
        if(clk'event and clk='1') then
                if ld_temp= '1' then
                        mtemp:= intdbus;
                else
                        mtemp:=mtemp;
                end if;
        end if;
                temp_hold <=mtemp;
        end process tempreg;
-- accumulator a; can be loaded from dbus and cleared
```

```
acca: process(clk)
        variable macca: bit_16;
begin
if(clk'event and clk='1') then
        if ld_a= '1' then
                macca:= intdbus;
        elsif clr_a='1' then
                macca:= "0000000000000000";
        else
                macca:=macca;
        end if;
end if;
        acca_hold <=macca;
end process acca;

-- accumulator b; can be loaded from dbus, cleared, incremented, decremented,
-- complemented

accb: process(clk)
        variable maccb: bit_16;
begin
if(clk'event and clk='1') then
        if ld_b= '1' then
                maccb:= intdbus;
        elsif clr_b='1' then
                maccb:= "0000000000000000";
        elsif inc_b='1' then
                maccb:=maccb+1;
        elsif dec_b='1' then
                maccb:=maccb-1;
        elsif com_b='1' then
                maccb:= not(maccb);
        else
                maccb:=maccb;
        end if;
end if;
        accb_hold <=maccb;
end process accb;

-- stack pointer register; can be initialised on start-up, loaded from dbus and
-- incremented and decremented
```

```
    spreg: process(clk)
            variable msp: bit_16;
    begin
    if(clk'event and clk='1') then
            if init_sp='1' then
                    msp:= "0000000011111111";
            elsif ld_sp= '1' then
                    msp:= intdbus;
            elsif inc_sp='1' then
                    msp:=msp+1;
            elsif dec_sp='1' then
                    msp:=msp-1;
            else
                    msp:=msp;
            end if;
    end if;
            sp_hold <=msp;
    end process spreg;


-- stack pointer auxiliary register; can be loaded from stack pointer and
-- initialised to the same value as stack  pointer

    streg: process(clk)
            variable mst: bit_16;
    begin
    if(clk'event and clk='1') then
            if init_st='1' then
                    mst:= "0000000011111111";
            elsif ld_st= '1' then
                    mst:= sp_hold;
            elsif inc_st='1' then
                    mst:=mst+1;
            else
                    mst:=mst;
            end if;
    end if;
            st_hold <=mst;
    end process streg;
-- multiplexers
-- data internal signal has the source either from internal memory block or from
-- external data lines depending on the value of effective address; lower 256 16-
```

```
-- bit locations implemented as internal memory block

        data <= dmemout when maddress(15 downto 8) = "00000000" else
                        datain;

-- implementation of dbus

        intdbus <= pc_hold when  dbsel = selpc2db else
                                 temp_hold when dbsel = seltemp2db else
                                 x_hold when dbsel = selx2db else
                                 data when dbsel = seldata2db else
                                 sp_hold when dbsel = selsp2db else
                                 acca_hold when dbsel=passa2db else
                                 accb_hold when dbsel=passb2db else
                                 aluout when dbsel = selalu2db else
                                 aluout;

-- implementation of abus

        intabus <= ar_hold when absel = selar else
                                 pc_hold when absel = selpc2ab else
                                 sp_hold when absel = selsp2ab else
                                 ir_hold when absel = selir2ab else
                                 x_hold when absel = selx2ab else
                                 ir_hold;

-- implementation of address bus

        maddress <= ar_hold when adsel = selar2adr else
                                 x_hold when adsel2adr = selx1 else
                                 sp_hold when adsel2adr = selsp1 else
                                 st_hold when adsel= selst2adr else
                                 ar_hold;


-- arithmetic logic unit; inputs acca, accb, result from adder/subtracter, and alu
-- operation select alusel; result on aluout lines

        alu: process(alusel, acca_hold, accb_hold, result)
        begin
                case alusel is
                        when addd =>
```

```
                    aluout<=result;
            when andd =>
                    aluout<= acca_hold and accb_hold;
            when subb =>
                    aluout <= result;
            when orr =>
                    aluout <= acca_hold or accb_hold;
            when others =>
                    aluout <= "0000000000000000";
        end case;

-- generation of z condition bit

        if aluout(15 downto 8)="00000000" and aluout(7 downto
        0)="00000000" then
                z1<= '1';
        else
                z1<= '0';
        end if;

    end process alu;

-- condition code bits stored in individual flip-flops

    ccr: process(clk)
            variable z, c, v: std_logic;
    begin
    if(clk'event and clk='1') then
            if clr_z='1' then
                    z:='0';
            elsif ld_z = '1' then
                    z:= z1;
            else
                    z:=z;
            end if;

            if clr_c='1' then
                    c:='0';
            elsif ld_c = '1' then
                    c:= c1;
            else
                    c:=c;
```

```
                end if;

                if clr_v='1' then
                        v:='0';
                elsif ld_v = '1' then
                        v:= v1;
                else
                        v:=v;
                end if;
        end if;

                zout <=z;  -- outputs from data path to control unit
                cout <=c;
                vout <=v;

        end process ccr;

-- external data output lines and address lines

        addressout <= maddress;
        dataout <= intdbus;
end combined;
```

### *16.5.3 Control Unit Implementation*

The control unit represents the core of the FLIX microcomputer. It provides proper timing and sequencing of all microoperations as required by the user program stored in the internal or external memory. It provides proper start-up of the microprocessor upon power-up or manual reset. The control unit is also responsible for interrupt sequences as shown in preceding sections. The global structure of the control unit is presented in Figure 16.10. It receives information from the data path both concerning the instructions that have to be executed and the results of ALU operations. It also accepts reset and interrupt request signals. Using these inputs it carries out all steps described in control flowcharts of Figures 16.5 and 16.6. Obviously, in order to carry out proper steps in the appropriate machine (clock) cycles, a pulse distributor which will produce four periodic non-overlapping clock phases is needed.

They are used in conjunction with the information decoded by the operation decoder to determine actions, register transfers and microoperations, which will be undertaken by the data path. The only exceptions occur in two cases presented by external hardware signals:

- When the system is powered-up or reset manually, the operation of the pulse distributor is seized for four machine cycles needed to initialize data path resources (program counter and stack pointer), as well as interrupt control circuitry.

When the interrupt request signal is activated and if the interrupt structure is enabled, the control unit provides interruption of the current program upon the completion of the current instruction and the jump to predefined starting address of an interrupt service routine; this is done by passing the control unit through the interrupt cycle.

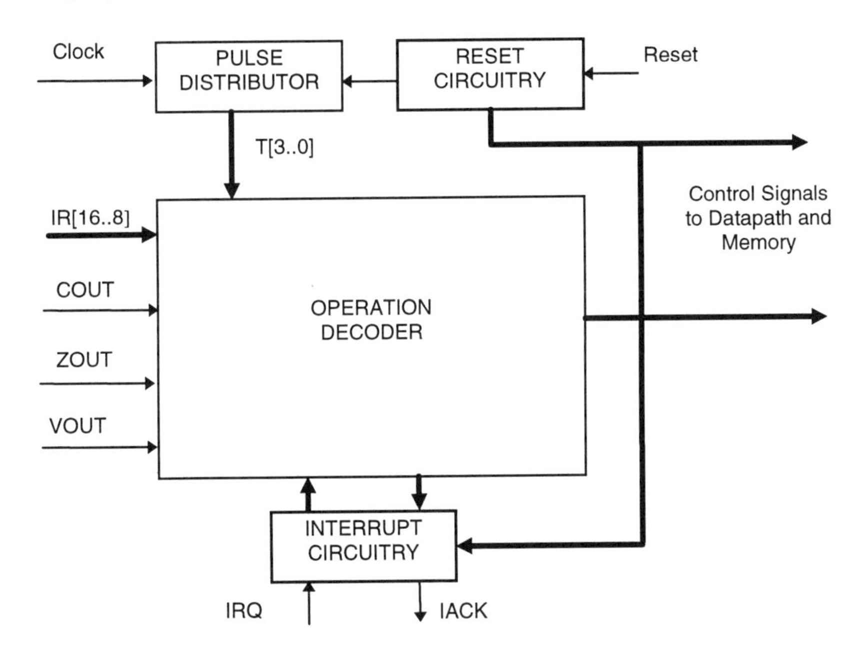


Figure 16.10 The FLIX Control Unit global view

## Pulse Distributor

Pulse distributor takes the system clock to provide four non-overlapping sequences called `T[3..0]`. It also has two input control lines as shown in Figure 16.11. The first, called "clear pulse distributor", or `CLR_PD`, is used to bring the pulse distributor to initial state T[3..0]=0001 which indicates that the `T0` machine cycle is present. The second, called "enable pulse distributor", or ENA_PD , is used to enable operation of the pulse distributor. The pulse distributor is specified as a VHDL process within the whole control unit below.

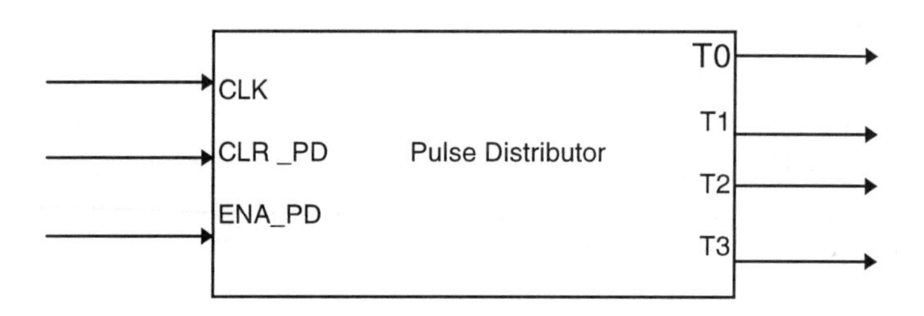


Figure 16.11 Pulse Distributor

## Operation Decoder

The Operation Decoder represents the combinational circuit that recognizes input signals to the control unit, as well as the current state of the control unit in order to provide the proper control signals. It is implemented as a single VHDL process that is shown in the VHDL design of the whole control unit.

## Reset Circuitry

Reset circuitry has the role of initializing the FLIX at power-up or manual external reset. The only input is a reset signal, but several outputs are activated in order to do proper initialization. The initialization consists of providing the initial values for the program counter and stack pointer, enabling the interrupt enable flip-flop IEN, and clearing internal IFF flip-flops. Upon initialization, external interrupts will be enabled and the control unit will automatically enter instruction execution cycle. This will happen as soon as pulse distributor is enabled and initialized.

The reset circuitry causes initialization to last exactly four system clock cycles. In the case of active RESET signal, an internal flip-flop is set. The output of this flip-flop represents the enable signal of an internal counter providing that the internal counter will count until reaches value 11. While this counter is counting, the initialization process is repeated in each machine cycle. When it stops, the initialization is also stopped, and the pulse distributor enabled to continue with its normal instruction execution cycle. The reset circuitry is represented by a single VHDL process within the whole control unit.

### Interrupt Circuitry

The interrupt circuitry has only one external input, the interrupt request IRQ, and one external output, the interrupt acknowledgment IACK. Upon interrupt request assertion, an interrupt request IRQ flip-flop is set producing the IRQA signal which is used by operation decoder circuitry. If the interrupts are enabled, and IRQA set, the operation decoder will set the interrupt flip-flop IFF to force the control unit to enter the interrupt cycle. In the interrupt cylce, the interrupt acknowledgment IACK flip-flop whose output will be available to the circuitry outside the FLIX will be set for two machine cycles. The interrupt enable flip-flop, IEN, can be set by operation decoder or reset circuitry, and the interrupt flip-flop IFF can be cleared by the operation decoder or reset circuitry. The interrupt circuitry is presented as a single VHDL process within the whole control unit design.

A more detailed illustration of the FLIX control unit and its constituent parts is given in Figure 16.12. It reflects the major parts of the overall design that are implemented as separate VHDL processes.

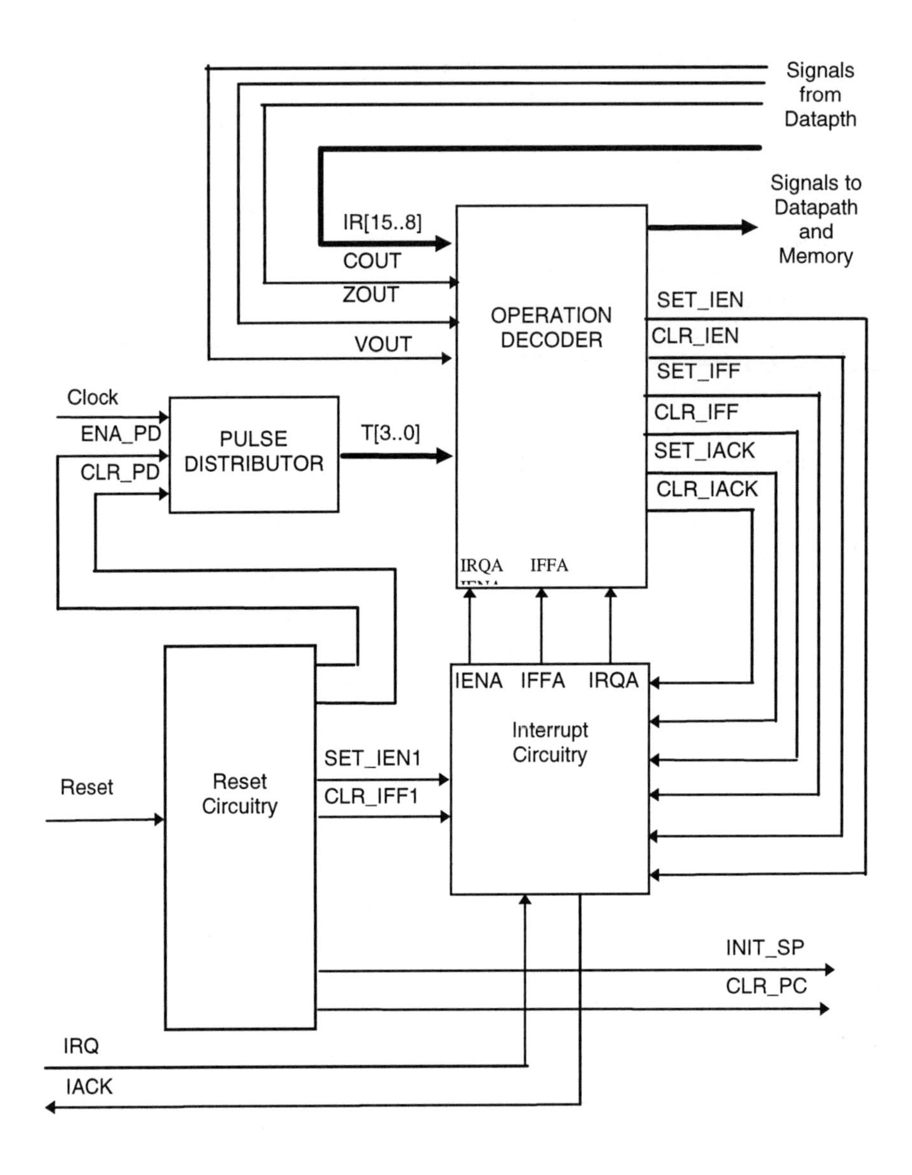


Figure 16.12 The FLIX Control Unit structure detailed view

The VHDL description of the FLIX control unit together with the comments of individual parts of the design is given below.

```
library ieee;
use ieee.std_logic_1164.all;
use ieee.std_logic_unsigned.all;
use ieee.std_logic_arith.all;
use work.flix_types.all;
use work.various_constants.all;
use work.opcodes.all;

entity controlunit is
        port(
        clk, reset: in std_logic;

-- interrupt related external signals
        irq: in std_logic;
        iack: out std_logic;


-- signals from data path

                ir: in byte_8;
                cout, zout, vout: in std_logic;


-- control of multiplexers and alu

                dbsel: out bit_3;
                absel: out bit_2;
                adsel: out bit_2;
                alusel: out bit_2;

-- control of data path

                init_pc, lda_pc, ldd_pc, inc_pc : out std_logic;  -- program
                                                          -- countercontrol
                ld_x, inc_x, dec_x: out std_logic;  -- adress (indirect) register
                                                                  -- control
                ld_arl, lda_arh, ldp_arh, init_ar: out std_logic; -- address
                                                          -- register control
                ld_ir: out std_logic;  -- instruction register control
                ld_temp: out std_logic; -- temp. register control
                ld_a, clr_a: out std_logic;  -- accumulator a control
                ld_b, inc_b, dec_b, clr_b, com_b: out std_logic;
```

```
                                                -- accumulator b control
            ld_pr: out std_logic; -- page register control
            init_st, inc_st, ld_st: out std_logic; -- auxiliary sp register
                                                -- control
            init_sp, inc_sp, dec_sp, ld_sp: out std_logic; -- stack pointer
                                                      -- control
            clr_c, clr_z, clr_v: out std_logic; -- control of condition bits
            ld_c, ld_z, ld_v: out std_logic;

            write: out std_logic – write signal
      );
end controlunit;

architecture combined of controlunit is

-- internal signals as described in preceding paragraphs

      signal   tim:  bit_4;  -- phases of the internal clock
      signal clr_pd, ena_pd: std_logic; -- control of pulse distributor

      signal   irqa, iffa, iena: std_logic;  -- interrupt related signals
      signal   set_ien, clr_ien: std_logic;
      signal   set_iff, clr_iff: std_logic;
      signal   set_iack, clr_iack: std_logic;
      signal   clr_irq: std_logic;
      signal set_ien1, clr_iff1: std_logic;


begin

-- pulse distributor process

pulsedistributor: process(clk)
      variable mpd: bit_2; -- bits of internal counter
begin

-- internal 2-bit counter

      if(clk'event and clk='1') then
            if clr_pd='1' then
                  mpd:="00";
            elsif ena_pd='1' then
```

```
                        mpd:=mpd+1;
                else
                        mpd:=mpd;
                end if;
        end if;

-- counter state decoder

        case mpd is
                when "00" =>
                        tim <="0001";
                when "01" =>
                        tim <="0010";
                when "10" =>
                        tim <="0100";
                when "11" =>
                        tim <="1000";
                when others =>
                        tim <="0001";
        end case;

end process pulsedistributor;

-- reset circuitry

resetcircuit: process(clk, reset)
        variable mrs: bit_2; -- bit of internal counter
        variable sr: std_logic;  -- internal state flip-flop; set when the start-up is
                                        -- on the way
begin
        if(clk'event and clk='1') then
                if reset ='1' then
                        sr:='1';
                        ena_pd <= '0';
                        clr_pd <= '1';
                elsif reset='0' or mrs = "11" then --
                        sr:='0';
                        ena_pd <='1';
                        clr_pd <='0';
                end if;

                if sr ='1' then
```

```
                    mrs:=mrs+1;
                    init_ar<='1';
                    init_sp<='1';
                    init_pc<='1';
                    set_ien1 <='1';
                    clr_iff1<='1';
          elsif sr ='0' then
                    mrs:=mrs;
                    init_sp<='0';
                    init_pc<='0';
                    set_ien1 <='0';
                    clr_iff1 <='0';
          end if;
     end if;
end process resetcircuit;

-- interrupt circuitry; maintains internal flip-flops

interrupt: process(clk)
     variable iff, irqff, ienff, iackff: std_logic; -- internal flip-flops
begin
if(clk'event and clk='1') then

          if set_ien='1' or set_ien1='1' then

                    ienff:= '1';
          elsif clr_ien='1' then
                    ienff:='0';
          else
                    ienff:=ienff;
          end if;

          if set_iff ='1' then
                    iff:='1';
          elsif clr_iff ='1' or clr_iff1='1' then
                    iff:='0';
          else
                    iff:=iff;
          end if;

          if set_iack ='1' then
                    iackff:='1';
```

```
                    elsif clr_iack ='1' then
                              iackff:='0';
                    else
                              iackff:=iackff;
                    end if;

                    if irq='1' then
                              irqff:='1';
                    elsif clr_irq='1' then
                              irqff:='0';
                    else
                              irqff:=irqff;
                    end if;

end if;

          irqa <= irqff;  -- signals to operation decoder
          iffa <= iff;
          iena <= ienff;
          iack <= iackff; -- external interrupt acknowledgement

end process interrupt;

-- operation decoder circuit

opdec: process(tim, ir, cout, zout, vout, irqa, iffa, iena)

begin

-- t0 cycle

if tim(0)='1' and iffa='0' then -- no interrrupts
          absel<= selpc2ab;  -- ar<-pc
          ld_arl<='1';
          lda_arh <= '1';

elsif tim(0) ='1' and iffa='1' then -- interrupt request processing
          dbsel <= selpc2db; --  m(sp) <-pc
          adsel <=selsp2adr;
          write<='1';
          clr_irq<='1';
end if;
```

```
-- t1 cycle - fetch instruction from memory

if tim(1) = '1' and iffa ='0' then -- no interrupts
          adsel<=selar2adr;  --ir <-m(ar)
          dbsel<=seldata2db;
          ld_ir<='1';
          absel <= selsp2ab; st<-sp
          ld_st<='1';
          inc_pc <='1'; pc<pc+1
elsif tim(1) ='1' and iffa ='1' then  -- interrupt cycle processing
          set_iack <='1';
          dec_sp <='1';
          init_st <='1'; --  st<-interrupt vector
end if;

-- t2 cycle - start decoding instruction

if tim(2)='1' and iffa='0' then

-- detect addressing mode and prepare for instruction execution

          case ir(7 downto 6) is

                    when funit => -- stack and functional unit addressing
                              absel<= selar2ab;  -- ar<-sp
                              ld_arl<='1'; -- lower part
                              lda_arh <='1'; -- uper part
                              dbsel<=selp2cdb; -- temp <- pc
                              ld_temp <='1';
                              inc_st <='1';

                    when indirect => -- register indirect addressing
                              absel <= selx2ab;  -- ar <-x
                              ld_arl <='1';
                              lda_arh<='1';
                              dbsel<=selpc2db; -- temp <- pc
                              ld_temp <='1';

                    when others =>
                              absel <= selir2ab; -- ar <- ir(9 downto 0)
                              ld_arl <='1';  -- address formed from pr and direct
                                                     -- address
                              ldp_arh <='1';
```

```
                        dbsel<= selpc2db; -- temp from pc
                        ld_temp <='1';
        end case;

elsif tim(2)='1' and iffa='1' then  -- interrupt cycle processing
                        adsel <= selst2adr;  -- pc ,- interrupt service routine
                                                  -- address
                        dbsel <= seldata2db;
                        ldd_pc <='1';
end if;
-- t3 cycle -- instruction execution

if tim(3)='1' and iffa ='0' then
        if ir(7 downto 6) = "00" then -- implicit
                case ir(5 downto 1) is

                when adab =>
                        ld_a <= '1';
                        dbsel <= selalu2db;
                        alusel <= addd;
                        ld_c <='1';
                        ld_z <= '1';
                        ld_v <= '1';

                when sbab =>
                        ld_a <= '1';
                        dbsel <= selalu2db;
                        alusel <= subb;
                        ld_c <='1';
                        ld_z <= '1';
                        ld_v <= '1';


                when anab =>
                        ld_a <= '1';
                        dbsel <= selalu2db;
                        alusel <= andd;
                        ld_z <= '1';

                when orab =>
                        ld_a <= '1';
                        dbsel <= selalu2db;
```

```
            alusel <= orr;
            ld_z <= '1';

when clra =>
            clr_a <='1';

when clrb =>
            clr_b <='1';

when comb =>
            com_b <='1';

when incb =>
            inc_b<='1';

when decb =>
            dec_b<='1';

when incx =>
            inc_x <='1';

when decx =>
            dec_x <='1';

when incs =>
            inc_sp<='1';

when decs =>
            dec_sp<='1';

when clrc =>
            clr_c <='1';

when clrz =>
            clr_z <='1';

when clrv =>
            clr_v <='1';

when ion =>
            set_ien <='1';
```

```
        when ioff =>
                clr_ien <='1';

        when skc =>
                if cout='1' then
                        inc_pc <='1';
                elsif cout = '0' then
                        absel <= selir2ab;
                end if;

        when skz =>
                if zout='1' then
                        inc_pc <='1';
                elsif zout = '0' then
                        absel <= selir2ab;
                end if;

        when skv =>
                if vout='1' then
                        inc_pc <='1';
                elsif vout = '0' then
                        absel <= selir2ab;
                end if;

        when ldpr =>     --the only immediate addr. mode instruction
                absel <= selir2ab;
                ld_pr <= '1';

        when noop =>
                ld_c<='1'; -- do nothing

        when others =>
                ld_z <='1'; -- should be invalid instruction code

        end case;

    elsif ir(7 downto 6) = "01" then -- stack or functional units
        case ir(5 downto 2) is

        when pula =>
                ld_a <='1';
                adsel <= selst2adr;
```

```
                    dbsel <= seldata2db;
                    inc_sp <= '1';

            when ret =>
                    adsel <= selst2adr;
                    dbsel <= seldata2db;
                    ldd_pc <='1';
                    inc_sp<='1';

            when psha =>
                    adsel <= selar2adr;
                    dbsel <= passa2db;
                    write <='1';

            when jsr =>
                    adsel <= selar2adr;
                    dbsel <= seltemp2db;
                    write <='1';
                    absel <= selir2ab;
                    lda_pc <='1';
                    dec_sp <='1';

            when others =>
                    ld_z<='1'; -- should be invalid instruction code

            end case;

    elsif ir(7) = '1' then – direct or register indirect addressing

            case ir(5 downto 2) is

            when lda =>
                    ld_a <='1';
                    adsel <= selar2adr;
                    dbsel <= seldata2db;

            when ldb =>
                    ld_b <='1';
                    adsel <= selar2adr;
                    dbsel <= seldata2db;

            when sta =>
```

```
                    adsel <= selar2adr;
                    dbsel <= passa2db;
                    write <='1';

          when stb =>
                    adsel <= selar2adr;
                    dbsel <= passb2db;
                    write <='1';

          when ldx =>
                    ld_x <='1';
                    adsel <= selar2adr;
                    dbsel <= seldata2db;

          when stox =>
                    adsel <= selar2adr;
                    dbsel <= selx2db;
                    write <='1';
          when ldsp =>
                    ld_sp <='1';
                    adsel <= selar2adr;
                    dbsel <= seldata2db;

          when jmp =>
                    lda_pc <='1';
                    absel <= selir2ab;

          when others =>
                    ld_v <='1'; -- should be invalid instruction code

          end case;

     end if;

     if iena = '1' and irqa ='1' then
          set_iff <='1';
          clr_ien <='1';
     end if;

elsif tim(3)='1' and iffa='1' then – end of interrupt cycle
     clr_iack <='1';
     clr_iff <='1';
```

```
end if;

end process opdec;

end combined;
```

Final integration of the data path and control unit is obvious. It can be performed using a pure structural model.

The FLIX microcomputer presented in this chapter represents an open base for further modifications and customization as required by specific applications. The details of customization with functional units are left to the reader as an open-ended exercise.

## 16.6 Questions and Problems

16.1. Assume that FLIX has to be extended with indexed indirect addressing mode with the post auto-incrementing of the index register. That means, after generation of effective address, index register is incremented by one.

What part of FLIX description should be changed and how? What should be done to enable auto-incrementing for an offset which can take values 2, 4, and 8? Make modifications of the FLIX VHDL description.

16.2. Is it possible to extend FLIX with indexed indirect addressing mode with pre auto-incrementing (effective address is generated after incrementing index register) without changing the number of machine cycles in the instruction?

16.3. An application can be solved with a much shorter program if the other index register is introduced. What are the modifications that have to be performed on the FLIX data path, and in the control unit?

Make modifications to both the FLIX data path and control unit VHDL descriptions that fulfill the above requirements.

# 17 APPLICATION-SPECIFIC IMAGE ENHANCEMENT COMPUTING

The aim of this Chapter is to demonstrate the application of FPLDs in a customizable, application-specific computing. The preceding Chapter showed the power of both VHDL and FPLDs to specify and implement a general-purpose microcomputer core that can be easily modified to application-specific requirements. A class of image processing applications is well suited to FPLD-based applications because of the repetitive operations that are performed on all, or groups of, image pixels which include various simple arithmetic operations, or small algorithms consisting of a number of such operations. Still, those operations are not suitable to be performed on general purpose computing platforms, because they usually require large numbers of data manipulations such as transfer of data between memory and registers, or between registers themselves. Even to perform relatively simple algorithms, the number of operations and transfers becomes excessive requiring a huge number of clock cycles. On the other hand, many of the algorithms applied in image processing inherently allow a high degree of parallelism, much higher than almost any other type of application. They also require a certain degree of flexibility and adaptability to a specific problem, which can not be achieved using a custom VLSI. These features of image processing algorithms land themselves naturally to FPLDs as implementation platforms. Also, if we combine high performance, which can be achieved by using specialized application-specific units to carry out time-critical parallel and repetitive tasks, with the flexibility of program-controlled algorithm execution, very good results can be achieved.

In this Chapter we consider an application-specific image enhancement microcomputer based on the FLIX core. The image enhancement microcomputer consists of a customized FLIX core and a number of functional units that perform time-critical operations. In this case we introduce two functional units which perform two image enhancement tasks each initiated by a single FLIX instruction. Those tasks are of high-boost filtering and histogram equalization. We first introduce briefly those two algorithms, then the concept of functional units which perform the operation in the framework of an image enhancement microcomputer. An intermediate step, the implementation of application-specific co-processors to existing computers like a standard PC, is also discussed and presented in this section. Finally, we show the implementation of those two functional units using VHDL and Altera's FPLDs.

## 17.1 High-boost Filtering and Histogram Equalization

Contrast enhancement is a digital image processing methodology, which has a wide range of applications in medical and non-medical areas. In this Chapter, we consider a way to improve the computational speed of contrast enhancement using low-cost FPLD-based hardware for implementation. In particular, we consider an enhancement method that consists of filtering followed by histogram modification which yields good results for scanned X-ray images. Filtering is done via the high boost filter (HBF) which is based on unsharp masking, and the histogram modification is based on global histogram equalization (GHE). Both the HBF and GHE (in a more popular form known as adaptive HE (AHE)) are well established methods and have been proposed in the literature for both medical and non-medical images. The hardware implementation that has low cost and yet is simple to design for both HBF and HE we follow in this Chapter is partly presented in [16].

The enhancement algorithm consists of filtering followed by histogram flattening. High boost filtering is a classical technique where the low (spatial) frequency components of an image are attenuated while, simultaneously, the high frequency details, such as edges, are emphasized. The basis for this enhancement technique lies in the human visual physiology which enables differing perceptions of the brightness of a region depending on the brightness of its background. The method considered in this Chapter uses subtracting a proportion of the low-pass filtered image $I_{LP}$ from the original image I:

$$I_{HB} = \alpha I - I_{LP}$$

The factor $\alpha$ determines the degree of high-frequency emphasis, with increasing $\alpha$ resulting in decreasing level of edge enhancement. The low pass filtered image is usually efficiently computed in the spatial domain by formulating the filtering as an averaging operation over some suitable neighborhood. Using a square neighborhood of size n x n, the time required for low pass filtering an input image of dimension N x N under sequential implementation is $O(n^2N^2)$. Computing the enhanced image requires further $N^2$ multiplication and $N^2$ additions.

Enhancement via histogram modification is a technique used to redistribute the pixel values in a given image. This process, aims to improve the contrast by increasing the dynamic range of pixel values. In histogram equalization, all pixel values of an input image are converted such that the histogram of the resulting image is distributed uniformly, i.e., flat. This is accomplished by using a mapping curve for the pixel conversion, which is the accumulation of the histogram of the input image. The method described above can be applied either globally, to the entire input image, or adaptively to smaller regions of the input image. In our case we show implementation of the GHE.

Digital images used for enhancement are, on average, of higher resolution in medical rather than non-medical applications. Medical images are of sizes 512x512 or higher and are digitized at 8 bits or higher resolution. Therefore, their enhancement using HBF and GHE poses a considerable computational load when implemented in software.

## 17.2 Image Enhancement Frameworks

Our approach to implementation of an image enhancement algorithm, as presented in the preceding section, is based on a combination of software and hardware techniques. The main goal is to improve performance using parallelisms and fast execution of some repetitive operations in FPLD-based hardware, and also to use flexibility of the software in decision making situations. Two approaches to the image enhancement co-processor are considered first:

1. Using a general-purpose FPLD prototyping board as a PC add-on board on which the functional units are implemented.
2. Using functional units as application-specific additions to the FLIX core.

The actual design of functional units that perform image enhancement algorithms is practically the same for these two approaches, except in the interface part.

### *17.2.1 PC Add-on Co-Processor for Image Enhancement*

The image enhancement co-processor, IMECO, uses standard PC as its host. The PC provides storage resources, programming facilities and flexibility, and controls the operation of IMECO. The hardware part of IMECO is implemented on a general-purpose FPLD prototyping RAPROS board developed for rapid prototyping of digital systems and hardware/software co-design in PC/FPLD environment presented in Chapter 12.

IMECO employs two key features of FPLDs:

1. Custom configurability of hardware resources present on the prototyping board, and
2. Dynamic reconfigurability using time-multiplexing of hardware functional units and implementation of virtual hardware.

IMECO represents an important intermediate step in the development of the full application-specific processor and demonstrates the feasibility of our approach with a fully application-specific microcomputer in the form of a functionally extended FLIX microcomputer core. On the other hand, it also can be considered as a powerful tool/environment for the development of hardware functional units that implement partial image enhancement algorithms and their integration using software. Besides the PC with its programming environment, and FPLD board with its software support, IMECO contains the library of hardware implemented algorithms that can be combined by software into complex image enhancement algorithms to adapt to specific goals of image enhancement and controlled by software. Examples of such partial algorithms are high-boost filter (HBF) and global histogram equalization (GHE). The HBF function and GHE function can be called from any programming language. IMECO enables downloading of the desired hardware version of the algorithm to the FPLD board and control of its operation from a program. Different versions of the algorithm represent different hardware designs that include computational units of various performance, and are more or less suitable for specific types of images. However, from the interfacing point of view each functional unit implementing a specific algorithm is the same. As is shown in the next section, the HBF unit can employ hardware units for arithmetic operations that require different amounts of hardware resources on the FPLD and have different performances. The user can select the implementation which suits best the required application. Then, a selected configuration, represented by a file on a PC hard disk, is downloaded to the board. The concept of hardware configurability is illustrated in the example of the HBF unit in Figure 17.1. The HBF hardware reads the input image from the on-board SRAM where it was stored by a program. The HBF function, invoked by a program, produces output image in the destination SRAM chips. The resulting image is available either to the GHE hardware or program for further processing. In the case when the resulting image is subsequently processed by the GHE hardware, the SRAM banks simply change their role. The output image from the HBF becomes the input image to the GHE, and memory used to store the original input image becomes the output memory into which the result of the GHE operation will be stored.

Program control is only needed to control DMA transfers of the original and the final images to/from SRAM chips on the board and the change of hardware configuration of the FPLDs. A user-friendly interface provides easy selection of the configurations that will be used in the algorithm, and subsequently loaded by a configuration loader. In the remaining text we refer to the hardware units that implement steps of the overall algorithm as to functional units. Due to the fact that functional units are not permanently present on the FPLD board, they, together with the FPLD board, make a sort of virtual hardware. The same FPLDs are used on a time-multiplexed basis to implement different functional units. The functional units that implement HBF and GHE algorithms are presented in sections 17.3 and 17.4.

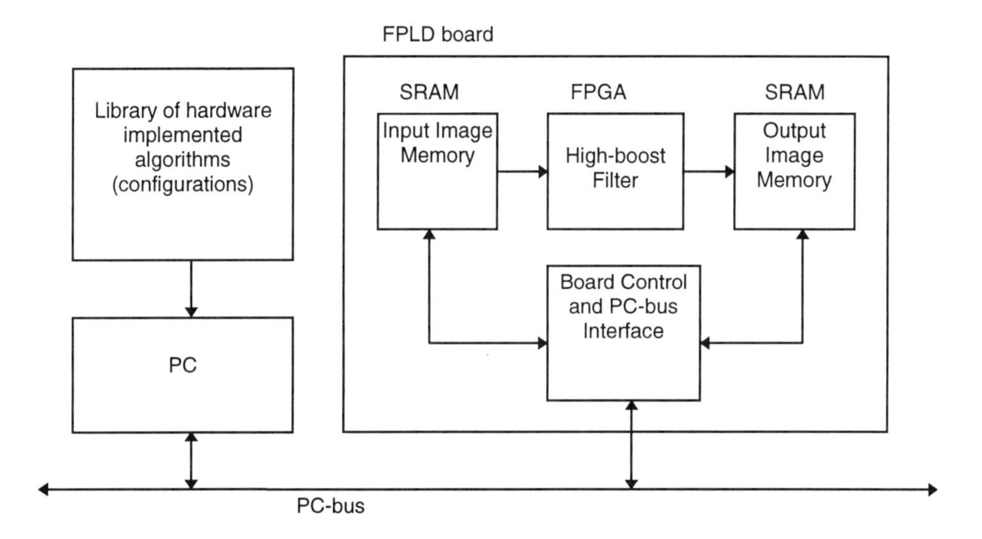


Figure 17.1. Framework for HBF implementation in IMECO environment

### *17.2.2 Image Enhancement Microcomputer*

Another approach to image enhancement application-specific platforms is to use customizable FLIX core and functional units that implement image enhancement algorithms. This approach provides a much tighter connection of processor core and basic instruction set to image enhancement functional units through image enhancement instructions, while the connection in the case of IMECO was provided by high-level subroutine-like calls implemented in the form of a dynamic link library.

The IMECOMP is an image enhancement microcomputer defined by the FLIX concept, which is briefly presented in this section. It consists of a general-purpose processor core with a simple data path and limited instruction set, and with facilities to extend it in an orderly way. FLIX allows two types of extensions: a) to add simple instructions that perform operations on the existing data path (simple register transfers and operations) not requiring any serious modification of either data path or control unit, and/or b) to add instructions that invoke more complex operations without limitation on the execution time; these operations are carried out in functional units added to existing core and controlled and synchronized by user-specific instructions. The number and the meaning of these instructions is practically unlimited. Generally, they support two modes of operations of functional units with respect to the core:

- synchronous operation, which presumes that the core halts program execution (and waits) until the functional unit completes the requested operation (instruction)
- asynchronous operation, which presumes that the core, after starting an instruction in a functional unit continues execution of the program and synchronizes with the functional unit by executing a wait instruction.

This enables writing of very flexible programs to implement and control the entire application. The designer of the functional unit only has to assign operation codes to the instructions which will be executed in the functional unit, and concentrate on the design of the functional unit itself. The IMECOMP framework is illustrated in Figure 17.2.

IMECOMP executes a program which consists of instructions from the FLIX core instruction set, which are executed in the FLIX data path under control of the FLIX control unit, and new instructions that represent commands for the HBF and GHE functional units. It should be noted that several functional units may be attached to the FLIX core simultaneously, as was presented in Chapter 16. The functional unit can execute one or more instructions from its own instruction set. Instruction execution in the functional unit is under the full control of its own control unit. Overall program execution control is transferred from the FLIX control unit (CU) to the image enhancement algorithm functional control unit (in our case HBF control unit or GHE control unit) and vice versa, as required by the program. The FLIX control unit transfers control to the functional unit control unit (FUCU) which, in turn, controls both its own and the core's data path as required. Upon completion of the required operations, the FUCU returns control to the FLIX control unit.

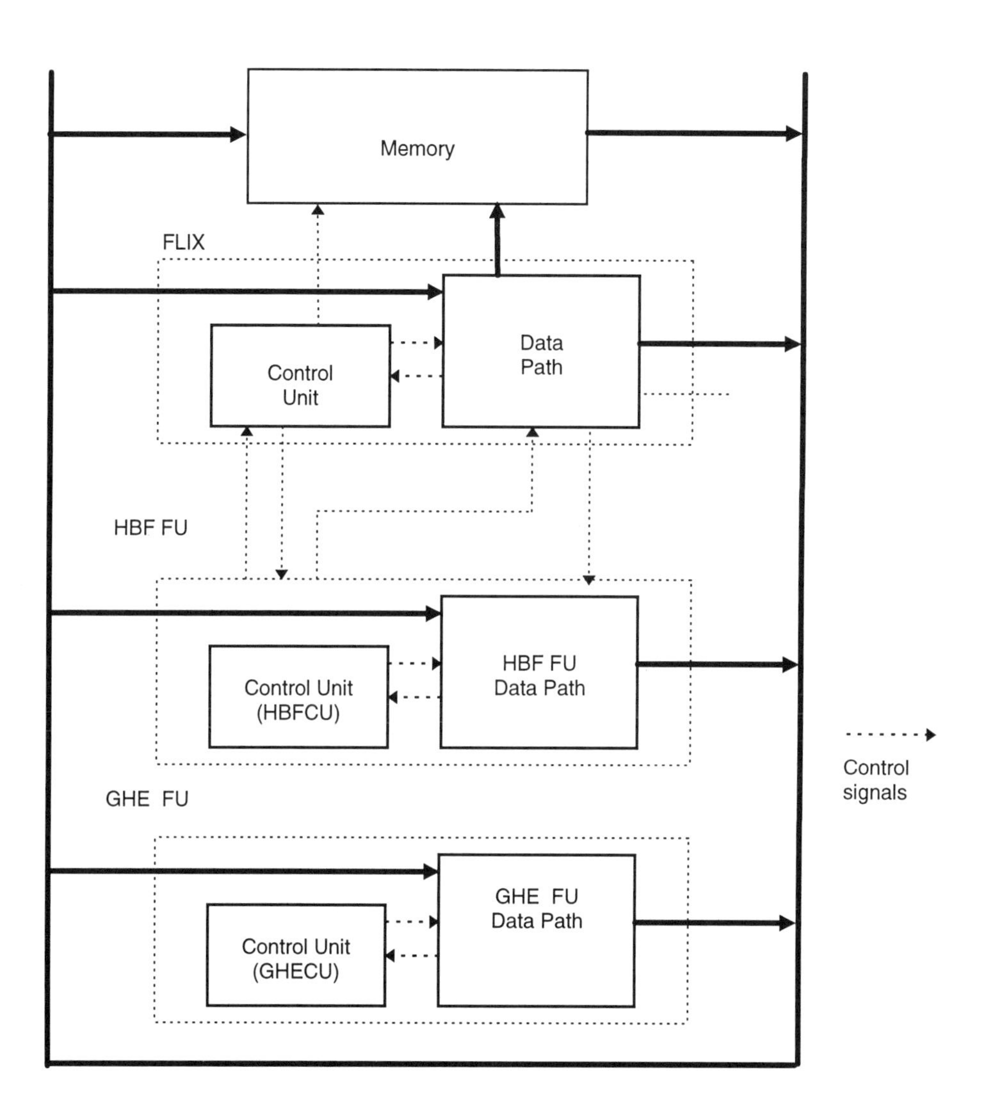


Figure 17.2 Overall structure of IMECOMP

In order to communicate with functional units, the instruction set required by the specific application has to be specified. A number of generic instructions are presented in Chapter 16. The FUCU can generally be in one of four major states:

- initialization, when the functional unit circuitry is initialized
- waiting on a new instruction (command) from the FLIX core
- executing an instruction synchronously with the FLIX operation, and
- executing an instruction asynchronously with the FLIX operation or waiting to synchronize with the FLIX core.

### *17.2.3 Functional Unit Architecture*

The global architecture of the functional unit, regardless of implementation on a FPLD add-on board as an IMECO co-processor, or as a functional unit of IMECOMP application-specific microcomputer, and its interfacing, is presented in Figure 17.3. It consists of interface to the processor bus, interface to the image memory, and three main parts:

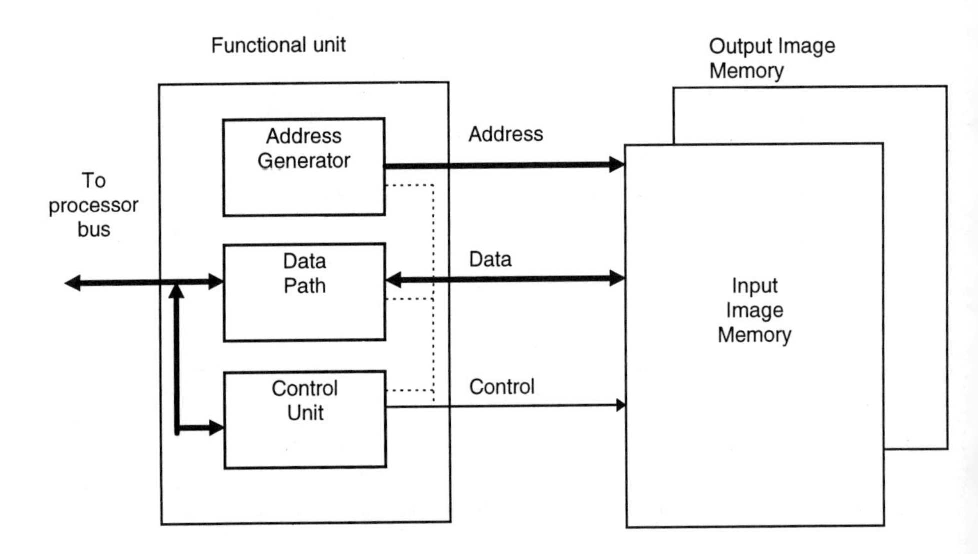


Figure 17.3 Functional unit organization – global view

- Address generator, which generates addresses of memory locations to access next pixel, either for reading or writing.

- Data path, which contains all processing and temporary storage elements that take part in algorithm implementation.
- Control unit, which coordinates and synchronizes all operations within data path and address generator. It also senses and provides control signals at the interfaces to PC-bus or FLIX bus and image memory.

In the following sections we concentrate on the design of individual functional units that implement HBF and GHE algorithms.

## 17.3 High-Boost Filter Functional Unit

The basic concepts of our approach for the HBF (used as an example) implementation are illustrated in Figure 17.6. The task of the HBF is to calculate for each pixel in the filtered image $P_{HB}(x, y)$, $(x, y = 0, 1, \ldots, N-1)$

$$P_{HB}(x,y) = \alpha P_{IN}(x,y) - P_{LP}(x,y)$$

where $P_{IN}(x, y)$ is the input image, $P_{LP}(x, y)$ is the low-pass filtered (LPF) image, and $\alpha$ is a constant that can take different values (>1). The LPF is defined as

$$P_{LP}(x,y) = \frac{1}{n^2} \sum_{(x,y)\in M} P_{IN}(x,y)$$

where $(x,y) \in M$ denotes pixels within a square mask M of size n x n. Therefore, the LPF operation is an averaging operation over a local neighborhood M. This operation depends on the size of the mask and requires $n^2$ additions and one division operation. Finding the high boost filtered version of an input image can be considered as a process in which each pixel of the input image must be processed in the same way: it becomes the center pixel of a square window, of size equal to the size of the filter mask, which moves (or slides) along the whole input image in an ordered way. A straightforward solution is to slide it along either rows or columns of the image, from top to down or left to right. The number of operations that have to be performed on an image is proportional to the size of image ($N^2$) and the size of the window ($n^2$).

The efficiency of software implementation of HBF depends on the programming language used. Even if we use a low-level programming language, the problems of memory allocation and address generation remain, but overall control of the number of operations is better than when using high-level languages. If a DSP processor is used, performance improvement is achieved because of efficient implementation of arithmetic operations, leading to a lower number of cycles to perform the algorithm, but the problems of memory allocation and address generation still remain. Generally, resource management is more efficient at the lower level of programming abstraction. Processing image boundaries requires additional checks thereby increasing the algorithm execution time.

In our algorithm, the window slides for one bit position downwards until its leading pixel reaches the last row of an image. This process is continued by moving the sliding window to the top of the next column, to process the next column's pixels, and repeated until the last pixel in the image is reached.

### *17.3.1 HBF Data Path*

We have reduced the number of memory read operations by recognizing and exploiting the fact that the only change when the sliding window moves to a new row is the addition of a new row. Therefore, only the sum of pixels of the new (leading) row is calculated. All these pixels occupy adjacent memory locations. The new mean window value, which is actually the low-pass filtered value of the center pixel, is calculated by subtracting the value of the row dropped from the sliding window (last trailing row) and adding the value of the new leading row. This is illustrated below for the case of a 5 x 5 ($n^2 = 25$) sliding window:

$$P_{LP}(x,y) = P_{LP}(x,y-1) + S_P$$

$$S_P = \frac{1}{25}[\text{Lead}(x,y+2) - \text{Trail}(x,y-3)]$$

Here Lead and Trail represent the partial sums of pixels of the new leading and previous trailing row of the sliding window. A graphical illustration of the main parts of the data path to implement the HBF is shown in Figure 17.4. The sums of sliding

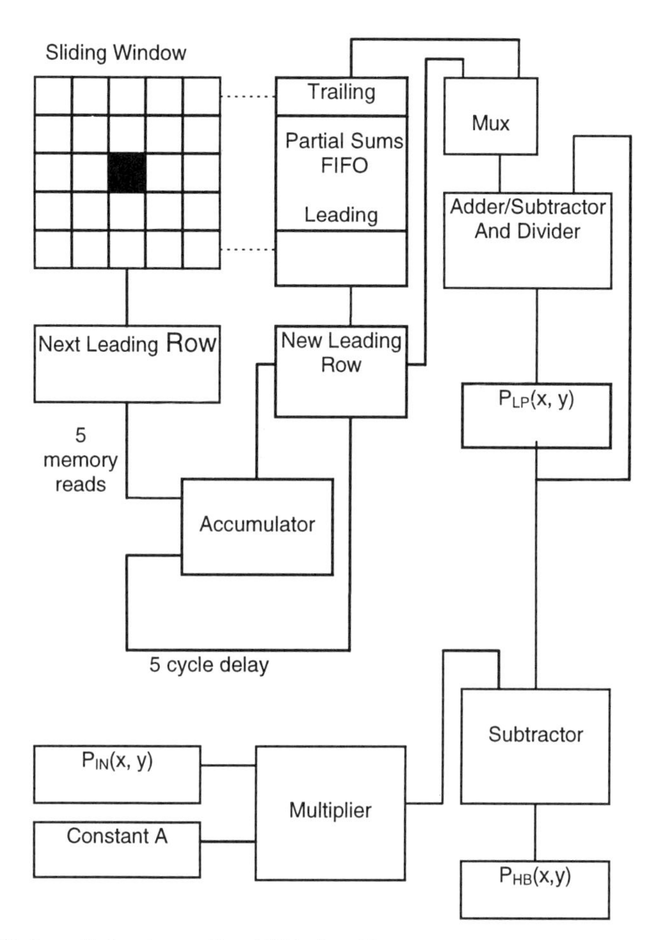


Figure 17.4 Data path implementing HBF algorithm

window rows are calculated and stored in a FIFO structure. The FIFO always contains sums of five current rows of the sliding window. When a new leading row is encountered, its sum is calculated in the accumulator. Next, the sum of the trailing row is subtracted and the sum of the new leading row is added, to the existing sum of all rows belonging to the sliding window. This sum is divided by $n^2$ to obtain the low-pass filtered pixel value. Finally, the high boost-filtered pixel value is calculated. The total number of operations in our approach is proportional to the size of the image ($N^2$) and the width (n) of the applied mask. The whole algorithm is implemented in CPLD hardware and employs parallelism. At the beginning of each new column, contents of the FIFO is cleared and partial sums of the rows are calculated to be stored in the FIFO as the window slides downwards.

The described algorithm requires just one memory read operation of each pixel value for calculation of HBF values of all pixels belonging to one column of original image, and also reduces number of additions compared to straightforward approach summarized in the beginning of this section. The total number of operations involved in our approach is proportional to the number of pixels in the image and the square root of number of pixels in the applied mask. This algorithm can be implemented in software, but requires complex management of FIFO and sliding window, with additional support operations. A software implementation also can not use inherent features of the algorithm that allow parallel execution of some of operations. Our approach is to implement the whole algorithm in hardware and employ the parallelism. At the beginning of each new column, contents of the FIFO is cleared. Then, partial sums of the rows are calculated and stored into the FIFO as it slides downwards. In this way the FIFO is passing an initialization procedure until central pixel reaches first row represented by x=0.

The implementation of the HBF data path as presented by the VHDL description below is slightly simplified version of the HBF. We suppose that the input image is multiplied by coefficient $\alpha$ equal to 2, and the low-pass filtered value is found by division of the mask sum by 32. This obviously simplifies arithmetic operations, and consequently the overall data path. Still, the HBF filter with these characteristics provides satisfactory results. The data path uses the following 11-bit register as component for implementation of 5 locations deep FIFO. The register length of 11 bits is needed to store the partial sums of the row pixels.

```
library ieee;
use ieee.std_logic_1164.all;
use ieee.std_logic_arith.all;

entity regg is
port(
        clk, ld_fiforeg, reset: in std_logic;
        datain: in std_logic_vector(10 downto 0);
        dataout: out std_logic_vector(10 downto 0));
end regg;

architecture beh of regg is
begin

reg: process(clk, reset)
        variable m1: std_logic_vector(10 downto 0);
begin
```

```
if reset='1' then
        m1:="00000000000";
elsif(clk'event and clk='1') then
                if ld_fiforeg='1' then
                        m1:=datain;
                else
                        m1:=m1;
                end if;
end if;
        dataout<=m1;
end process reg;
end beh;
```

The HBF data path VHDL description follows:

```
library ieee;
use ieee.std_logic_1164.all;

use ieee.std_logic_arith.all;

entity hbfdatapath is
port(clk, reset: in std_logic;
-- control signals that initiate and synchronize operations
        ld_indr: in std_logic;
        ld_inpix: in std_logic;
        clr_acc: in std_logic;
        ld_fiforeg, clr_fiforeg: in std_logic;
        inpsel, ena_sum: in std_logic;
        ld_hbreg: in std_logic;
        datain: in std_logic_vector(7 downto 0);      -- input 8-bit data
        dataout: out std_logic_vector(7 downto 0)); -- output 8-bit data
end hbfdatapath;

architecture combined of hbfdatapath is
component regg                          - register used to form FIFO
        port (
        clk, ld_fiforeg, reset: in std_logic;
        datain :in std_logic_vector(10 downto 0);
        dataout :out std_logic_vector(10 downto 0));
end component;
-- internal signals used to interconnect components and processes
        signal rowin: std_logic_vector(10 downto 0);
```

```
        signal rowout: std_logic_vector(10 downto 0);
        signal dout: std_logic_vector(7 downto 0);
        signal reg1_hold, reg2in, reg2_hold, reg3_hold,
                reg4_hold: std_logic_vector(10 downto 0);
        signal indr_hold: std_logic_vector(7 downto 0);
        signal pixout: std_logic_vector(8 downto 0);

begin

-- input data register stores the value read from input memory before
-- accumulation

        indr: process(clk)
                variable m: std_logic_vector(7 downto 0);
        begin
        if (clk'event and clk='1') then
                if ld_indr= '1' then
                        m:= datain;
                else
                        m:="00000000";
                end if;
        end if;
                indr_hold<=m;
        end process indr;

-- input center pixel stored in separate register

        inpix: process(clk)
                variable n: std_logic_vector(7 downto 0);
        begin
        if (clk'event and clk='1') then
                if ld_inpix= '1' then
                        n:= indr_hold;
                else
                        n:=n;
                end if;
        end if;
                pixout(8 downto 1)<=n; -- output multiplied by 2 –
                        -- shifted left one bit
                pixout(0) <= '0';

        end process inpix;
```

```
-- process that accumulates pixels of a row

        acc: process(clk)

                variable p: std_logic_vector(10 downto 0);

        begin
        if reset='1' then
                p:="00000000000";
        elsif(clk'event and clk='1') then
                if clr_acc='1' then
                        p:="00000000000";
                else
                        p:= p + indr_hold;
                end if;
        end if;
                rowin <= p;

        end process acc;

-- fifo structure represented by 5 cascaded 11-bit registers

u1: regg port map (clk, ld_fiforeg, clr_fiforeg, rowin, reg1_hold);
u2: regg port map (clk, ld_fiforeg, clr_fiforeg, reg1_hold, reg2_hold);
u3: regg port map (clk, ld_fiforeg, clr_fiforeg, reg2_hold, reg3_hold);
u4: regg port map (clk, ld_fiforeg, clr_fiforeg, reg3_hold, reg4_hold);
u5: regg port map (clk, ld_fiforeg, clr_fiforeg, reg4_hold, rowout);


-- calculation of the low pass filtered value of the center pixel

        lpf: process(clk)
                variable m: std_logic_vector(12 downto 0); -- 13-bit
                                                    -- pass filtered value
        begin low
        if reset='1' then
                m:="0000000000000";
        elsif (clk'event and clk='1') then
                if ena_sum='1' then
                        if inpsel='0' then
                                m:=m-rowout; -- subtract trailing row
                        elsif inpsel='1' then
```

```
                        m:=m+rowin; -- add leading row
                    end if;
            end if;
    else
            m:=m;
    end if;
            dout<=m(12 downto 5); -- output divided by 32 –
                    -- shifted right 5 bits
    end process lpf;

    hbreg: process(clk)
            variable r: std_logic_vector(8 downto 0);
    begin
    if (clk'event and clk='1') then
            if ld_hbreg='1' then
                    r:=pixout - dout; -- high boost filtered value
                                      -- calculated
            else
                    r:="000000000";
            end if;
    end if;
            dataout<=r(7 downto 0);
    end process hbreg;

end combined;
```

### *17.3.2 Address Generator*

An important part of the whole HBF is the address generator that is used to generate addresses of both input pixels' and output pixels' memory locations. In this example we assume that the input image is of the size 256 x 256 pixels represented by bytes stored in memory in the row major order, as it is illustrated in Figure 17.5.

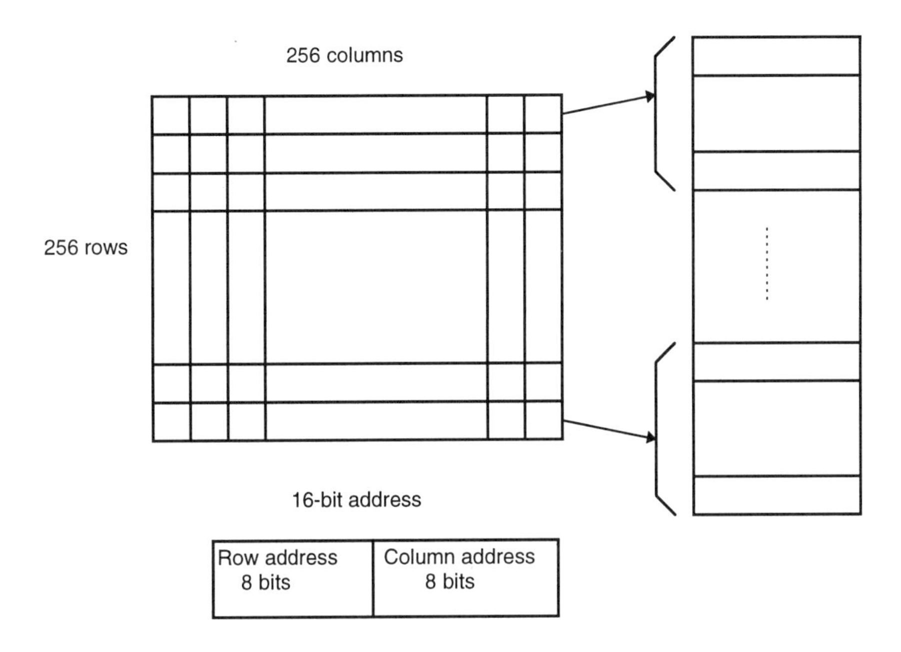


Figure 17.5 Memory allocation of image pixels

Address generator provides three addresses at any time:

- The address of the current row which is accumulated (leading row) to the row sum. When the end pixel within the mask is encountered, the sum is calculated and this address must be incremented to encounter the next leading row.
- The address of the current column. This address represents an internal information within the address generator. When the last row is processed, this address is incremented by one, and used as the initial address for calculation of sums of rows' pixels for that column.
- The address of a pixel which is currently accumulated to the row sum (of the new leading row). The pixels that belong to a row within the mask are accessed from the current column pixel to the right-hand side pixel within the row by incrementing the value of the pixel register.

These three addresses are sufficient to form two output addresses:

- An input memory address representing next pixel which will be read from the input image memory, and
- An output memory address representing location into which the high boost filtered value will be stored.

The address generator structure (data path) is represented in Figure 17.6. The HBF control unit provides control signals that initialize and increment values of the address registers, and monitors whether the boundary has been reached.

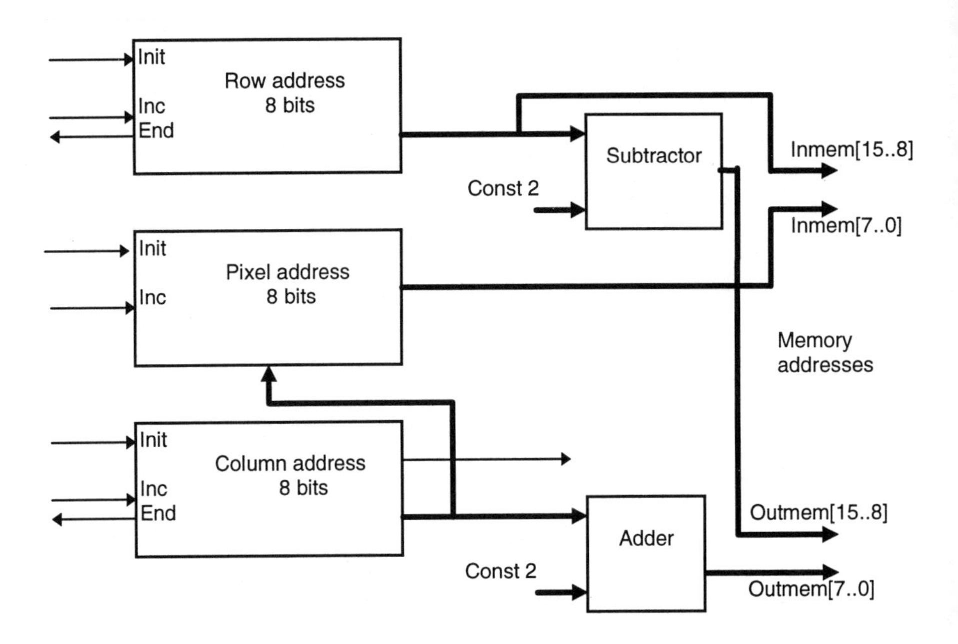


Figure 17.6 Address generator structure

The VHDL description of the HBF address generator is given below:

```
library ieee;
use ieee.std_logic_1164.all;
use ieee.std_logic_arith.all;

entity addrgen is
port(clk, reset: in std_logic;
```

```
        init_rowaddr, init_coladdr, init_pixaddr: in std_logic;
        inc_rowaddr, inc_coladdr, inc_pixaddr: in std_logic;
        end_row, end_col: out std_logic;
        inmem: out std_logic_vector(17 downto 0);
        outmemory: out std_logic_vector(17 downto 0)
        );
end addrgen;
architecture integral of addrgen is
        signal coladdress: std_logic_vector(7 downto 0);
        constant rowmax: std_logic_vector(7 downto 0):= "11111111"; -

                -- boundary row address
        constant colmax: std_logic_vector(7 downto 0):= "11111010"; --
                -- boundary column address
        signal outmlow: std_logic_vector(7 downto 0);
        signal outmhigh: std_logic_vector(7 downto 0);

begin

        rowaddr: process(clk)
                variable m: std_logic_vector(7 downto 0);
        begin
        if reset='1' then
                m:="00000000";
        elsif(clk'event and clk='1') then
                if init_rowaddr ='1' then
                        m:="00000000";
                elsif inc_rowaddr = '1' then
                        m:=m+1;
                end if;
        end if;
        if m = rowmax then
                end_row <= '1';
        else
                end_row <= '0';
        end if;
                inmem(17 downto 8) <= m;
                outmhigh(7 downto 0) <= m - "00000010";
        end process rowaddr;


        pixaddr: process(clk)
```

```
        variable n: std_logic_vector(7 downto 0);
begin
if reset='1' then
        n:="00000000";
elsif(clk'event and clk='1') then
        if init_pixaddr ='1' then
                n:=coladdress;  -- move mask to the next
                                -- column
        elsif inc_pixaddr = '1' then
                n:=n +1;
        end if;
end if;
        inmem(7 downto 0) <= n;

end process pixaddr;

coladdr: process(clk)
        variable p: std_logic_vector(7 downto 0);
begin
if reset='1' then
        p:="00000000";
elsif(clk'event and clk='1') then
        if init_coladdr ='1' then
                p:="00000000";
        elsif inc_coladdr = '1' then
                p:=p+1;
        end if;
end if;
if p=colmax then
        end_col <= '1';
else
        end_col <= '0';
end if;

        coladdress(7 downto 0) <= p;
        outmlow(7 downto 0) <= p + "00000010";

end process coladdr;
        outmemory(17 downto 8)<=outmhigh(7 downto 0);
        outmemory(7 downto 0)<=outmlow(7 downto 0);

end integral;
```

### *17.3.3 HBF Control Unit*

Control flow of the hardware implemented HBF is represented by flow diagram in Figure 17.7 The HBF control is transferred from the processor to HBF functional unit by a single instruction. After initialization of the address registers within the address generator and registers within the data path that take part in calculations of intermediate and final results, the control unit clears the FIFO register and positions the mask to the leftmost position so that the leading row encounters pixels of the first row of the image (row with address 0). As long as the pixels belong to the leading row they are accumulated and the sum of the leading row pixels is calculated. In parallel, center pixel of the current mask is stored in inpix register "multiplied" by 2 waiting to calculate the low pass filtered value of the current mask. The low pass filtered value is found by subtracting the current trailing and by adding the current leading row sums. New leading row sum is pushed to the FIFO, and the previous trailing row sum is lost. Finally, the high boost filtered sum is found and stored into the hbf register, and in the next clock cycle it is stored into output image memory. Using parallelisms of execution of the microoperations, it becomes possible to calculate the next pixel's high boost filtered value every 8 clock cycles except in the case of boundaries of the image, when an additional clock cycle is required.

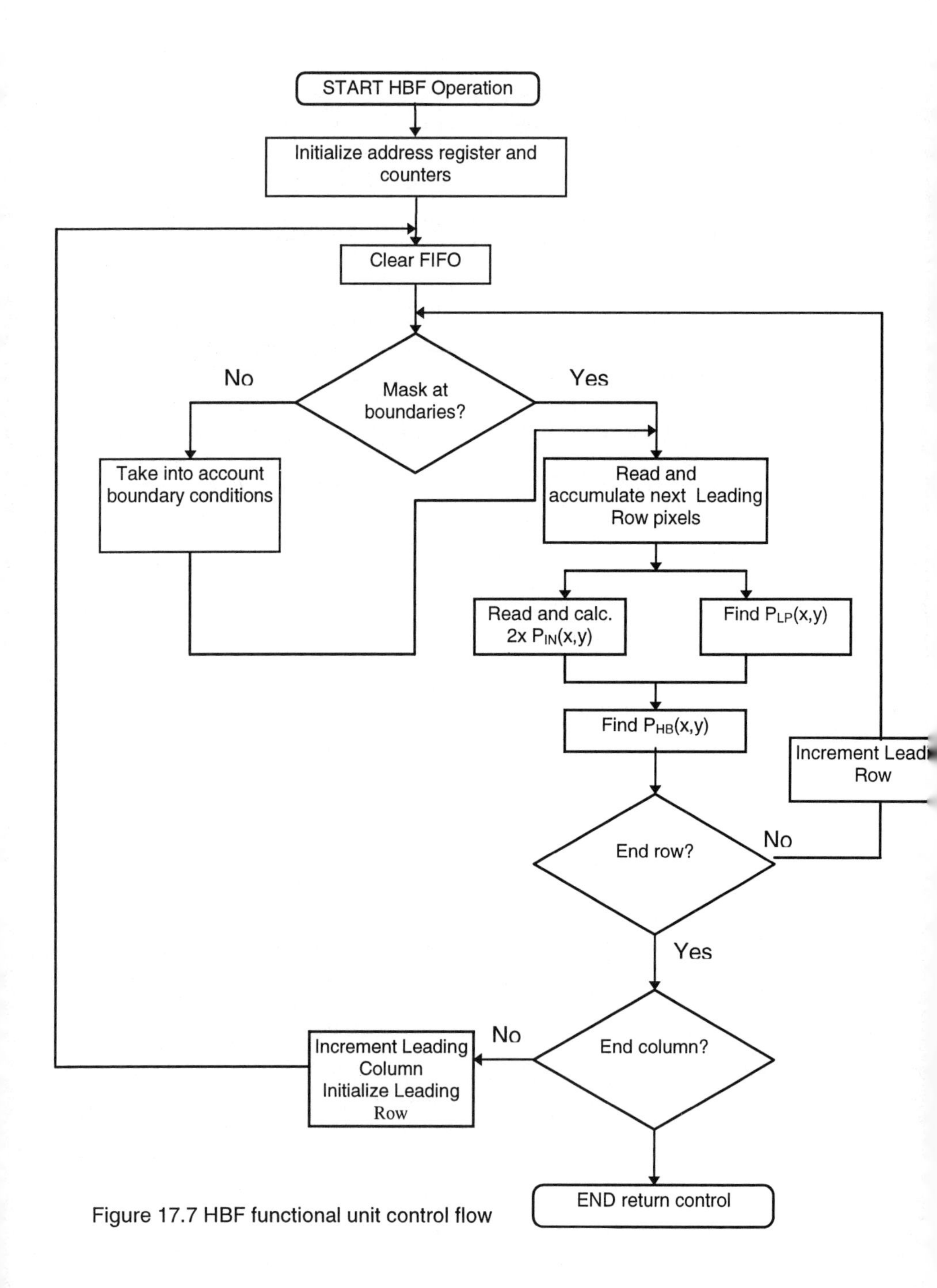


Figure 17.7 HBF functional unit control flow

The VHDL description of the HBF control unit is given below.

```
library ieee;
use ieee.std_logic_1164.all;
use ieee.std_logic_arith.all;

entity hbfcontr is
port (
        clk, start: in std_logic;
        end_row, end_col: in std_logic;
        clr_acc: out std_logic;
        ld_fiforeg, clr_fiforeg, inpsel, ena_sum: out std_logic;
        ld_inpix: out std_logic;
        ld_indr: out std_logic;
        ld_hbreg: out std_logic;
        inc_rowaddr, inc_pixaddr, inc_coladdr: out std_logic;
        init_rowaddr, init_pixaddr, init_coladdr: out std_logic;
        finished: out std_logic
                        );
end hbfcontr;

architecture beh of hbfcontr is
type state_type is (init, init1, s0, s1, s2, s3, s4, s5, s6, nextrow, nextcol,
                finish);
        signal state: state_type;
begin
        process (clk)
        begin
        if start = '1' then
                state <= init;
        elsif (clk'event and clk = '1') then
                case state is

                        when init =>
                                init_rowaddr <= '1';
                                init_coladdr <= '1';
                                init_pixaddr <='1';
                                state <= init1;
                        when init1 =>
                                init_rowaddr <= '0';
                                init_coladdr <= '0';
                                init_pixaddr <='0';
```

```
                inc_coladdr <='0';
                state <= s0;
        when s0 =>
                ld_indr <='1';
                inc_pixaddr<= '1';
                inc_rowaddr <='0';
                init_pixaddr<='0';
                ld_hbreg <= '1';
                ena_sum <= '0';
                state <= s1;
        when s1 =>
                ld_indr <='1';
                inc_pixaddr <= '1';
                ena_sum <= '0';
                ld_hbreg <= '0';
                state <= s2;
        when s2 =>
                ld_indr <='1';
                inc_pixaddr <= '1';
                ena_sum <= '0';
                state <= s3;
        when s3 =>
                ld_indr <='1';
                inc_pixaddr <= '1';
                ena_sum <= '0';
                state <= s4;
        when s4 =>
                ld_indr <='1';
                inc_pixaddr <= '1';
                ena_sum <= '0';
                state <= s5;
        when s5 =>
                ld_indr <='0';
                ena_sum <= '1';
                inc_pixaddr <= '0';
                state <= s6;
        when s6 =>
                ld_indr <='0';
                ena_sum <= '1';
                inc_pixaddr <= '0';
                if end_row ='0' then
                        state <= nextrow;
```

```
                        elsif end_row='1' then
                                state <=nextcol;
                        end if;

                when nextrow =>
                        ld_indr <='0';
                        ena_sum <= '0';
                        inc_pixaddr <= '0';
                        inc_rowaddr <='1';
                        init_pixaddr<='1';
                        state <= s0;
                when nextcol=>
                        if end_col = '0' then
                                inc_coladdr <='1';
                                init_rowaddr <='1';
                                ld_indr <='0';
                                ena_sum <= '0';
                                init_pixaddr<='1';
                                state <= init1;
                        else
                                state<=finish;
                        end if;
                when finish =>
                                inc_rowaddr <='0';
                                inc_pixaddr <='0';
                                ena_sum<='0';
                                ld_hbreg <='0';
                                state <= init;
                                state <=finish;
            end case;
        end if;
end process;

siggen: process(state)
begin

        if state=init1 then
                clr_fiforeg <= '1';
        else
                clr_fiforeg <= '0';
        end if;
```

```
        if state = s3 then
                ld_inpix <= '1';
        else
                ld_inpix <= '0';
        end if;

        if state = s6 then
                ld_fiforeg <= '1';
        else
                ld_fiforeg <= '0';
        end if;

        if state=nextrow then
                inpsel <= '1';
                clr_acc <= '1';
        else
                inpsel <= '0';
                clr_acc <= '0';
        end if;

        if state = finish then
                finished <= '1';
        else
                finished <= '0';
        end if;
    end process siggen;
    end beh;
```

The HBF control unit is implemented using two processes; the first process implements the single state machine that provides transitions through the states of the control unit, and the second process generates a number of output control signals that are activated in just a single state.

## 17.4 Histogram Equalization Functional Unit

The GHE functional unit performs histogram equalization of the entire image. The algorithm for histogram equalization that is implemented as a functional unit can be described as follows:

1. Compute histogram $H_A(k)$ of given image A of the size N x N, which is

the result of HBF operation,

$H_A(k) = n_k$

where, $n_k$ is the total number of pixels in the image at the kth gray level, with $k = 0,1,...L-1$.

2. Compute the equalisation value $s_k$ for each gray level k as

$$s_k = Int[\frac{L}{N^2}\sum_{j=0}^{K-1} H_A(j)]$$

where *Int* is an integer part of the calculated number.

3. Equalise and compute new image pixels $P_{HE}(x,y)$ as

If $A(x,y) = k$

Then $P_{HE}(x,y) = s_k$ for every x and y.

### *17.4.1 GHE Functional Unit Data path*

The GHE functional unit data path is presented in Figure 17.8. It finds input image in the memory block into which it was stored by the HBF functional unit. Then, it reads all pixel values and accordingly increments the histogram value of the corresponding gray level. The histogram values are stored in a separate memory called histogram memory. The gray level of a pixel fetched from the image memory represents the memory address of the histogram value that has to be incremented. An incrementer is used to do this operation and return a new histogram value to the corresponding histogram memory location. As soon as all pixels from the input image are processed, the second step of the algorithm takes place. It calculates equalization values for each gray level by reading the values from histogram memory and accumulating them to the previously found values. Only one pass of histogram memory read operations and memory write operations is needed to calculate equalization values which are now stored in histogram memory. This part of the algorithm is performed by an accumulator and arithmetic block that carries out multiplication of the accumulated value by L/M and rounding to the integer part.

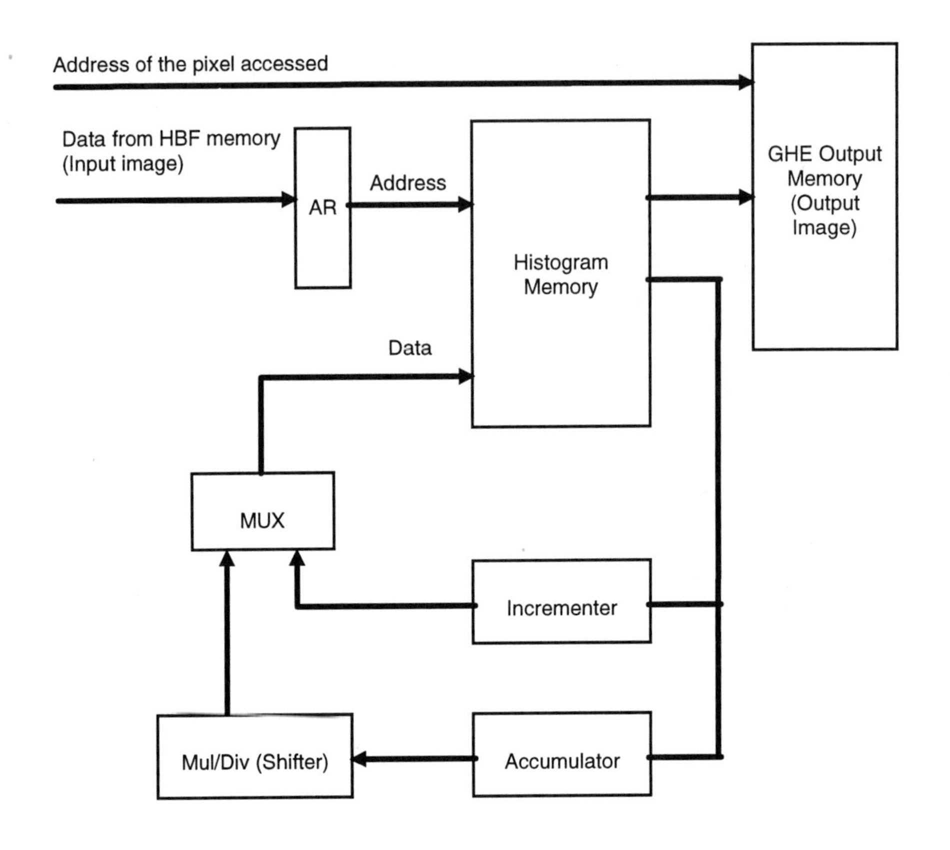


Figure 17.8 GHE functional unit data path

If we consider expression L/M, we can notice that usually we have $L=2^b$, where b is the number of bits which represent gray levels, and $M = N \times N = 2^{2w}$, where $2w > b$. Therefore, the expression L/M will produce result in the form $2^{-q}$, where q is an integer and $q>0$. This further means that the resulting accumulated sum must be divided by $2^q$ reducing multiplication and division operations to a simple shift right operation for q bits. The final result is stored in GHE memory block, actually being the block in which original image was stored. Number of operations involved in the GHE algorithm operation is proportional to $N^2$. The Altera FLEX 10K CPLDs provide internal embedded array blocks of the SRAM, which are in this case used to implement histogram memory. Two such blocks are used to implement 256 x 17 histogram memory, and a library parametrized module with address width and data width as parameters. The values of these two parameters are declared in a separate package in the beginning of the VHDL description, and can be easily changed if necessary. The rest of the VHDL description is not parametrized, although it can be done in order to make the whole design more flexible.

The VHDL description of the GHE functional unit data path follows:

```
package ram_constants is
constant addr_width: integer:= 8;
        constant data_width: integer:= 16;
end ram_constants;

library ieee;
use ieee.std_logic_1164.all;
use ieee.std_logic_arith.all;
use ieee.std_logic_unsigned.all;
library lpm;
use lpm.lpm_components.all;
library work;
use work.ram_constants.all;

entity ghedatapath is
        port(
        we, clk: in std_logic;
        hbfdataout: in std_logic_vector (7 downto 0); -- data input to the
                                                      -- GHE unit
        gheaddress: out std_logic_vector (15 downto 0); -- address of the
                                      -- pixel in input or output memory
        hmout: out std_logic_vector (7 downto 0); -- data output to the
                                                      -- GHE memory
        hmdatasel: in std_logic; -- selector of the input data to HM
                                      -- memory
        init_adr, inc_adr: in std_logic; -- control signals to internal
                                 -- registers
        ena_inc, ld_inc: in std_logic;
        ena_acc, clr_acc: in std_logic;
        ld_hmadr, inc_hmadr, init_hmadr: in std_logic;
        end_hm: out std_logic; -- indications of the end of HM
        end_hbf: out std_logic  -- or HBF memory
        );
end ghedatapath;

architecture combined of ghedatapath is
```

```
        signal hmdata: std_logic_vector (data_width-1 downto 0); --
                                              -- internal signals
        signal hmq: std_logic_vector (data_width-1 downto 0); -- used to
                              -- connect lpm_ram_dq
        signal hmaddress: std_logic_vector (addr_width-1 downto 0);
        signal ghevalue: std_logic_vector (15 downto 0); -- internal
                                        -- signal/output from accumulator
        signal inc_hold: std_logic_vector (15 downto 0); -- internal
                                        -- signal/output from incrementer

    begin

-- histogram memory instance
    hm_ram: lpm_ram_dq
    generic map (lpm_widthad => addr_width, lpm_width => data_width,
            lpm_outdata=> "unregistered")
    port map (data => hmdata, address => hmaddress,
            inclock => clk, we => we, q=> hmq);

-- incrementer process used to calculate histogram of the input image

    incr: process(clk)
            variable m: std_logic_vector(15 downto 0);

    begin
            if(clk'event and clk='1') then
                    if ld_inc= '1' then
                            m:= hmq;
                    elsif ena_inc = '1' then
                            m:= m + 1;
                    else
                            m:=m;
                    end if;
                    inc_hold <=m;

    end if;

    end process incr;

-- accumulator process calculates partial sums of histogram memory locations

    acc: process(clk)
```

```
            variable p: std_logic_vector(23 downto 0);

        begin
            if(clk'event and clk='1') then
                if clr_acc='1' then
                    p:="000000000000000000000000";
                elsif ena_acc = '1' then
                    p:= p + hmq;
                end if;
            end if;
                ghevalue(15 downto 0) <= p(23 downto 8);
        end process acc;

-- multiplexer to select the source of data input to histogram memory

        hmdata <= ghevalue when  hmdatasel = '0' else
                inc_hold;

-- generation of the address of location of histogram memory

        hmadr: process (clk)
            variable n: std_logic_vector(7 downto 0);

        begin
            if(clk'event and clk='1') then
                if ld_hmadr='1' then
                    n:= hbfdataout;
                elsif inc_hmadr='1' then
                    n:=n+1;
                elsif init_hmadr='1' then
                    n:="00000000";
                end if;
            end if;
            hmaddress <=n;

        end process hmadr;

-- indication that the end of histogram memory has been reached

        process(hmaddress)
        begin
            if hmaddress = "11111111" then
```

```
                    end_hm <='1';
          else
                    end_hm <='0';
          end if;
     end process;

-- generation of the address of HBF (input) or GHE (output) memory

     hbadr: process(clk)
          variable s: std_logic_vector(15 downto 0);
     begin
          if(clk'event and clk='1') then
                    if init_adr= '1' then
                              s:= "0000000000000000";
                    elsif inc_adr = '1' then
                              s:= s + 1;
                    end if;
                              gheaddress <=l;
                    end if;
                    if s="1111111111111111" then
                              end_hbf <='1'; -- indication that the last
                                        -- location has been reached
                    else
                              end_hbf <='0';
                    end if;
     end process hbadr;
          hmout<= hmq(7 downto 0);

     end combined;
```

### *17.4.2 GHE Functional Unit Control Unit*

The GHE functional unit control unit accepts a command from the host processor, starts the GHE operation and after its completion returns a signal to the host processor that the operation is completed. The control flow of the GHE functional unit control is presented by the control flow diagram in Figure 17.9.

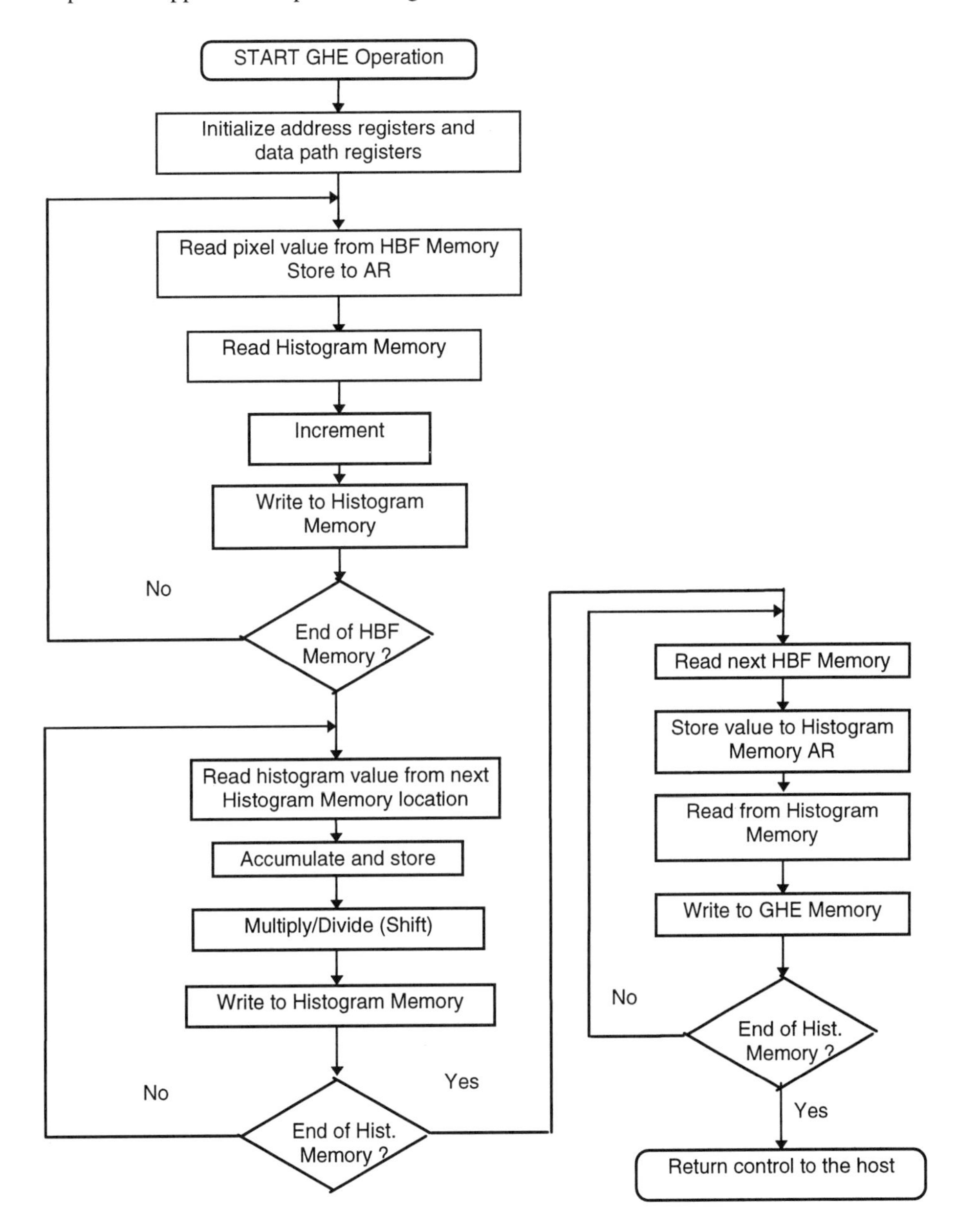


Figure 17.9 GHE functional unit control flow

The VHDL description of the GHE control unit operation is given below:

```
library ieee;
use ieee.std_logic_1164.all;
use ieee.std_logic_unsigned.all;
use ieee.std_logic_arith.all;

entity ghecontrol is
        port (
                clk, startghe: in std_logic;
                end_hm: in std_logic;
                hmdatasel: out std_logic;
                ld_hmadr, inc_hmadr, init_hmadr: out std_logic;
                ena_inc, ld_inc: out std_logic;
                ena_acc, clr_acc: out std_logic;
                init_adr, inc_adr: out std_logic;
                end_hbf: in std_logic;
                we, endghe: out std_logic;
                readhbf: out std_logic;
                writeghe : out std_logic
                        );
end ghecontrol;

architecture beh of ghecontrol is
type state_type is (init, readhbf1, hmadr1, hmadr2, wait1, wait2, wait3,
wait4, readhm1, inc, writehm1, writehm2, readhm2, init1, init2, readhbf2,
writeghe1, writeghe2, writeghe3, finish);
signal state: state_type;

begin

sm: process (clk)
begin
        if startghe = '1' then
                state <= init;
        elsif (clk'event and clk = '1') then

                case state is

                when init =>
                        readhbf <= '1';
                        state <= readhbf1;
```

```
when readhbf1 =>
        ld_hmadr<='1';
        readhbf <= '0';
        state <= hmadr1;

when hmadr1 =>
        ld_hmadr<='0';
        ld_inc <='1';
        state <= hmadr2;

when hmadr2 =>
        state <= wait1;

when wait1 =>
        ld_inc<='0';
        ena_inc <='1';
        state <= readhm1;

when readhm1 =>
        ena_inc<='0';
        inc_adr <= '1';
        state <= inc;


when inc =>
        hmdatasel <= '1';
        we <='1';
        inc_adr <='0';
        state <= writehm1;

when writehm1 =>
        we <='0';
        hmdatasel <= '0';

        if end_hbf = '1' then
                init_hmadr<='1';
                clr_acc<= '1';
                state <= init1;
        else
                readhbf <= '1';
                state <= readhbf1;
        end if;
```

```
when init1 =>
        we<='0';
        init_adr<='1';
        clr_acc <='0';
        init_hmadr<='0';
        state <= readhm2;

when readhm2 =>
        init_adr<='0';
        inc_hmadr<='0';
        state <= wait2;

when wait2 =>
        ena_acc <= '1';
        state <= writehm2;

when writehm2=>
        ena_acc <= '0';
        we <= '1';
        if end_hm = '1' then
                init_adr <='1';
                state <= init2;
        else
                state <= wait3;
        end if;

when wait3 =>
        we<='0';
        state<= wait4;

when wait4 =>
        inc_hmadr<='1';
        state <= readhm2;

when init2 =>
        we<='0';
        readhbf<='1';
        init_adr <='0';
        state <= readhbf2;

when readhbf2 =>
        readhbf<='0';
```

```
                    ld_hmadr<='1';
                    inc_adr <='0';
                    writeghe <='0';
                    state <= writeghe1;

            when writeghe1 =>
                    ld_hmadr<='0';
                    state <= writeghe2;

            when writeghe2 =>
                    inc_adr <='1';
                    state <= writeghe3;

            when writeghe3 =>
                    writeghe <='1';
                    if end_hbf='1' then
                            state <= finish;
                    else
                            readhbf<='1';
                            state <= readhbf2;
                    end if;
            when finish =>
                    writeghe <='0';
                    endghe <= '1';
            end case;
        end if;
        end process sm;

end beh;
```

Obviously, the GHE functional unit data path and control unit should be connected in an integral structural VHDL description to implement the whole GHE functional unit. That description is omitted from this book and left to the reader to implement the whole GHE functional unit.

## 17.5 Questions and Problems

17.1. Analyze the implications of using a general multiplier and divider for the high-boost filter implementation. Propose solution and implement both circuits using

VHDL. Analyze at least two possible approaches for both of these circuits: a) without using EABs, and b) with using EABs. Assume the accuracy of both circuits taking into account that the image pixels are presented with 8 bits.

17.2. Try to further improve the speed at which the HBF values are calculated by storing the partial sums of rows in a separate memory for the entire column, and using them for calculation of row sums for the next column. Assuming the same size of input image (256 x 256), use internal EABs of FLEX10K device to implement block that stores these partial sums. Describe the solution using VHDL. What performance improvement can be achieved in this way? Simulate your solution using the Max+Plus II Simulator, and implement it using UP prototyping board.

17.3. What changes are needed to implement the high-boost filter for an input image represented by 10-, 12- and 16-bit pixels. Modify solution presented in this Chapter, compile it and simulate using Max+Plus II Simulator and implement it on UP prototyping board.

17.4. Analyze the details of a customization of the FLIX core to accommodate both the HBF and GHE. Describe all details of the customization of the FLIX core and the interfacing details for both filters using VHDL.

17.5. Analyze the use of dynamic reconfigurability to implement the FLIX core with image enhancement filters (both the HBF and GHE). Assume that you are allowed to use two FLEX10K devices, one to implement the FLIX core and the other to implement one of the filters at the time. Specify conceptually permanent links between the FLIX core and filter FPLD to enable dynamic reconfigurability of the filter FPLD in order to implement either of the filters at the request from the program running in the FLIX core. Take into account the interfacing with memories used to store the input and output images.

# GLOSSARY

**Access Type** A data type analogous to a pointer that provides a form of indirection.

**Active-high (-low) node** A node that is activated when it is assigned a value one (zero) or Vcc (Gnd). In AHDL design files, an active-low node should be assigned a default value of Vcc with the Defaults statement.

**Aggregate** A form of expression used to denote the value of a composite type. An aggregate value is specified by listing the value of each element of the aggregate using either positional or named notation.

**AHDL** Acronym for Altera Hardware Description Language. Design entry language which supports Boolean equation, state machine, conditional, and decode logic. It also provides access to all Altera and user-defined macrofunctions.

**Alias** Statement used to declare an alternate name for an object.

**Antifuse** Any of the programmable interconnect technologies forming electrical connection between two circuit points rather than making open connections.

**Architecture** Describes the behaviour, dataflow, and/or structure of a VHDL entity. An architecture is created with an architecture body. A single entity can have more than one architecture. Configuration declarations are used to specify which architectures to use for each entity.

**Array** A collection of one or more elements of the same type that are accessed using one or more indices depending on dimension of array. Array data types are declared with an array range and array element type.

**ASIC** Acronym for Application-Specific Integrated Circuit. A circuit whose only final photographic mask process is design dependent.

**Assert** A statement that checks whether a specified condition is true. If the condition is not true, a report is generated during simulation.

**Assignment** In VHDL, assignment refers to the transfer of a value to a symbolic name or group, usually through a Boolean equation. The value on the right side of an assignment statement is assigned to the symbolic name or group on the left.

**Asynchronous input** An input signal that is not synchronized to the device Clock.

**Attribute** A special identifier used to return or specify information about a named entity. Predefined attributes are prefixed with ' character.

**Back annotation** Process of incorporating time delay values into a design netlist reflecting the interconnect capacitance obtained from a completed design. Also, in Altera's case, the process of copying device and resource assignments made by the Compiler into Assignment and Configuration File for a project. This process preserves the current fit in future compilations.

**Block** A feature that allows partitioning of the design description within an architecture.

**Cell** A logic function. It may be a gate, a flip-flop, or some other structure. Usually, a cell is small compared to other circuit building blocks.

**Cell library** The collective name for a set of logic functions defined by the manufacturer of an FPLD or ASIC. Simulation and synthesis tools use cell library when simulating and synthesizing a model.

**CLB** Acronym for Configurable Logic Block. This element is the basic building block of the Xilinx LCA product family.

**Clock** A signal that triggers registers. In a flip-flop or state machine, the clock is an edge-sensitive signal. The output of the clock can change only on the clock edge.

**Clock enable** The level-sensitive signal on a flip-flop with E suffix, e.g., DFFE. When the Clock enable is low, clock transitions on the clock input of the flip-flop are ignored.

**Component** Specifies the ports of a primitive or macrofunction in VHDL. A component consists of the name of the primitive or macrofunction, and a list of its inputs and outputs. Components are specified in the Component declaration

**Component instantiation** A concurrent statement that references a declared component and creates one unique instance of that component.

**Composite type** A data type that includes more than one constituent element (for instance, array or record).

**Concurrent statements** Statements that are executed in parallel.

**Configuration** It maps instances of VHDL components to design entities and describes how design entities are combined to form a complete design. Configuration declarations are used to specify which architectures to use for each entity.

**Configuration scheme** The method used to load configuration data into an FPGA.

**Constant** An object that has a constant value and cannot be changed.

**Control path** The path of signals generated by control unit used to control data path.

**CPLD** Acronym for Complex Programmable Logic Device. CPLDs include an array of functionally complete or universal logic cells in an interconnection framework that has foldback connection to central programming regions.

**Data path** The path which provides processing and transfer of information in the circuit through the blocks of combinational and sequential logic.

**Design entity** The combination of an entity and its corresponding architecture.

**Design file** A file that contains description of the logic for a project and is compiled by the Compiler.

**Design library** Stores VHDL units that have already been compiled. These units can be referenced in VHDL designs. Design libraries can contain one or more of the following units:

- Entity declarations
- Architecture declarations
- Configuration declarations
- Package declarations
- Package body declarations

**Design unit** A section of VHDL description that can be compiled separately. Each design unit must have a unique name within the project.

**Driver** Contains the projected output waveform for a data object. Each scheduled value is a driver.

**Dual-purpose pins** Pins used to configure an FPGA device that can be used as I/O pins after initialization.

**Dynamic reconfigurability** Capability of an FPLD to change its function "on-the-fly" without interruption of system operation.
**EDIF** Acronym for Electronic Design Interchange Format. An industry-standard format for the transmission of design files.

**Entity** See Design entity.

**Enumeration type** A symbolic data type that is declared with an enumerated type name, and one or more enumeration values.

**EPLD** Acronym for EPROM Programmable Logic Devices. This is a PLD that uses EPROM cells to internally configure the logic function. Also, Erasable Programmable Logic Device.

**Event** The change of value of a signal. Usually refers to simulation.

**Event scheduling** The process of scheduling of signal values to occur at some simulated time.

**Excitation function** Boolean function that specifies logic that directs state transitions in a state machine.

**Exit condition** An expression that specifies a condition under which a loop should be terminated.

**Expander** Section in the MAX LAB containing an array of foldback NAND functions. The expander is used to increase the logical inputs to the LAB macrocell section or to make other logic and storage functions in the LAB.

**Fan-in** The number of input signals that feed all the input equations of a logic cell.

**Fan-out** The number of output signals that can be driven by the output of a logic cell.

**FastTrack interconnect** Dedicated connection paths that span the entire width and height of a FLEX 8000 device. These connection paths allow the signals to travel between all LABs in a device.

**Field name** An identifier that provides access to one element of a record data type.

**File type** A data type used to represent an arbitrary-length sequence of values of a given type.

**For loop** A loop construct in which an iteration scheme is a for statement.

**Finite state machine** The model of a sequential circuit that cycles through a predefined sequence of states.

**Fitting** Process of making a design fit into a specific architecture. Fitting involves technology mapping, placement, optimization, and partitioning among other operations.

**Flip-flop** An edge-sensitive memory device (cell) that stores a single bit of data.

**Floorplan** Physical arrangement of functions within a design relative to the other.

**FPGA** Acronym for Field Programmable Gate Array. A regular array of cells that is either functionally complete or universal within a connection framework of signal routing channels.

**FPLD** An integrated circuit used for implementing digital hardware that allows the end user to configure the chip to realize different designs. Configuring such a device is done using either a special programming unit or by doing it "in system".

**Function prototype** Specifies the ports of a primitive or macrofunction in AHDL. It consists of the name of the primitive or macrofunction, and a list of its inputs and outputs in exact order in which they are used. An instance of the primitive or macrofunction can be inserted with an Instance declaration or an in-line reference.

**Functional simulation** A simulation mode that allows to simulate the logical performance of a project without timing information.

**Functional test vector** The input stimulus used during simulation to verify a VHDL model operates functionally as intended.

**Functionally complete** Property of some Boolean logic functions permitting them to make any logic function by using only that function. The properties include making the AND function with an invert or the OR function with an invert.

**Fuse** A metallic interconnect point that can be electrically changed from short circuit to an open circuit by applying electrical current.

**Gate** An electronic structure, built from transistors, that performs a function.

**Gate array** Array of transistors interconnected to form gates. The gates in turn are configured to form larger functions.

**Gated clock** A clock configuration in which the output of an AND or OR gate drives a clock.

**Generic** A parameter passed to an entity, component or block that describes additional, instance-specific information about that entity, component or block.

**Glitch or spike** A signal value pulse that occurs when a logic level changes two or more times over a short period.

**Global signal** A signal from a dedicated input pin that does not pass through the logic array before performing its specified function. Clock, Preset, Clear, and Output Enable signals can be global signals.

**GND** A low-level input voltage. It is the default inactive node value.

**Hierarchy** The structure of a design description, expressed as a tree of related components.

**Identifier** A sequence of characters that uniquely identify a named entity in a design description.

**Index** A scalar value that specifies an element or range of elements within an array.

**Input vectors** Time-ordered binary numbers representing input values sequences to a simulation program.

**Instance** The use of a primitive or macrofunction in a design file.

**I/O cell register** A register on the periphery of a FLEX 8000 device or a fast input-type logic cell that is associated with an I/O pin.

**I/O feedback** Feedback from the output pin on an Altera device that allows an output pin to be also used as an input pin.

**LAB** Acronym for Logic Array Block. The LAB is the basic building block of the Altera MAX family. Each LAB contains at least one macrocell and an I/O block and an expander product term array.

**Latch** A level-sensitive clocked memory device (cell) that stores a single bit of data. A high-to-low transition on the Latch Enable signal fixes the contents of the latch at the value of the data input until the next low-to-high transition on Latch Enable.

**Latch enable** A level-sensitive signal that controls a latch. When it is high, the input flows through the output; when it is low, the output holds its last value.

**Library** In VHDL denotes facility to store analyzed design units.

**Literal** A value that can be applied to an object of some type.

**Logic element** A basic building block of an Altera FLEX 8000 device. It consists of a look-up table i.e., a function generator that quickly computes any function of four variables, and a programmable flip-flop to support sequential functions.

**Long line** Mechanism inside an LCA where a signal is passed through repeating amplifier to drive a larger interconnect line. Long lines are less sensitive to metal delays.

**LPM** Acronym for Library of Parametrized Modules. Denotes the library of design units that contain one or more changeable parts, parameters, that are used to customize design unit as application requires.

**Macro** When used with FPGAs, a cell configuration that can be repeated as needed. It can be Hard and Soft macro.

**Macrocell** In FPGAs, a portion of the FPGA that is smallest indivisible building block. In MAX devices it consists of two parts: combinatorial logic and a configurable register.

**MAX** Acronym for Multiple Array MatriX, which is an Altera product family. It is usually considered to be a CPLD.

**MAX+PLUS II** Acronym for Multiple Array MatriX Programmable Logic User System II. A set of tools that allow design and implementation of custom logic circuits with Altera's MAX and FLEX devices.

**Mode** A direction of signal (either in, out, inout or buffer) used as subprogram parameter or port.

**Model** A representation that behaves similarly to the operation of some digital circuit.

**MPLD** Acronym for Mask-Programmed Logic Device.

**Netlist** A text file that describes a design. Minimal requirements are identification of function elements, inputs and outputs and connections.

**Netlist synthesis** Process of deriving a netlist from an abstract representation, usually from a hardware description language.

**NRE** Acronym for Non-Recurring Engineering expense. It reefers to one-time charge covering the use of design facilities, masks and overhead for test development.

**Objcct** A named entity of a specific type that can be assigned a value. Object in VHDL include signals, constants, variables and files.

**One Hot Encoding** A design technique used more with FPGAs than CPLDs. It assigns a single flip-flop to hold a logical one representing a state, with the rest of flip-flops being held at zeros.

**Package** A collection of commonly used VHDL constructs that can be shared by more than one design unit.

**PAL** (Programmable Array Logic) a relatively small FPLD containing a programmable AND plane followed by a fixed-OR plane.

**Parameter** An object or literal passed into a subprogram via that subprogram's parameter list.

**Partitioning** Setting boundaries within functions of a system.

**Physical types** A data type used to represents measurements.

**PLA** (Programmable Logic Array) a relatively small FPLD that contains two levels of programmable logic - an AND plane and an OR plane.

**Placement** Physical assignment of a logical function to a specific location within an FPGA. Once logic function is placed, its interconnection is made by routing.

**PLD** Acronym for Programmable Logic Device. This class of devices comprise PALs, PLAs, FPGAs and CPLDs.

**Port** A symbolic name that represents an input or output of a primitive or of a macrofunction design file.

**Primitive** One of the basic functional blocks used to design circuits with Max+Plus II software. Primitives include buffers, flip-flops, latch, logical operators, ports, etc. Functional prototypes for AHDL primitives are built into the Max+Plus II software. Component declarations for VHDL primitives are provided in the `maxplus2` package.

**Process** A basic concurrent statement represented by a collection of sequential statements that are executed whenever there is an event on any signal that appears in the process sensitivity list, or whenever an event occurs that satisfies condition of a wait statement within the process.

**Programmable switch** A user programmable switch that can connect a logic element or input /output element to an interconnect wire or one interconnect wire to another.

**Project** A project consists of all files that are associated with a particular design, including all subdesign files and ancillary files created by the user or by Max+Plus II software. The project name is the same as the name of the top-level design file without extension.

**Propagation delay** The time required for any signal transition to travel between pins and/or nodes in a device.

**Range** A subset of the possible values of a scalar type.

**Record** A composite data type that includes more than one of differing types. Record elements are identified by field names.

**Register** A memory device that contains more than one latch or flip-flop that are clocked from the same source clock signal.

**Resource** A resource is a portion of a device that performs a specific, user-defined task (e.g., pins, logic cells).

**Retargetting** A process of translating a design from one FPGA or other technology to another. Retargetting involves technology mapping and optimization.

**Routing** Process of interconnecting previously placed logic functions.

**RTL** Acronym for Register Transfer Level. The model of circuit described in VHDL that infers memory devices to store results of processing or data transfers. Sometimes it is referred to as dataflow-style model.

**Scalar** A data type that has a distinct order of its values, allowing two objects or literals of that type to be compared using relational operators.

**Semicustom** General category of integrated circuits that can be configured directly by the user of IC. It includes gate array, PLD, FPGA, PROM and EPROM devices.

**Signal** In VHDL a data object that has a current value and scheduled future values at simulation times. In RTL models signals denote direct hardware connections.

**Simulation** Process of modeling a logical design and its stimuli in which the simulator calculates output signal models.

**Slew rate** Time rate of change of voltage. Some FPGAs permit a fast or slow slew rate to be programmed for an output pin.

**Slice** A one-dimensional, contiguous array created as a result of constraining a larger one-dimensional array.

**Speed performance** The maximum speed of a circuit implemented in an FPLD. It is set by the longest delay through any path for combinational circuits, and by maximum clock frequency at which the circuit operates properly for sequential circuits.

**State transition diagram** A graphical representation of the operation of a finite state machine using directed graphs.

**Structural-type architecture** The level at which VHDL describes a circuit as an arrangement of interconnected components.

**Subprogram** A function or procedure. It can be declared globally or locally.

**Synthesis** The process of converting the model of a design described in VHDL from one level of abstraction to another, lower and more detailed level.

**Technology mapping** Process of translating the function of a design from one technology to another. All versions of the design would have the same function, but the cell used would be very different.

**Test bench** A VHDL model used to verify the correct behavior of another VHDL model, commonly known as unit under test.

**Type** A declared name and its corresponding set of declared values representing the possible values the type. Four general categories of types are used: scalar types, composite types, file types and access types.

**Type declaration** A declaration statement that creates a new data type. A type declaration must include a type name and a description of the entire set of possible values for that type.

**Universal logic cell** A logic cell capable of forming any combinational logic function of the number of inputs to the cell. RAM, ROM and multiplexers have been used to form universal logic cells. Sometimes they are also called look-up tables or function generators.

**Usable gates** Term used to denote the fact that not all gates on an FPGA may be accessible and used for application purposes.

**Variable** In VHDL a data object that has only current value that can be changed in variable assignment statement.

**VCC** A high-level input voltage represented as a high (1) logic level in binary group values. It is a default active node value in AHDL.

**VHDL** Acronym for VHSIC (Very High Speed Integrated Circuits) Hardware Description Language. VHDL is used to describe function, interconnect and modeling.

# REFERENCES AND SELECTED READING

We are witnesses to a high interest in hardware description languages and filed-programmable logic, and huge numbers of references and material created as a part of educational or research efforts throughout the world. We can list only a small portion, mainly those which influenced the contents of this book.


[1] Armstrong, J.R., Gray F.G. *Structured Logic Design With VHDL*, Prentice-Hall, 1993

[2] Ashenden P., *The Designer's Guide to VHDL*, Morgan Kaufman Publishers, 1993

[3] Bolton M., "Digital Systems Design with Programmable Logic", Addison-Wesley Publishing Co., 1990

[4] Brown S. et al., *Field-Programmable Gate Arrays*, Kluwer Academic Publishers, 1992

[5] Brown S., and Rose J., "FPGA and CPLD Archhitectures: A Tutorial", IEEE Design and Test of Computers, Summer 1996

[6] Mazor S., Langstraat P., *A Guide to VHDL*, Kluwer Academic Publishers, 1989

[7] Pellerin D., *VHDL Made Easy*, Prentice-Hall 1997

[8] Perry D., *VHDL*, Second Edition, McGraw-Hill, 1994

[9] Rose J., El Gamal A., and Sangiovanni-Vincentelli A., "Architecture of Field-Programmable Gate Arrays", Proc. IEEE, Vol. 81, No.7, July 1993

[10] Salcic Z., "FLIX - A FLexible Instruction eXecution Microcomputer", Tech. Report , Auckland University, Department of Electrical and Electronic Engineering, September 1997

[11] Salcic Z., "PROTOS- A Microcontroller/FPGA-based prototyping System for Embedded Applications", *Elsevier Science Journal on Microprocessors and Microsystems*, (in press)

[12] Salcic Z., "SimP-A Simple Custom-Configurable Processor Implemented in FPGA", Tech. Report no.567/96 , Auckland University, Department of Electrical and Electronic Engineering, July 1996

[13] Salcic Z., Cheng M.S., " RAPROS - A Rapid Prototyping System for PC-compatible Hardware/Software Solutions", *World Manufacturing Congress '97, Proceedings of World Manufacturing Congress - Proceedings of the International Symposium on Manufacturing Technology*, Auckland, N.Zealand, November, S. Nahavandi & M. Saadat Eds., Academic Press, 1997, pp. 74-80

[14] Salcic Z., Maunder B., "SimP - a Core for FPLD-based Custom-Configurable Processors", Proceedings of Internatonal Conference on ASICS - ASICON '96, Shanghai, 1996

[15] Salcic Z., Maunder B., "CCSimP - An Instruction-Level Custom-Configurable Processor for FPLDs", Field-Programmable Logic 96, R.Hartenstein, M.Glesner (Eds), Lecture Notes in Computer Science 1142, Springer 1996

[16] Salcic Z., Sivaswamy J., "High-speed Image Enhancement Using FPGA Customised Hardware", Proceedings of the *Fifth ANZ International Conference on Intelligent Information Processing Systems ICONIP'97, Dunedin*, Springer, 1997, pp. 1157-1160

[17] Salcic Z., Smailagic A., *Digital Systems Design and Prototyping Using Field Programmable Logic*, Kluwer Academic Publishers, 1997

[18] Smith D.J., *HDL Chip Design*, Doone Publications, 1996

[19] Trimberger S., ed., *Field-Programmable Gate Array Technology*, Kluwer Academic Publishers, 1994

[20] Proc.IEEE Symposium FPGAs for Custom-Computing Machines, IEEE Computer Society Press, Los Alamitos, 1993-1996

[21] Field-Programmable Logic, FPL '94 Prague, Springer Verlag, 1994

[22] Field-Programmable Logic, FPL '96 Darmstadt, Springer Verlag, 1996

[23] Field-Programmable Logic, FPL '97 London, Springer Verlag, 1997

[24] *Max+Plus II Programmable Logic Development System: AHDL*, Altera Corporation, 1995

[25] *VHDL - Language Reference*, IEEE Press, 1994

[26] Various data sheets, application notes and application briefs by Altera Co.and Xilinx Co.

# PACKAGE TEXTIO

**package** TEXTIO **is**

-- Type Definitions for Text I/O
**type** LINE **is access** STRING; -- a LINE is a pointer to a STRING value
**type** TEXT **is file of** STRING; -- a file of variable-length ASCII records
**type** SIDE **is** (RIGHT, LEFT); -- for justifying output data w/in fields
**subtype** WIDTH **is** NATURAL; -- for specifying widths of output fields

-- Standard Text Files

**file** INPUT: TEXT **is in** "STD_INPUT";
**file** OUTPUT: TEXT **is out** "STD_OUTPUT";

-- Input Routines for Standard Types

**procedure** READLINE (**variable** F: **in** TEXT; L: **inout** LINE); -- made F a
-- variable; L an inout

**procedure** READ (L: **inout** LINE; VALUE: **out** BIT;GOOD: **out** BOOLEAN);
**procedure** READ (L: **inout** LINE; VALUE: **out** BIT);
**procedure** READ (L: **inout** LINE; VALUE: **out** BIT_VECTOR; GOOD: **out**
BOOLEAN);
**procedure** READ (L: **inout** LINE; VALUE: **out** BIT_VECTOR);
**procedure** READ (L: **inout** LINE; VALUE: **out** BOOLEAN; GOOD: **out**
BOOLEAN);
**procedure** READ (L: **inout** LINE; VALUE: **out** BOOLEAN);
**procedure** READ (L: **inout** LINE; VALUE: **out** CHARACTER; GOOD: **out**
BOOLEAN);
**procedure** READ (L: **inout** LINE; VALUE: **out** CHARACTER);

**procedure** READ (L: **inout** LINE; VALUE: **out** INTEGER; GOOD: **out**
BOOLEAN);
**procedure** READ (L: **inout** LINE; VALUE: **out** INTEGER);

**procedure** READ (L: **inout** LINE; VALUE: **out** REAL; GOOD: **out** BOOLEAN);

**procedure** READ (L: **inout** LINE; VALUE: **out** REAL);
**procedure** READ (L: **inout** LINE; VALUE: out STRING; GOOD: **out** BOOLEAN);
**procedure** READ (L: **inout** LINE; VALUE: **out** STRING);
**procedure** READ (L: **inout** LINE; VALUE: **out** TIME; GOOD: **out** BOOLEAN);
**procedure** READ (L: **inout** LINE; VALUE: **out** TIME);

-- Output Routines for Standard Types

**procedure** WRITELINE (**variable** F: **out** TEXT; L: **inout** LINE); -- explicitly
-- made F a variable; L an inout
**procedure** WRITE (L: **inout** LINE; VALUE: **in** BIT; JUSTIFIED: **in** SIDE:= RIGHT; FIELD: **in** WIDTH := 0);
**procedure** WRITE (L: **inout** LINE; VALUE: **in** BIT_VECTOR; JUSTIFIED: **in** SIDE:= RIGHT; FIELD: **in** WIDTH := 0);
**procedure** WRITE (L: **inout** LINE; VALUE: **in** BOOLEAN; JUSTIFIED: **in** SIDE:= RIGHT; FIELD: **in** WIDTH := 0);
**procedure** WRITE (L: **inout** LINE; VALUE: **in** CHARACTER; JUSTIFIED: **in** SIDE:= RIGHT; FIELD: **in** WIDTH := 0);
**procedure** WRITE (L: **inout** LINE; VALUE: **in** INTEGER; JUSTIFIED: **in** SIDE:= RIGHT; FIELD: **in** WIDTH := 0);
**procedure** WRITE (L: **inout** LINE; VALUE: **in** REAL; JUSTIFIED: **in** SIDE:= RIGHT; FIELD: **in** WIDTH := 0; DIGITS: **in** NATURAL:= 0);
**procedure** WRITE (L: **inout** LINE; VALUE: **in** STRING; JUSTIFIED: **in** SIDE:= RIGHT; FIELD: **in** WIDTH := 0);
**procedure** WRITE (L: **inout** LINE; VALUE: **in** TIME; JUSTIFIED: **in** SIDE:= RIGHT; FIELD: **in** WIDTH := 0; UNIT: **in** TIME:= ns);

**end** TEXTIO;

# INDEX

## A



## B



## C



## D

## M



## N



## O



## P



## R

## S



## T



## U



## V



## W



## X

# About the Accompanying CD-ROM

***VHDL and FPLDs in Digital Systems Design, Prototyping and Customization,*** First Edition includes a CD-ROM that contains Altera's MAX+PLUS II 7.21 Student Edition programmable logic development software. MAX+PLUS II is a fully integrated design environment that offers unmatched flexibility and performance. The intuitive graphical interface is complemented by complete and instantly accessible on-line documentation, which makes learning and using MAX+PLUS II quick and easy. MAX+PLUS II version 7.21 Student Edition offers the following features:

- ✓ Operates on PCS running Windows 3.1, Windows 95, and Windows NT 3.51 and 4.0.
- ✓ Graphical and text-based design entry, including Altera Hardware Description Language (AHDL) and VHDL.
- ✓ Design compilation for product-term (MAX 7000S) and look-up table (FLEX 10K) device architectures.
- ✓ Design verification with full timing simulation.

**Installing with Windows 3.1 and Windows NT 3.51**

Insert the MAX+PLUS II CD-ROM in your CD-ROM drive. In the Windows Program Manager, choose **Run** and type: *<CD-ROM drive>*: `\pc\maxplus2\install` in the *Command Line* box. You are guided through the installation procedure.

**Installing with Windows 95 and Windows NT 4.0**

Insert the MAX+PLUS II CD-ROM in your CD-ROM drive. In the Start menu, choose **Run** and type: *<CD-ROM drive>*: `\pc\maxplus2\install` in the *Open* box. You are guided through the installation procedure.

**Registration & Additional Information**

To register and obtain an authorization code to use the MAX+PLUS II software, to **http://www.altera.com/maxplus2-student**. For complete installation instructions, refer to the **read.me** file on the CD-ROM or to the ***MAX+PLUS II Getting Started Manual***, available on the Altera world-wide web site (**http://www.altera.com**).

This CD-ROM is distributed by Kluwer Academic Publishers with *ABSOLUTELY NO SUPPORT* and *NO WARRANTY* from Kluwer Academic Publishers.

Kluwer Academic Publishers shall not be liable for damages in connection with, or arising out of, the furnishing, performance or use of this CD-ROM.